NEW FOUNDATION OF ARTIFICIAL INTELLIGENCE

Other Related Titles from World Scientific

Handbook of Machine Learning
Volume 1: Foundation of Artificial Intelligence
by Tshilidzi Marwala
ISBN: 978-981-3271-22-7

Handbook of Machine Learning
Volume 2: Optimization and Decision Making
by Tshilidzi Marwala and Collins Achepsah Leke
ISBN: 978-981-120-566-8

What is Artificial Intelligence?
A Conversation between an AI Engineer and a Humanities Researcher
by Suman Gupta and Peter H Tu
ISBN: 978-1-78634-863-0

NEW FOUNDATION OF ARTIFICIAL INTELLIGENCE

Ming Xie
Nanyang Technological University, Singapore

Hui Chen
Tongji University, China

Zhencheng Hu
Kumamoto University, Japan

World Scientific

NEW JERSEY • LONDON • SINGAPORE • BEIJING • SHANGHAI • HONG KONG • TAIPEI • CHENNAI • TOKYO

Published by

World Scientific Publishing Co. Pte. Ltd.
5 Toh Tuck Link, Singapore 596224
USA office: 27 Warren Street, Suite 401-402, Hackensack, NJ 07601
UK office: 57 Shelton Street, Covent Garden, London WC2H 9HE

Library of Congress Control Number: 2020948736

British Library Cataloguing-in-Publication Data
A catalogue record for this book is available from the British Library.

NEW FOUNDATION OF ARTIFICIAL INTELLIGENCE

ISBN 978-981-4271-63-9 (hardcover)
ISBN 978-981-4271-64-6 (ebook for institutions)
ISBN 978-981-123-222-0 (ebook for individuals)

For any available supplementary material, please visit
https://www.worldscientific.com/worldscibooks/10.1142/7264#t=suppl

Typeset by Stallion Press
Email: enquiries@stallionpress.com

Printed in Singapore

This book is dedicated to my family

and all those who have worked with me in one way or another

Dedication also goes to the blessings from God and Buddhas

Preface

The founding event of artificial intelligence was the Dartmouth summer workshop held in 1956, which was financially supported by Rockefeller Foundation. The organizers of the Dartmouth summer workshop include John McCarthy, Marvin Minsky, Nathaniel Rochester and Claude Shannon. It was John McCarthy who chose the name of artificial intelligence for this new field at that time. The Dartmouth summer workshop triggered the first wave of research on artificial intelligence which focused on developing machine's mental capabilities of thinking, reasoning and inference. The second wave of research on artificial intelligence was marked by the Michigan State University (MSU) workshop in 2000, which advocated the special focus on machine's mental capabilities of autonomous development and incremental learning. The organizers of the MSU workshop include John Weng, James McCelland and Alex Pentland. Today, the research on artificial intelligence has already entered its third wave, which directs the focus towards developing machine's mental capabilities of learning human languages and mastering cognition. The focus of this third wave should be the truthful foundation of artificial intelligence.

It goes without saying that human beings are living inside an ocean of signals. These signals are the inputs to the sensory systems of human beings. Undoubtedly, the series of transformation from signals to cognitive states of understanding the meanings behind these signals are the most powerful, as well as the most mysterious, ability of human beings' minds. Therefore, the fundamental question of human intelligence should be: what are the physical principles behind the series of transformation from signals to cognitive states of knowing the meanings behind these signals by human beings?

Similarly, the creatures of human beings are also living inside the same ocean of signals. In particular, it is highly expected that future machines, such as intelligent vehicles and smart robots, could possess human-like mental capability in understanding the meanings behind various types of surrounding signals. Hence, the fundamental question of artificial intelligence, or machine intelligence, should be: what are the physical principles behind the series of transformation from signals to cognitive states of knowing the meanings behind these signals by machines themselves?

Therefore, this book aims at promoting the paradigm of artificial intelligence as being a scientific discipline with the focus on the study of physical principles which enable machines to self-transform signals into the cognitive states of knowing the meanings behind these signals. This paradigm is fundamentally different from the existing paradigms such as the one which considers artificial intelligence as the study of computerizing human intelligence, or translating human ways of dealing with knowledge into computer programs (or software). In other words, this book advocates machine's self-intelligence, instead of computer-aided human intelligence (or computational intelligence).

This book consists of four parts, which are:

- Knowledge Representation: This book defines knowledge as being sets of properties and constraints which govern the entities in both the physical world and the conceptual worlds. Most importantly, this book admits that the best way to represent knowledge is to use human languages with extensions to both the technical language (i.e. mathematics) and programming languages.
- Knowledge Acquisition: Knowledge exist in both the physical world and the conceptual worlds. And, the most challenging, or the most mysterious, task is to transform physical quantities in the physical world into the conceptual quantities in the conceptual worlds. This is the classical problem of symbol grounding in artificial intelligence. In this part of the book, solutions addressing this challenge will be presented.
- Knowledge Computation: Machines could learn from human ways of computing new knowledge from given knowledge. And, the best practice could be to computerize human ways of inferring new knowledge from given knowledge. In this part of the book, some examples will be given in order to illustrate the way of making use of knowledge already discovered by human beings for the purpose of determining new knowledge.
- Knowledge Application: The status of self-intelligence will enable machines to achieve autonomy in learning, designing, conversational dialogue, decision-making, and action-taking, etc. In this part of the book, applications of knowledge to achieve autonomous behaviors such as knowledge-based physical activities and knowledge-based mental activities will be explained.

The methodologies described in this book will be applicable to all kinds of future machines with self-intelligence. However, special focus is put on the case of designing self-intelligence of future intelligent vehicles such as city-cars. It is interesting to take note that an intelligent vehicle could actually be renamed as a high-speed mobile robot.

In terms of mobility, a vehicle has three degrees of freedom. If we miniaturize a vehicle into a device (i.e. a head) which could be mounted on the shoulder of a robot, the self-intelligence of a vehicle could be readily implanted into smart robots. On the other hand, the outcome of autonomous driving scenarios could also be achieved with the combination of normal vehicles and intelligent robot drivers. In practice, a robot driver could be a humanoid robot which possesses self-intelligence. Hence, this book will illustrate principles and concepts in using either intelligent vehicles or smart humanoid robots as alternative examples.

Writing a book in general is a huge effort, which requires many years of preparations, writings and proof-readings. Writing a book about artificial intelligence in particular is even more demanding. This is because artificial intelligence is an emerging discipline with many schools of thoughts. Hopefully, this book could contribute to the research community as well as the education in science and engineering at the undergraduate and/or postgraduate levels.

Obviously, without support from organizations and people, it will be hard for a book to become a product of high quality. Such support or help must be acknowledged.

First of all, the author would like to thank the Future System Directorate under the Ministry of Defense (Singapore) which has provided a grant to the author's project on outstanding sense-making mind (OSMIND) in 2007. This project has helped the author to significantly advance a new foundation of artificial intelligence which has been nurtured during a preceding period of three years from 2003 to 2006. Most importantly, the research findings have encouraged the author to announce the third wave of research on artificial intelligence in 2009, which should focus on developing machine's mental capabilities of learning human languages and transforming signals into cognitive states of knowing the meanings behind signals.

Secondly, the author would like to thank Nanyang Technological University (Singapore) for assigning him to be in charge of revamping the course under the name of MP4G06 Artificial Intelligence and Machines in 2012. This educational task has motivated the author to prepare all the new materials devoted to the writing of this book. Also, these new materials have been put into practice of teaching, which in return has helped fine tune some important topics in this book. Thirdly, the author would like to thank World Scientific for providing the full editorial support which make the proof-reading and production of this book happy moments. Finally, thanks also go to those who have worked with the author in one way or another in the past decade.

Last but not the least, the author would like to express his deep gratitude toward all the members of his family for their daily love and unconditional care.

M. Xie
2020

Contents

Preface vii

Part 1 Knowledge Representation **1**

1. Basics of Knowledge 3

1.1. Introduction 3
1.2. Physical World 3
1.2.1. Entity 4
1.2.2. Properties of Entity 5
1.2.3. Constraints of Entity 6
1.3. Conceptual Worlds 7
1.3.1. Entity with Self-Intelligence 8
1.3.2. Role of Human Language 9
1.3.3. Conceptual World and Its Definition 10
1.4. Knowledge and Its Definition 11
1.5. Artificial Self-Intelligence and Its Roadmap 12
1.6. Summary 13
1.7. Exercises 14

2. Basics of Human Language 15

2.1. Introduction 15
2.2. Use of Human Language 15
2.3. Invention of Nouns 16
2.4. Invention of Pronouns 17
2.5. Invention of Verbs 18
2.6. Invention of Adverbs 20
2.7. Invention of Adjectives 20
2.8. Invention of Prepositions 21
2.9. Invention of Grammar 21

2.9.1. Language Models of Phrases . . . 22
2.9.2. Language Models of Sentences . . . 23
2.10. Summary . . . 24
2.11. Exercises . . . 25

3. Basics of Technical Language 29

3.1. Introduction . . . 29
3.2. Use of Technical Language . . . 29
3.3. Point . . . 30
3.4. Vector . . . 31
3.5. Matrix . . . 33
3.6. Polynomials . . . 35
3.7. Integration . . . 35
3.8. Derivative . . . 36
3.9. Logic . . . 37
3.10. Probability . . . 41
3.11. Possibility . . . 45
3.12. Fourier Transform . . . 48
3.13. Laplace Transform . . . 50
3.14. Summary . . . 54
3.15. Exercises . . . 55

4. Basics of Programming Language 57

4.1. Introduction . . . 57
4.2. Use of Programming Language . . . 57
4.3. Data Structure . . . 59
4.4. Instruction Set . . . 61
4.5. Function . . . 62
4.6. Library . . . 67
4.7. Dynamic Class . . . 68
4.8. Programmable Script . . . 73
4.9. Summary . . . 74
4.10. Exercises . . . 74

Part 2 Knowledge Acquisition **75**

5. Knowledge Acquisition from Text 77

5.1. Introduction . . . 77
5.2. Representation of Human Language . . . 77
5.2.1. Representation of Physical Meanings . . . 78
5.2.2. Representation of Conceptual Meanings . . . 81
5.2.3. Representation of Concept-physical Meanings . . . 84

5.3. Transformation From Text to Speech 86
5.4. Transformation From Speech to Text 90
5.4.1. Sensing of Analogue Sound Waves 90
5.4.2. Sampling of Analogue Sound Signals 91
5.4.3. ADC of Discrete-time Sound Signals 91
5.4.4. Segmentation of Digital Sound Signals 91
5.4.5. Fourier Transform of Digital Sound Signals 92
5.4.6. Symbol Grounding of Digital Sound Unit 94
5.4.7. Composition of Words 95
5.4.8. Composition of Sentences 96
5.5. Transformation From Image to Text 96
5.5.1. Digitization of Input Image 96
5.5.2. Identification of Background Color 97
5.5.3. Segmentation of Text Image 98
5.5.4. Alignment of Text Image 98
5.5.5. Segmentation of Word Image 98
5.5.6. Symbol Grounding of Word Image 99
5.6. Transformation From Text to Meanings 102
5.7. Summary 103
5.8. Exercises 104

6. Basics of Visual Perception 105

6.1. Nature of Human Vision 105
6.1.1. Human Vision Is Composite 106
6.1.2. Human Vision Is Parallel 106
6.1.3. Human Vision is Quantitative 107
6.1.4. Human Vision Is Qualitative 108
6.1.5. Human Vision Is Situated 109
6.1.6. Human Vision Is Developmental 111
6.1.7. Human Vision Is Cognitive 113
6.1.8. Human Vision Is Attentive 119
6.2. Nature of Machine Vision 122
6.2.1. What Is Machine Vision 122
6.2.2. Machine Vision Is Computational 125
6.2.3. Machine Vision Is Cognitive 126
6.3. Summary 136
6.4. Exercises 136

7. Knowledge Acquisition from Color 139

7.1. Introduction 139
7.2. Transformation From Light to Color Image 140
7.2.1. Light Gathering 140
7.2.2. Color Sensing 142

7.2.3. Color Synthesis 144
7.2.4. Color Image Formation 145
7.3. Representation of Color 146
7.3.1. Color's Properties in Physical World 146
7.3.2. Color's Properties in Conceptual Worlds 160
7.3.3. Data Structure of Organized Meanings 162
7.4. Cognition of Color 163
7.4.1. Visual Computing 163
7.4.2. Visual Learning 168
7.5. Recognition of Color 173
7.5.1. Use of Symbol Grounding Network 174
7.5.2. Use of Fuzzy Logic 175
7.5.3. Use of Multi-scale Color Indices 176
7.6. Summary 178
7.7. Exercises 181

8. Knowledge Acquisition from Shape 183

8.1. Introduction 183
8.2. Transformation from Image to Shape 183
8.2.1. Definition of Shape 185
8.2.2. Use of Color Histogram 186
8.2.3. Use of Color Region 189
8.2.4. Use of Color Edge 196
8.2.5. Use of Color Texture 200
8.3. Representation of Shape 207
8.3.1. Properties of Shape in Physical World 207
8.3.2. Representation of Shape's Physical Properties 209
8.3.3. Representation of Shape's Conceptual Properties 225
8.3.4. Data Structure of Organized Meanings 228
8.4. Cognition of Shape 229
8.4.1. Visual Computing 230
8.4.2. Visual Learning 235
8.5. Recognition of Shape 237
8.5.1. Open-loop Approach to Shape Recognition 237
8.5.2. Closed-loop Approach to Shape Recognition 240
8.6. Summary 241
8.7. Exercises 242

9. Knowledge Acquisition from Depth 245

9.1. Introduction 245
9.2. Transformation From Depth to Image 247
9.2.1. Properties of Image Plane 248

9.2.2. Properties of Camera 252
9.2.3. Relative Pose Between Camera's Coordinate Systems . . 255
9.2.4. Relative Motion Between Camera's Coordinate Systems . 259
9.2.5. Baseline Between Camera's Coordinate Systems 261
9.2.6. Projection of Points from Scene to Image 265
9.2.7. Projection of Velocities from Scene to Image 266
9.3. Transformation from Image to Apparent Motion 268
9.3.1. Formation of Apparent Motion 268
9.3.2. Registration of Image Sequence 269
9.3.3. Use of Conglomerated Appearance 270
9.3.4. Matching of Conglomerated Appearances 275
9.3.5. Match Search Space 278
9.4. Representation of Depth 282
9.4.1. Use of Depth-map 282
9.4.2. Continuous Depth-map in Camera Frame 284
9.4.3. Discrete Depth-map in Camera Frame 284
9.4.4. Discrete Scene in Viewer Frame 286
9.4.5. Generalized Depth-map in Viewer Frame 288
9.4.6. Concept of Look-through Scene 289
9.4.7. Concept of Depth Redundancy 289
9.4.8. Conceptual Meaning of Depth 291
9.4.9. Data Structure of Organized Meanings 291
9.5. Cognition of Depth 293
9.5.1. Depth from Visual Computing 293
9.5.2. Visual Learning of Depth 307
9.6. Recognition of Depth 309
9.6.1. Problem Statement 310
9.6.2. Use of Predication and Verification Loop 311
9.6.3. Estimation of Recognized View's Pose 313
9.7. Summary 314
9.8. Exercises 315

Part 3 Knowledge Computation 317

10. Computing Statics of Generic Data 319

10.1. Introduction 319
10.2. Computation of Data's Center 319
10.3. Computation of Data's Spread 320
10.4. Computation of Data's Occurrence 320
10.5. Computation of Data's Similarity 321
10.6. Computation of Data's Dissimilarity 321
10.7. Computation of Data's Correlation 322

10.8. Computation of Data's Principal Components 324
10.9. Summary 326
10.10. Exercises 326

11. Computing Dynamics of Mechanical Systems 327

11.1. Introduction 327
11.2. Computation of Mechanism's Motion 327
11.2.1. Equation of Path 328
11.2.2. Equation of Trajectory 329
11.3. Computation of Mechanism's Kinematics 334
11.3.1. Equation of Forward Kinematics 334
11.3.2. Equation of Inverse Kinematics 336
11.4. Computation of Mechanism's Dynamics 337
11.4.1. Equation of Forward Dynamics 337
11.4.2. Equation of Inverse Dynamics 340
11.5. Summary 340
11.6. Exercises 341

Part 4 Knowledge Application 343

12. Knowledge-enabled Applications 345

12.1. Introduction 345
12.2. Vision-centric Intelligence and Its Applications 345
12.2.1. Vision-based Measurement 346
12.2.2. Vision-based Guidance 350
12.2.3. Visual Recognition 353
12.2.4. Visual Understanding 356
12.3. Speech-centric Intelligence and Its Applications 359
12.3.1. Conversational Dialogue 360
12.3.2. Voice-controlled Action 360
12.3.3. Massive Sense-making 361
12.3.4. Meaning-based Machine Translation 361
12.3.5. Voice-based Inquiry 362
12.4. Planning-centric Intelligence and Its Applications 363
12.4.1. Task Planning 363
12.4.2. Action Planning 364
12.4.3. Motion Planning 365
12.5. Control-centric Intelligence and Its Applications 365
12.5.1. Achieving Stability 366
12.5.2. Achieving Smallest Delay in Time 366
12.5.3. Achieving Smallest Error in Output 366

12.6. Data-centric Intelligence and Its Applications 367
12.6.1. Strategy for Market Reward and Market Acceptance . . . 369
12.6.2. Development of Internal Capabilities 370
12.6.3. Optimization of Internal Processes 371
12.7. Design-centric Intelligence and Its Applications 371
12.7.1. Designing Products 371
12.7.2. Designing Factories 373
12.7.3. Designing Universities 374
12.7.4. Designing Intelligent Transportation Systems 375
12.8. Summary 376
12.9. Exercises 376

Bibliography 377

Index 381

Part 1
Knowledge Representation

In Artificial Intelligence, one of the major issues is knowledge representation. However, it is a matter of fact that the best tool to represent knowledge is the existing set of human languages (i.e. natural languages invented by human beings) with extensions to both the technical language (i.e. mathematics) and programming languages. Hence, it is important to start this book with the part which is focused on knowledge representation. More specifically, this first part of the book contains chapters related to knowledge definition and knowledge representation with the use of human languages, technical language and programming languages. The focus is on designing self-intelligence for future vehicles or drivers in the form of humanoid robots.

Chapter 1

Basics of Knowledge

1.1 Introduction

We are living on earth. We are part of the universe. We act and interact with other entities on earth and in the universe. What are the inner abilities for us to act? What are the common enabling factors which enable entities to interact? This chapter will reveal answers to these questions. These answers will lead to a better definition of knowledge. In addition, on the basis of knowledge, a roadmap toward achieving artificial self-intelligence in connection with human intelligence will be discussed.

1.2 Physical World

Here, the physical world means the universe. Our universe consists of entities which could act and interact among themselves. By definition, a set of elements or entities which act or interact for achieving common goals is called as a system. Therefore, the universe is actually organized into the inner hierarchy of clusters of systems at various levels.

Refer to Fig. 1.1. The systems at the smallest scale are energies. At a higher level, we have particles which are systems as well. And, we know that atoms are systems with particles such as protons, neutrons and electrons. When relevant atoms tightly interact together, the result will be the formation of molecules, which in turn form systems of materials. What are mostly interesting to us are the systems at the level of entities and beyond. For example, a vehicle itself is a system. A fleet of vehicles could also form a system at a higher scale.

Refer to Fig. 1.1 again. Intrinsically, there are vibrations behind energies. Vibrations emit signals, which in turn show the manifestation of properties and constraints. Among the properties and constraints, there are governing rules called as causality. For example, an apple drops from a tree and falls onto the ground due to masses and gravity. When causality occurs among the entities inside a system, we will see actions and reactions. Since interactions inside a system could evolve over time and in space, what we observe is called the dynamics of a system.

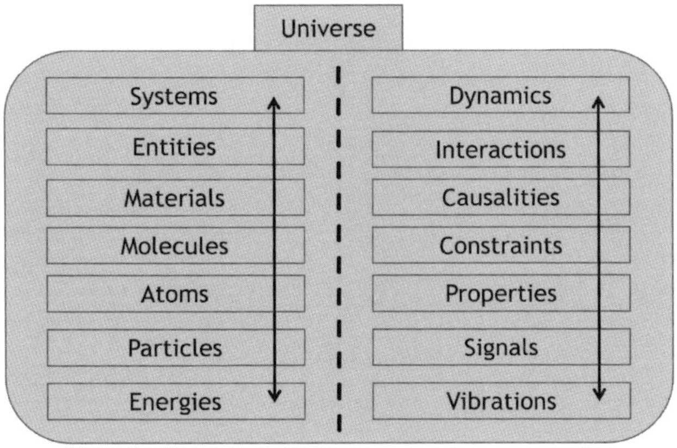

Fig. 1.1 The universe consists of the inner hierarchy of clusters of systems and the inner hierarchy of clusters of knowledge. We will define knowledge later on.

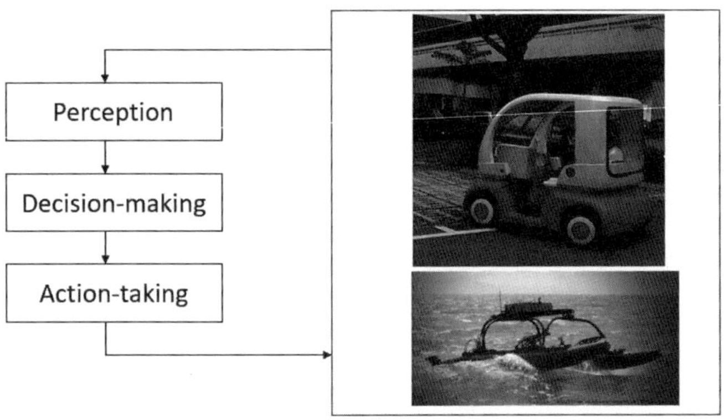

Fig. 1.2 Examples of Entity: A land vehicle is an entity, a surface vehicle is also an entity, both of them could include the constituent elements such as perception module, decision-making module and action-taking module.

1.2.1 *Entity*

An entity is any existence in the universe at any level of scale. An entity is a system with its own constituent elements at a lower scale.

In Fig. 1.2, we show two examples (i.e. one land vehicle and one surface vehicle) of entity, each of which may have the constituent elements such as perception module, decision-making module and action-taking module.

In general, an entity could be in one of the following four states, namely: solid state, liquid state, gas state, and plasma state. For example, water could exist in the form of ice, liquid and vapor.

Since an entity could be treated as a system, we can talk about open-loop type of systems and closed-loop type of systems. In Fig. 1.3, we give one example of open-loop type of systems. In this example, the output voltage is a function of the

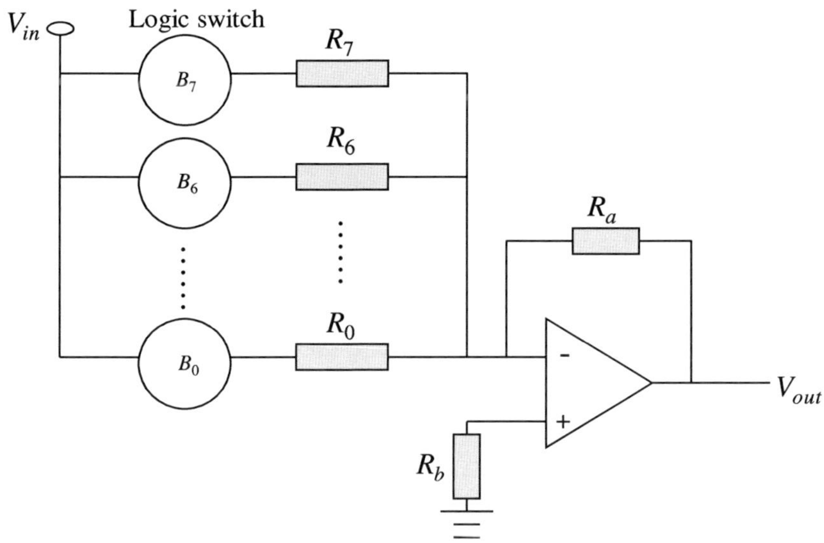

Fig. 1.3 An example of open-loop type of systems. (photo courtesy of free source in Internet).

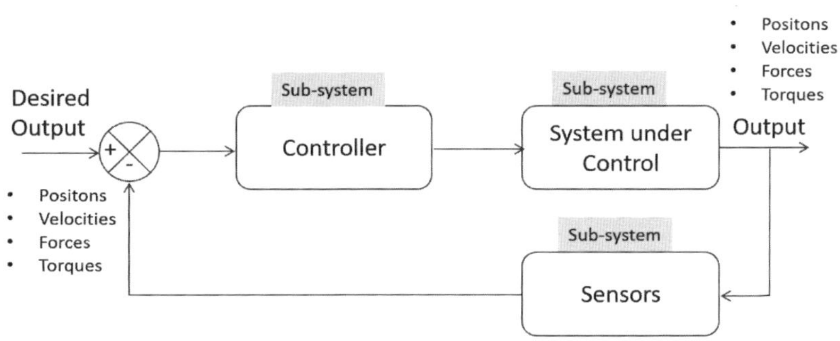

Fig. 1.4 An example of closed-loop type of systems.

input voltage multiplied by the logic (or switch) values (i.e. either 0 or 1) of bits $\{B_7, B_6, B_5, B_4, B_3, B_2, B_1, B_0\}$.

In Fig. 1.4, an example of closed-loop type of systems is given. The three basic building modules of a closed-loop system are controller, system under control and sensors. With a properly designed controller, a closed-loop system could make the actual values of output, such as positions, velocities, forces, and torques, to be exactly equal to the desired output values.

1.2.2 *Properties of Entity*

An entity is an existence in the universe, which could be observed by an observer such as a human being. This truth implies that all entities have something called as properties which show evidences of an entity's existence.

Fig. 1.5 Example of properties in a system with a fleet of tanks. (photo courtesy of free source in Internet with modification).

For example, a rigid vehicle has mass, dimension, shape, color, volume, etc. In Fig. 1.5, we show an example, in which we have highlighted properties such as length, color, temperature, speed, weight, pressure and friction.

In fact, the journey of scientific research is to discover the properties of all entities in the universe as much as possible. Such journey is being undertaken by different groups of scientists in different directions. And, these directions are commonly labeled as physics, biology, chemistry, economics, political science, material science and engineering, electrical engineering, mechanical engineering, civil engineering, computer science and engineering, automotive engineering, aerospace engineering, control science and engineering, etc. Hence, it is not the purpose of this book to list all the properties of entities discovered by our scientists during the long history of mankind. Interested readers are referred to the textbooks published in science and engineering so as to get the detailed information about all the possible properties of entities.

1.2.3 *Constraints of Entity*

The existence of an entity imposes at least two constraints, which are space and time. For example, the property of dimension or volume of an entity implies that the entity will occupy a space. This gives rise to the existence of space constraint. On the other hand, the life span of an entity implies that the entity will evolve over time from its appearance in space to its disappearance from space.

Besides the space-time constraints, there are constraints in other domains such as physics (e.g. gravity), biology, chemistry, economics, political science, material science and engineering, electrical engineering (e.g. electrical force), mechanical engineering (e.g. Newton's second law), civil engineering, computer science

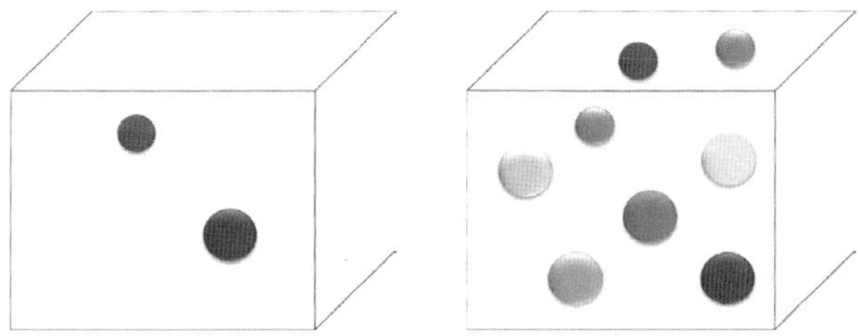

(a) (b)

Fig. 1.6 When there are more than one entity inside a space such as in the cases (a) and (b), interactions among these entities will occur due to various constraints.

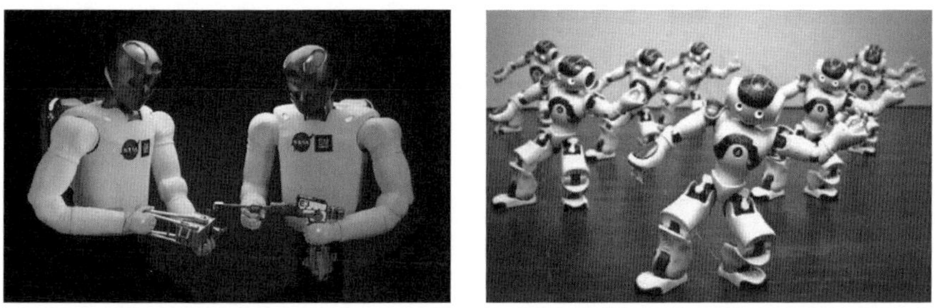

(a) (b)

Fig. 1.7 (a) Interactions between robots, and (b) Interactions among multiple robots. (photo courtesy of free source in Internet).

and engineering, automotive engineering (e.g. kinematics and dynamics), aerospace engineering, control science and engineering, etc. Again, it is not the purpose of this book to list all the constraints of entities discovered by our scientists during the long history of mankind. Interested readers are referred to the textbooks published in science and engineering so as to get the detailed information about all the possible constraints of entities.

It is interesting to take note that constraints play key roles inside the games of causality or interaction among the entities in the universe. As shown in Fig. 1.6, interactions may occur when there are more than one entity inside a small space.

In our daily life, we could easily encounter many scenarios of interactions, which demonstrate the inter-play of imposed constraints on entities such as human beings, machines, vehicles, or robots. In Fig. 1.7, we give two real examples of interactions. One is between two robots, and the other is among robots in a team.

1.3 Conceptual Worlds

Among the entities in the universe, there is a special category of entities, which have the ability to sense, to perceive, to understand, and to describe the properties and constraints. Such special entities are named as the entities with self-intelligence.

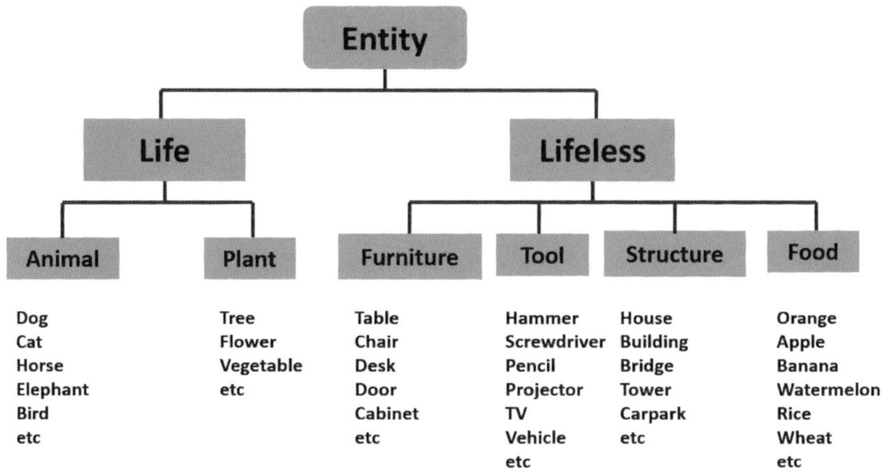

Fig. 1.8 Example of classifying entities into two broad categories: those with life and those without life.

Most importantly, the outcomes of sensing, perception, understanding and description by these entities with self-intelligence will form a set of inner worlds which are very special. Hence, this book calls these special inner worlds the conceptual worlds.

1.3.1 *Entity with Self-Intelligence*

There are a huge amount of entities in the universe or the physical world. For example, on earth, there are human beings, animals, plants, entities in solid state, entities in liquid state, and entities in gas state, etc. Broadly speaking, these entities could be categorized into two classes, namely: those with life and those without life. Fig. 1.8 shows one example of classifying entities into two broad categories: the ones with life and the ones without life.

Furthermore, among the entities with life, we could distinguish the ones with self-intelligence (e.g. human beings) from the ones without self-intelligence (e.g. trees). If we examine the key constituent elements of a human being, we will agree that a human being is broadly composed of three sub-entities, namely: body, brain and mind, as shown in Fig. 1.9. This book intentionally separates mind from brain. This is because the blueprint of a mind is different from the blueprint of a brain. Most importantly, it is the functionalities of a mind, which condition the level of self-intelligence.

Refer to Fig. 1.9. The primary function of body is to act and react. The knowledge computation and data storage are supported by the brain of an entity with self-intelligence. And, it is the mind of an entity with self-intelligence, which will demonstrate the ability of undertaking cognition (including learning) and re-cognition (including understanding). Therefore, this book advocates the

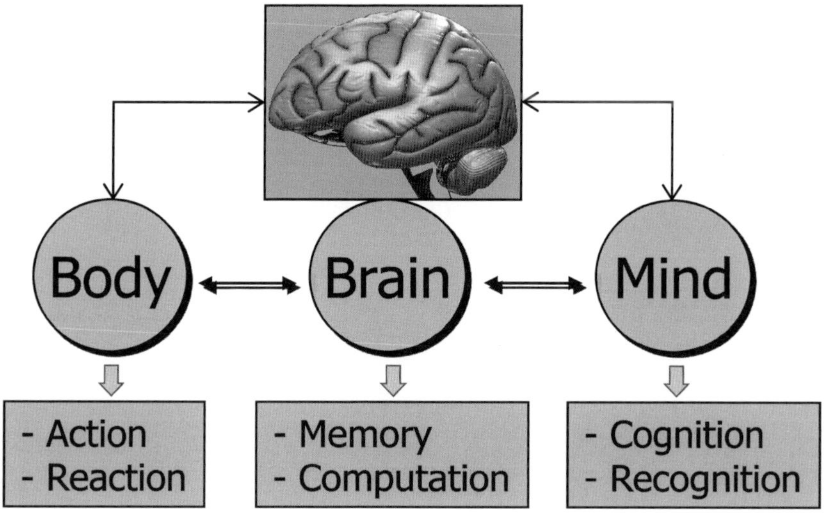

Fig. 1.9 An entity with self-intelligence is composed of body, brain and mind.

statement that intelligence is specifically the ability and outcome of mind, instead of brain. For example, an elephant may have a very big brain (which is about three times heavier than a human being's brain), but a very small mind.

1.3.2 *Role of Human Language*

Human beings co-exist with other entities on earth and in the universe. Most of these entities are sources, which emit signals. Therefore, human beings are surrounded by all kinds of signals. Among them, there are two particular sources of signals, which are extremely important to human beings' mental capabilities. The first ones are the visual signals (e.g. lights) and the second ones are audio signals (e.g. voices). Refer to Fig. 1.10. From the viewpoint of signals, the first two fundamental issues in the study of artificial intelligence are most likely to be:

- What are the physical principles behind the series of transformations from visual signals to the cognitive state of knowing the meanings behind these visual signals?
- What are the physical principles behind the series of transformations from audio signals (e.g. human voices) to the cognitive state of knowing the meanings behind these audio signals?

The above two issues will trigger a third fundamental question in the study of artificial intelligence, namely: what is the best way of representing the meanings extracted from signals? Interestingly enough, the answer to the third issue is the use of human languages with extensions to both the technical language (i.e. mathematics) and programming languages (including flowcharts). The evidence behind this answer is the existence of libraries, in which knowledge discovered by human

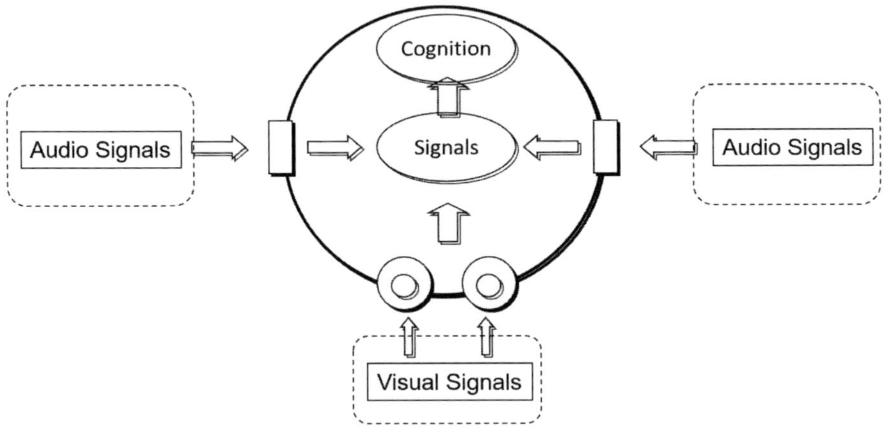

Fig. 1.10 The first two fundamental issues in the study of artificial intelligence: (a) What are the physical principles behind the series of transformations from visual signals to the cognitive state of knowing the meanings behind these visual signals?, and (b) What are the physical principles behind the series of transformations from audio signals (e.g. human voices) to the cognitive state of knowing the meanings behind these audio signals?

beings in all domains are all written into the texts of human languages with possible extensions to both the technical language (i.e. mathematics) and programming languages (including flowcharts).

Hence, the primary role of human language is to represent knowledge perceived by entities with self-intelligence. Knowledge stored in the form of texts written in human languages could also be retrieved by entities with self-intelligence. These mental capabilities will not only allow entities with self-intelligence to discover knowledge, but also to share, transfer, and communicate knowledge among themselves.

1.3.3 *Conceptual World and Its Definition*

Through the process of sensing, perception, understanding and description, an entity with self-intelligence is capable of building up an inner version of the outer physical world. And, the description of such inner version is in the form of texts written in a specific human language with possible extensions to both the technical language and programming languages.

Since there are many human languages on earth, the physical world could produce many inner versions, which could be described in different human languages such as English, French, Germany, Spanish, Italian, Japanese and Chinese, etc. Therefore, we could now put forward the following definition of conceptual worlds.

Definition 1.1. A conceptual world is a perceived version of the meanings in the physical world, which is written in the form of texts of a human language with possible extensions to both the technical language and programming languages.

A conceptual world is manifested in the form of texts in a specific human language. And, a text is a collection of words. Hence, each word in a specific human language could be treated as an entity with its own properties and constraints (i.e. grammatical properties and constraints). As a result, a conceptual world also contains a set of entities (i.e. words) which could act and interact together in order to form descriptions of various actions and interactions which occur in the physical world.

1.4 Knowledge and Its Definition

In layman terms, knowledge vaguely refers to information, truths, news, meanings, statements, etc. However, it is important to have a commonly agreeable definition of knowledge in the study of artificial intelligence for machines such as vehicles and robots.

Refer to 1.1. Simply speaking, the universes consists of systems on one hand and knowledge on the other hand. The knowledge of the universe includes the components such as vibrations, signals, properties, constraints, causality, interactions, and dynamics. Among these components of knowledge, we could divide them into three levels of complexity, for example:

- Primitive knowledge: vibrations and signals.
- Primary knowledge: properties and constraints.
- Composite knowledge: causality, interactions and dynamics.

Since the properties and constraints are the dominant features of all the entities in the universe as well as in the conceptual worlds, we would like to advocate the following definition of knowledge in general:

Definition 1.2. Knowledge consists of properties and constraints possessed by entities in the physical world as well as in the conceptual worlds.

It goes without saying that the research and innovation in science and engineering are the primary force which enriches knowledge discovered by human beings. In addition, the application of knowledge for the purpose of inventing technologies also contributes to the growth of knowledge created by human beings.

As illustrated by Fig. 1.11, knowledge is the most important content which motivates human beings to act and interact (e.g. to speak, to listen, to read, to write, etc). In addition, human languages are the most important tools which facilitate the sharing of knowledge as well as knowledge-enabled interactions among human beings. Similarly, machines with self-intelligence (e.g. intelligent vehicles or smart robots) could also join in such knowledge-centric games of sharing, actions and interactions. For example, it is expected that vehicles, or robots, of future could be engaged in conversational dialogues with human beings.

1.5 Artificial Self-Intelligence and Its Roadmap

Refer to Fig. 1.11. Human beings have the astonishing mental capability of transforming input signals into the cognitive state of knowing the meanings behind these signals. In addition, the mind of a human being is capable of incrementally accumulating knowledge into the construction of an inner world which is a perceived version of the outer physical world. Inside the inner world or conceptual world, the mind of a human being could undertake various types of mental activities such as the discovery of new knowledge, the creation of new knowledge, and the invention of new technologies, etc. Only with these mental capabilities, an entity could claim to have reached the level of having self-intelligence. Now, the next two questions will be:

- What are the physical principles behind the series of transformations from signals to cognitive state of knowing the meanings of these signals?
- What should be the roadmap for a machine's mind to be developed into the level of having artificial self-intelligence?

The primary focus of this book is to address the above two questions. In particular, the content of this book will follow the roadmap as shown in Fig. 1.12.

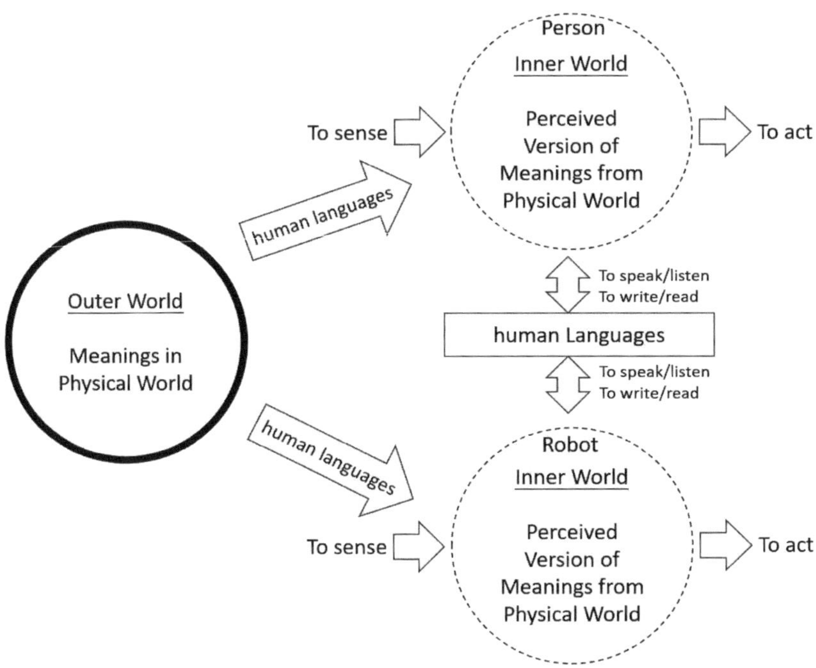

Fig. 1.11 Relationship between outer world and inner world: There is a common outer world which is the physical world. But, there are many versions of inner worlds or conceptual worlds.

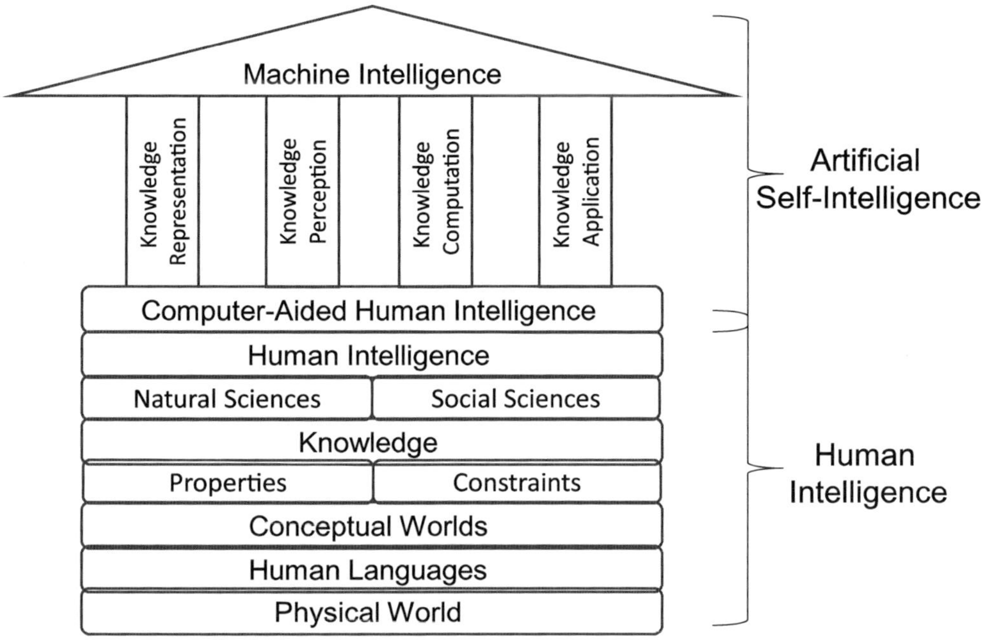

Fig. 1.12 Roadmap toward the development of artificial self-intelligence, in which we see the relationship among human intelligence, computer-aided human intelligence and artificial self-intelligence.

Refer to Fig. 1.12. One of the major contributions from human beings' mental activities is the creation of a huge body of knowledge in science and engineering. These knowledge cover the properties and constraints of most entities in the physical world. And, the invention and extensions to both the human languages have greatly facilitated the representation, sharing and communication (including knowledge transfer or education) of knowledge in human societies. Today, everyone of us appreciates very much the power of human intelligence, and has also enjoyed the wonderful outcome of human intelligence, in terms of better ecosystems, better services and better products.

The next journey of human beings' mental activities will be to construct the four pillars (i.e. knowledge representation, knowledge perception, knowledge computation, and knowledge application) which will lead to the achievement of artificial self-intelligence. And, it is the intention of this book to lay down the details inside these four pillars of artificial self-intelligence.

1.6 Summary

In this chapter, we have studied the nature of knowledge in both the physical world and the conceptual worlds. Also, we understand the importance of human languages for the representation of knowledge discovered and invented by human beings.

1.7 Exercises

(1) What is computer-aided human intelligence?
(2) What is intelligence by computer?
(3) What are the most important issues in the study of artificial self-intelligence?
(4) Does a software of computer game (e.g. chest game, go game) have its own intelligence?
(5) Mass is a property of a rigid body. Describe a scenario in which a rigid body's mass becomes a constraint.
(6) Provide a list of constraints in mechanical domain.
(7) Provide a list of constraints in electrical domain.
(8) A human being could learn any human language. Explain why it is true.
(9) An animal could never master any human language. Explain why it is true.

Chapter 2

Basics of Human Language

2.1 Introduction

Probably, the greatest invention on earth is human language. And, it is important to recognize that the major force, which drives the invention of human language, is the need or demand of representing knowledge discovered or created by human beings. In general, a human language consists of vocabulary (i.e. a set of words) and grammar (i.e. a set of syntax). The intention of this chapter is to highlight the relationship between knowledge and invention of vocabulary as well as grammar. The understanding of such relationship has two benefits. The first one is for educators in literature to rethink of better ways of teaching human languages from the viewpoint of language invention. The second benefit is for inventors of intelligent machines (e.g. vehicles, robots) to think of possible ways of making machines of future to master human languages.

2.2 Use of Human Language

Human languages are the most important invention of human beings' minds. In return, human languages empower human beings' minds to represent knowledge in the physical world as well as in conceptual worlds. In analogy to video cameras, we could imagine that the use of human languages actually play a similar function of projecting knowledge from the physical world into texts in conceptual worlds. Such process is illustrated in Fig. 2.1.

We know that primary knowledge come from properties of entities as well as constraints of entities. In practice, we name property-related knowledge as propositional knowledge. For constraint-related knowledge, they are called as procedural knowledge. Below are two examples:

Example 2.1. *Propositional Knowledge*: Acceleration of a mass is proportional to the applied net force on it, and inversely proportional to the mass itself.

<><><><><>

Fig. 2.1 In analogy to video cameras, the use of human languages play a similar function of projecting knowledge from the physical world into texts in conceptual worlds.

Example 2.2. *Procedural Knowledge*: When you see a traffic light to turn on red, you must brake and stop. When you see a slow moving vehicle in front, you should follow it if you cannot change the lane, or by-pass it if you can change the lane. When you are at a junction, you must give priority to cars coming from your right side (or your left side).

<><><><><>

Therefore, the use of human languages is to represent, to store, to share, and to communicate knowledge among people. Hopefully, human languages will also be able to empower machines of future to develop or grow their self-intelligence.

2.3 Invention of Nouns

When a new entity appears in the universe, it normally will exist for a certain period of time. Hence, when a new entity is discovered by human beings for the first time, this is a need for human beings to repeatedly name this newly discovered entity. For example, a horse is named with the label "cheval" in French language or "ma" in Chinese language. Another familiar example is the list of names attributed to nine planets in our solar system as shown in Fig. 2.2.

Therefore, the invention of nouns is motivated by the need of naming a newly discovered knowledge such as the identify of entity, a property of entity, or a constraint of entity, etc. In the vocabulary of a human language, a noun is a word. Such a word is considered to be an entity in a conceptual world because it carries the property of "being a noun" and must follow grammatical constraints.

For example, in English language, there are the following types of nouns (note: the list is not exhaustive):

- Common nouns, such as: David, Jack, John, Smith, etc.
- Collective nouns, such as : team, committee, army, etc.
- Compound nouns, such as: sister-in-law, hard-disk, etc.
- Singular nouns, such as: USA, UK, France, Germany, Italy, Spain, Singapore, China, Japan, etc.
- Plural nouns, such as: cars, cows, rooms, buildings, etc.

Figure 2.3 gives another example with familiar names attributed to fruits. For example, watermelon is a compound noun. And, fruits is a plural noun.

Fig. 2.2 The names of nine planets in our solar system. (photo courtesy of free source in Internet).

Fig. 2.3 Example of familiar names attributed to fruits. (photo courtesy of free source in Internet with modification).

2.4 Invention of Pronouns

In order to facilitate the tracking of reference to nouns during the description of knowledge, people have further invented an extension to nouns, which are called as pronouns.

For example, in English language, there are the following types of pronouns (note: the list is not exhaustive):

- Personal pronouns, for example: I, you, he, she, me, us, them, etc.
- Possessive pronouns, for example: mine, yours, ours, hers, his, etc.
- Interrogative pronouns, for example: which, whose, what, how much, etc.
- Relative pronouns, for example: which, that, whose, who, whom, etc.
- Demonstrative pronouns, for example: this, that, these, those, etc.
- Reflexive pronouns, for example: myself, yourself, themselves, etc.
- Indefinite pronouns, for example: all, any, anybody, nothing, none, each, etc.
- Reciprocal pronouns, for example: each other, one another, etc.

2.5 Invention of Verbs

The physical world has many entities. Most importantly, the constraints of entities, which are present in a common space, are the multiple forces which drive entities to act and react. When the interactions among entities evolve over time and in space, we will observe the dynamic scenario which is called as event. When multiple related events occur over time and in space, the resulting scenario is called as episode. And, the continuous sequence of episodes will form a much more complex dynamic scenario which is called as story. It goes without saying that the physical world, in which we are living in, is full of dynamics-driven stories. Figure 2.4 gives an illustration of such a hierarchy of dynamic scenarios which range from interactions, events, episodes, to stories.

Now, we are ready to state that the motivation behind the invention of verbs is to facilitate the description and communication of the dynamics underlying actions, reactions, interactions, events, episodes and stories. Verbs are words inside the vocabulary of a human language.

For the example of robotic game as shown in Fig. 2.5, the description of this type of dynamic stories requires the use of invented verbs in English, such as: move, kick, stop, avoid, push, collide, etc.

As we know, the physical capabilities of entities with self-intelligence, such as human beings and animals, are different. Hence, there are many types of invented verbs.

For example, in English, the invented verbs could be divided into the following types (note: the list is not exhaustive):

- Action verbs, for example: jump, fly, hit, push, swing, pull, write, close, walk, etc.
- Saying verbs, for example: say, speak, talk, argue, explain, announce, beg, etc.

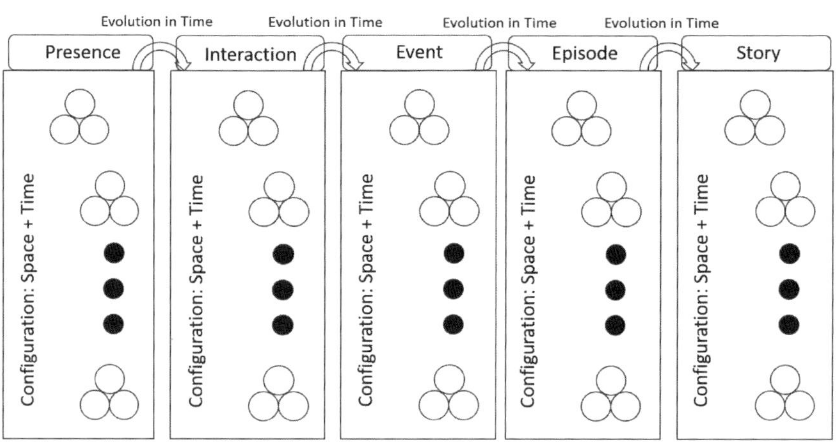

Fig. 2.4 When entities with the ability to act and to react are present within a common space, dynamic scenarios will occur. And, the dynamics of a physical scene will consist of a hierarchy of dynamic scenarios which range from interactions, events, episodes, to stories.

Fig. 2.5 Example of robotic game in which the description of actions and reactions requires the use of inverted verbs in English, such as: move, kick, stop, avoid, push, collide, etc. (photo courtesy of free source in Internet with modification).

Fig. 2.6 Example of actions which could be described by the verb "fly". (photo courtesy of free source in Internet).

- Sensing verbs, for example: smell, feel, hear, touch, observe, perceive, read, etc.
- Mental verbs, for example: think, infer, reason, solve, calculate, decide, etc.
- Linking verbs, for example: be, seem, become, appear, have, etc.
- Modal verbs, for example: will, must, might, would, may, can, could, etc.

Figure 2.6 shows an example, in which the actions could be described by the verb "fly". For instance, we could say that many air-planes are flying closely in the sky.

2.6 Invention of Adverbs

Normally, actions and reactions will take place over time and in space. In addition, actions and reactions could be undertaken in different manners and with different degrees of efforts. Hence, on top of verbs, people have further invented words which are called as adverbs.

For example, in English language, we have the following types of adverbs (note: the list is not exhaustive):

- Adverbs of manner, for example:
  - He easily understands your talk.
  - She writes beautifully.
  - The robot seats there quietly.
- Adverbs of degree, for example:
  - I completely agree with you.
  - The robot acts so powerfully.
  - He almost fails the exam.
  - He is very happy.
- Adverbs of time, for example:
  - He will come tomorrow.
  - I saw him two days ago.
- Adverbs of place, for example:
  - A car moves down a hill.
  - A robot walks forward, turns left, and then walks upstairs.
- Adverbs of connection, for example:
  - Firstly, I read this article.
  - Then, I write a summary of it.
  - Finally, I present the summary to you.

2.7 Invention of Adjectives

One interesting characteristics of using human languages is the ability to undertake both crisp description of knowledge (e.g. mass of 1.2 kg) and fuzzy description of knowledge (e.g. blue sky). The crisp description of knowledge will depend on the use of numerical values or symbols. However, the fuzzy description of knowledge will make use of invented words which are called as adjectives.

For example, in English language, we have the following types of adjectives (note: the list is not exhaustive):

- Adjectives of size, for example: tiny, small, big, huge, etc.
- Adjectives of shape, for example: triangular, rectangular, circular, polygonal, etc.
- Adjectives of age, for example: infant, young, old, etc.

- Adjectives of color, for example: red, green, blue, yellow, orange, purple, etc.
- Adjectives of origin, for example: near, far, close, etc.
- Adjectives of material, for example: solid, liquid, gaze, etc.
- Adjectives of behavior, for example: slow, fast, violent, gentle, etc.
- Adjectives of event, for example: terrible, wonderful, etc.
- Adjectives of process, for example: random, stationary, deterministic, etc
- Adjectives of episode, for example: unthinkable, unbelievable, exciting, interesting, wonderful, etc.

2.8 Invention of Prepositions

Adverbs allow people to describe general constraints in terms of space and time. However, for achieving crisp description of more precise, or specific, constraints in space and time, we must have words which provide references in space, time and direction. As a result of these requirements, prepositions have been invented.

For example, in English, we have the following types of prepositions (note: the list is not exhaustive):

- Prepositions of time reference, for example:
  - The lectures on Tuesday start at 4.30 pm.
  - You have to submit your project report before 18 April 2016.
- Prepositions of place reference, for example:
  - The robot puts its hands on top of a table.
  - The main campus of NTU is located in Jurong West.
- Prepositions of direction reference, for example:
  - The car moves along a road.
  - The robot walks out through a door.
  - The bus heads toward the main campus of NTU.
  - I seats in his left side.
- Prepositions of general reference, for example:
  - He gives a book to her.
  - He is playing outdoor.
  - I will teach Part 1 which is about knowledge representation.

2.9 Invention of Grammar

The vocabulary of a human language consists of a set of words. Each word has its own property such as being a noun, being a verb, etc. Individual words are not good enough to describe the knowledge underlying the dynamics of entities which act and react over time and in space. Words of different types must be able to be put together in order to form structures of words, which provide better representation of knowledge from the physical world.

In human languages, the basic structures of words are phrases and sentences. In order to standardize the formation of phrases and sentences, grammatical rules or syntax must be invented in advance. In this section, we simply highlight language models which will guide the construction of phrases and sentences in English.

2.9.1 *Language Models of Phrases*

In English language, the two typical phrases are noun-phrases and verb-phrases.

Noun phrases could be formed by following one of these two models:

- "Pre-modifier + Noun" model, such as:
  - "Demonstrative determiner + Noun", for example: that car.
  - "Possessive determiner + Noun", for example: our car.
  - "Interrogative determiner + Noun", for example: whose car?
  - "Adjective + Noun", for example: red car.
  - "Noun + Noun", such as: metal bar, school uniform.
  - "Quantifier + Noun", such as: three cars, many books, a lot of robots.
- "Noun + Post-modifier" model, such as:
  - "Noun + Preposition" model, for example:
    - the teacher in red hat
    - the book about robotics
    - the building with green windows
  - "Noun + Relative Clause" model, for example:
    - the student whose brother is soldier
    - the worker whose wife is doctor
  - "Noun + Non-finite Clause" model, for example:
    - the researcher to give a talk
    - the driver wearing a dark glasses
    - the robot kicking the ball

And, verb-phrases could be formed according to one of the following three models:

- Present tense, for example:
  - He likes study.
  - She is doing a project in the lab.
  - The lecture has finished.
  - The robot has started walking.
- Past tense, for example:
  - I sent you an email yesterday.
  - The robot was working last night.

  - I saw you on TV this morning.
  - He had been in Japan in year 2000.

- Future tense, for example:
  - We will study lecture 11 next week.
  - I shall travel to Europe soon.
  - She will be doing the project tomorrow.
  - He is going to submit the report next week.

Noun-phrase and verb-phrase are two basic models. The combination of them allows us to obtain composite models of noun-verb phrases or verb-noun phrases, for example:

- A lady wearing a hat is dancing.
- Eating a red apple per day is a good habit.

2.9.2 *Language Models of Sentences*

The description of the knowledge in the physical world will form a conceptual world which is a collection of texts in a human language. And, a text itself is a collection of sentences. Therefore, sentence models, which are invented by human beings and for human languages, are very important for the purpose of producing texts which describe knowledge from the physical world.

In English language, a simple sentence could be formed by applying the invented model of "Noun phrase + Verb Phrase + Noun Phrase". For example, the "Noun phrase + Verb Phrase + Noun Phrase" model, which is an invention in English language, could allow us to produce the following types of simple sentences:

- Simple sentences in the format of "Subject + Verb + Object 1 + Object 2 + ...", for example:
  - He asks teacher for helps.
  - He drives a sport car on road.
- Simple sentences in the format of "Subject + Verb + Indirect Object + Direct Object", for example:
  - The teacher gives us three assignments.
  - The robot brings me a drink.
- Simple sentences in the format of "Subject + Verb + Object + Adverb/Preposition", for example:
  - We complete the assignments timely.
  - He finished all the lectures happily yesterday.
  - The robot picks up a key from a table.

The combination of simple sentences could produce more complex sentences such as:

- Compound sentences, for example:
  - Robot A is doing assembly and robot B is doing painting.
- Composite sentences, for example:
  - I look at the robot which is doing material handling.

In addition, in English language, there are models which are invented for the purpose of constructing special sentences such as:

- Yes/No questions constructed by using the invented model of "Verb + Subject", for example:
  - Is this machine a robot?
  - Could a robot protect human beings?
- Who-What-Where-How questions (or WH questions in short) constructed by using the invented model of "Verb + Subject + Verb", for example:
  - Who give you that robot?
  - How many hours can this robot work?
- Tag questions constructed by using the invented model of "Declarative sentence + Question Tag", for example:
  - A robot can do difficult and dangerous jobs, isn't it?
  - He has not yet bought any robotics textbook, has he?
- Requests constructed by using the invented model of "Verb + Object + ...", for example:
  - Enjoy your study!
  - Have a nice day!
  - Hand in the reports by 2 pm.
- Commands constructed by using the invented model of "Verb + Object + ...", for example:
  - Finish the works by 5 pm!
  - Enter the building! Switch off the light!
- Exclamations constructed by using the invented model of "Pronoun + Noun-phrase", for example:
  - What a wonderful car!
  - What a crazy robot!

2.10 Summary

In this chapter, we have studied the nature of human languages from the viewpoint of inventions. In particular, we have studied the cases of inventing nouns, pronouns, verbs, adverbs, adjectives, and prepositions. Also, we have studied models of sentences in English language.

2.11 Exercises

(1) Describe the knowledge seen from the image as shown in Fig. 2.7, in the form of texts in English.

Fig. 2.7 Describe the knowledge seen from this image.

Fig. 2.8 Describe the knowledge seen from this image (photo courtesy of robotx.org).

(2) Describe the knowledge seen from the image as shown in Fig. 2.8, in the form of texts in English.
(3) Describe the knowledge seen from the image as shown in Fig. 2.9, in the form of texts in English.

Fig. 2.9 Describe the knowledge seen from this image.

(4) Describe the knowledge seen from the image as shown in Fig. 2.10, in the form of texts in English.

Fig. 2.10 Describe the knowledge seen from this image (photo courtesy of free source in Internet).

(5) Describe the knowledge seen from the image as shown in Fig. 2.11, in the form of texts in English.

Fig. 2.11 Describe the knowledge seen from this image (photo courtesy of free source in Internet).

Chapter 3

Basics of Technical Language

3.1 Introduction

As we have discussed in Chapter 1, an entity in the physical world has properties and is subject to constraints. Property-related knowledge are called as propositional knowledge while constraint-related knowledge are named as procedural knowledge. These knowledge could be sensed and perceived by entities with self-intelligence.

However, human beings' sensory systems have innate limitations. For example, our eyes can only see visible lights and our ears can only hear audible signals. And, without physical contact, our skin cannot sense the temperature of another entity. In addition, our skin cannot tell us the exact value of the temperature of a surface with which our skin is in contact.

As a result, some knowledge in the physical world could be directed sensed and perceived by human beings' sensory systems while some knowledge must reply on the use of scientific instruments as well as the analysis of experimental data in order to be understood by human beings. Therefore, in practice, the knowledge which could be directly perceived by human beings' sensory systems are called as explicit knowledge and those knowledge which could not be perceived directly are called as implicit knowledge.

It is interesting to take note that the invention of technical language (i.e. mathematics) occurs during the process of discovery of implicit or deep knowledge. For example, the knowledge behind Newton's second law is implicit and is better described by the equation: $f = m \bullet a$ (i.e. a net force f applied to a mass m will produce an acceleration a on it). In this chapter, a coverage about the basic part of technical language is provided. Here, technical language specifically means mathematics.

3.2 Use of Technical Language

The use of technical language is to describe implicit knowledge discovered by human beings. In return, the process of discovering new implicit knowledge will enrich technical language. Here, we show two familiar examples of using technical language to better describe implicit knowledge.

Example 3.1. A vehicle is moving along a circular path with radius R and an uniform circular velocity v. A coordinate system OXY is placed at the center of the circular path. If the vehicle is initially at position $(R, 0)$, what are the coordinates (x, y) of the vehicle at time t with respect to coordinate system OXY?

Solution:

$$x(t) = R\cos\left(\frac{v}{R}t\right) \tag{3.1}$$

$$y(t) = R\sin\left(\frac{v}{R}t\right) \tag{3.2}$$

<><><><><>

Example 3.2. A player of 2.0 m tall is launching a basket ball at initial speed v_0 and in a direction which is θ degrees above the horizon. The ball will travel in a vertical plane. If a coordinate system OXY is placed at the center of one foot of the player, what are the coordinates (x, y) of the ball at time t?

Solution:

$$x(t) = v_0 \cos(\theta)t \tag{3.3}$$

$$y(t) = 2.0 + v_0 \sin(\theta)t + \frac{1}{2}(-g)t^2 \tag{3.4}$$

where g is the gravitational acceleration on earth.

<><><><><>

3.3 Point

The existence of an entity in the physical world will occupy a space. Hence, the position of an entity is a spatial constraint. Such a spatial constraint could be represented by a point, which is an entity of the technical language. More specifically, a point inside a geometrical space with a reference (i.e. a coordinate system) represents a position of an entity. In mathematics, a coordinate system is a reference which consists of one or more axes. Each axis has an origin, a division (i.e scale) and a direction.

When we study the motion of rigid bodies in a space, a common practice is to define a global coordinate system. Then, within the space referenced by the global coordinate system, each entity is assigned with a local coordinate system. In this way, the point of a local coordinate system's origin represents the position of the entity to which the local coordinate has been attached. Figure 3.1 shows an example with one global coordinate system and three local coordinate systems which are attached to three entities. The origins of these three local coordinate

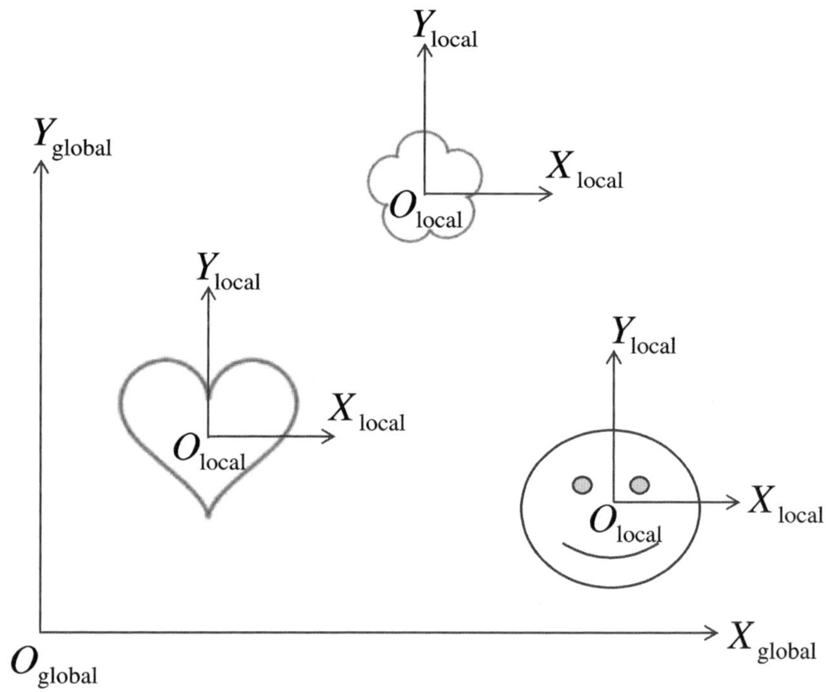

Fig. 3.1 Example of a space referenced by a global coordinate system in which there are three local coordinate systems attached to three entities, respectively.

systems represent the respective positions of the three entities with respect to the global coordinate system.

It is interesting to take note that a series of consecutively connected points will form a geometrical entity which is called as a curve in a two-dimensional space or a surface in a three-dimensional space. Also, the use of point is applicable to non-geometrical entities. For example, a color is a point in a color space. A feature of any kind is a point in a feature space.

3.4 Vector

When an entity changes its state of motion, its position will change. Such change results in displacement. This knowledge is the motivation behind the invention of vector. Hence, vector is an entity of the technical language.

In mathematics, a vector consists of a pair of tail and head. A vector has a length (which is the distance between its tail and head), a direction (which is the angle between the vector's line and the horizontal axis of the coordinate system attached to the vector's space), and a dimension (which is the number of elements contained by a vector). When a vector's tail is always at the origin of a coordinate system, such vector is also called as position vector.

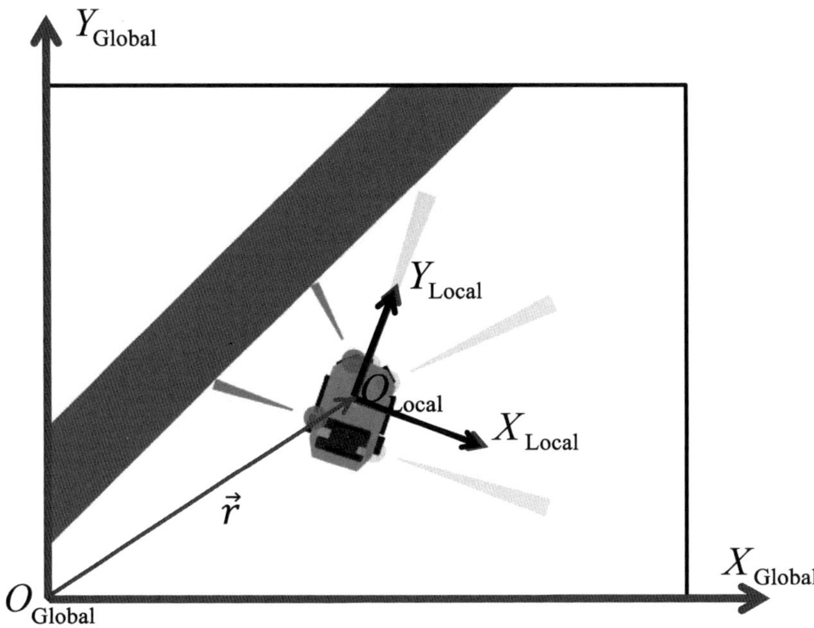

Fig. 3.2 Use of position vector to indicate the position of a mobile robot with respect to a global coordinate system OXY. (photo courtesy of free source in Internet with modification).

Figure 3.2 shows an example of using position vector $\vec{r}$ to represent the position of a mobile robot with respect to a global coordinate system. In this case, the mobile robot's position vector has two elements, which are coordinate $x(t)$ and coordinate $y(t)$. Therefore, the dimension of the mobile robot's position vector is 2. And, we can express $\vec{r}(t)$ with: $\vec{r}(t) = (x(t), y(t))$.

When the length of a vector is equal to 1, it is called an unit vector. One important use of unit vectors is to represent a coordinate system itself. For example, in Fig. 3.2, the global coordinate system could be represented by the following two unit vectors:

$$\vec{i} = (1, 0) \tag{3.5}$$

$$\vec{j} = (0, 1) \tag{3.6}$$

With the notion of position vector, we can easily define displacement vector which is the difference between two position vectors. For example, in Fig. 3.2, if the mobile robot's position vector changes from $\vec{r}(t-1)$ at time $t-1$ to $\vec{r}(t)$ at time t, the mobile robot's displacement vector will be $\vec{d}(t) = \vec{r}(t) - \vec{r}(t-1)$.

In the study of rigid body's motion, a displacement vector is also called as a translation vector, which represents the linear motion undertaken by an entity when it moves from the displacement vector's tail to its head within a certain period of time. In other words, a translation vector describes a translational motion in a physical space.

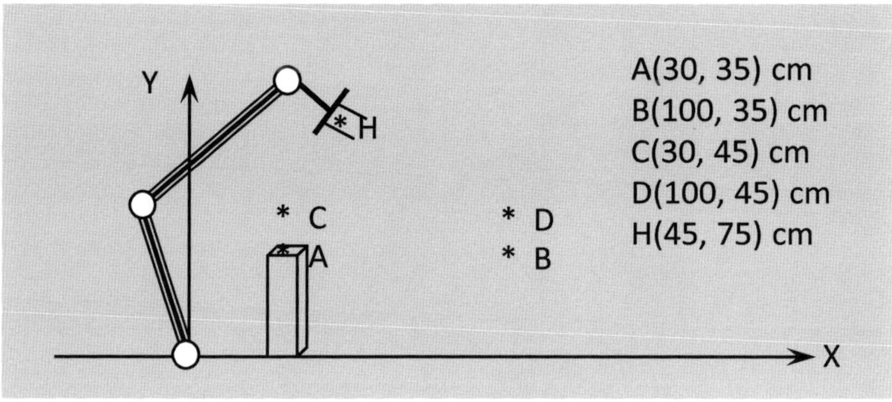

Fig. 3.3 A robot is doing pick-and-place operation. Positions A, B, C and D are predefined with specific coordinates.

In addition, the change rate of displacement vector in time is called as velocity vector. And, the change rate of velocity vector in time is called as acceleration vector. For example, in Fig. 3.2, the mobile robot's velocity vector will be $\vec{v}(t) = \frac{d\vec{r}(t)}{dt}$ and its acceleration vector will be $\vec{a}(t) = \frac{d\vec{v}(t)}{dt}$.

3.5 Matrix

Matrix is an entity in the technical language. Matrix could be considered as the extension of vector. Since a vector is a list of elements, each element could be a numeric value, an equation, or another vector. When the elements inside a vector are themselves vectors of a same dimension, such vector is called a matrix.

For example, in Fig. 3.3, a robot is doing pick-and-place operation. The robot's gripper is initially at a home position H. The robot moves its gripper from position H to position C and then to position A to pick an object. Subsequently, the robot moves the picked object from position A back to position C, then goes to position D, and finally places the object at position B. If we consider the sequence of positions H, C, A, C, D, B as a set of way-points, the path of the robot's gripper for the pick-and-place operation could be represented by a matrix as follows:

$$\begin{pmatrix} 45 & 30 & 30 & 30 & 100 & 100 \\ 75 & 45 & 35 & 45 & 45 & 35 \end{pmatrix} \tag{3.7}$$

where the first column comes from position $H(45, 75)$ cm.

From the above discussions, it is clear that a matrix has at least two dimensions. A matrix is an extension of a vector. In other words, a vector could be extended into a two-dimensional matrix. A two-dimensional matrix could be extended into a three-dimensional matrix, and so on.

On the other hand, we could also consider that a matrix is an invention motivated by the demand or need in representing the knowledge related to a rigid body's

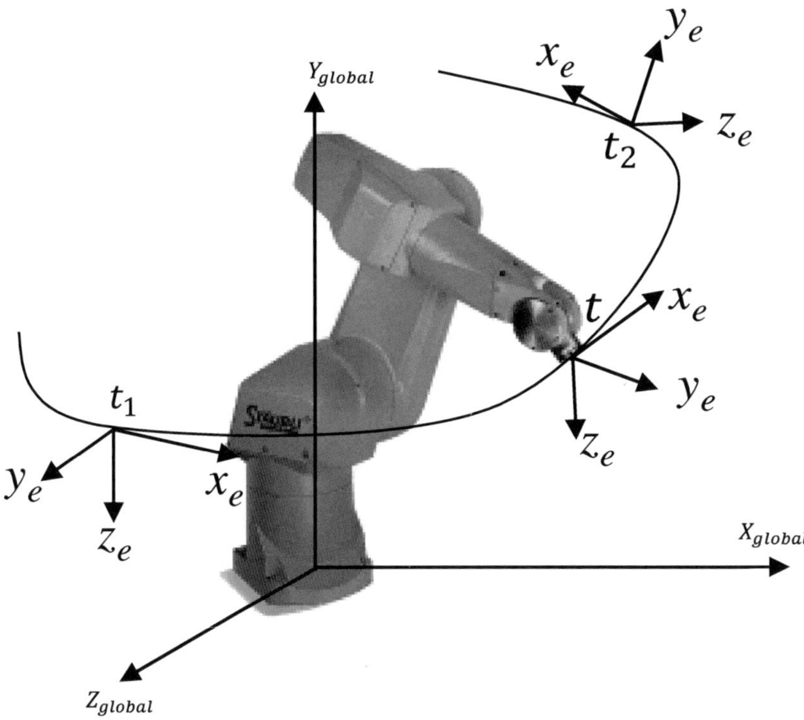

Fig. 3.4 The orientation and rotational motion of an end-effector attached to a robotic arm could be represented by a matrix which is composed of the three unit vectors of the local coordinate system $OXYZ|_e$'s three axes, with respect to the global coordinate system $OXYZ|_{global}$. (photo courtesy of free source in Internet with modification).

orientation and its rotational motion. For example, in Fig. 3.4, there is a robotic arm which is placed at the origin of a global coordinate system $OXYZ|_{global}$ and has an end-effector mounted on its wrist. In order to represent the orientation and rotational motion of the end-effector, a local coordinate system $OXYZ|_e$ is assigned to it. If the three unit vectors of the local coordinate system are $\vec{i}_e = [x_i(t), y_i(t), z_i(t)]$, $\vec{j}_e = [x_j(t), y_j(t), z_j(t)]$, and $\vec{k}_e = [x_k(t), y_k(t), z_k(t)]$, then the end-effector's orientation at time t could be represented by rotation matrix $R(t)$ as follows:

$$R(t) = \begin{bmatrix} x_i(t) & x_j(t) & x_k(t) \\ y_i(t) & y_j(t) & y_k(t) \\ z_i(t) & z_j(t) & z_k(t) \end{bmatrix} \tag{3.8}$$

And, the change of orientation R within time interval $[t_1, t]$ is a rotational motion, which could be represented by $\triangle R(t_1, t)$ as follows:

$$\triangle R(t_1, t) = R^{-1}(t_1) \bullet R(t) \tag{3.9}$$

where $R^{-1}(t_1)$ means the inverse of matrix $R(t_1)$.

Another motivation behind the invention of matrix is the demand or need in representing images, which are the output of eyes or cameras. For example, a gray-scale image is represented by a two-dimensional matrix of numeric values. Each numeric value represents the intensity of an incoming light which enters a camera.

However, a color image must be represented by a two-dimensional matrix of color vectors, or a three-dimensional matrix of numeric values. In a color space, a color vector has three elements. For example, in a *RGB* color space, the three elements of a color vector are red component value, green component value, and blue component value.

3.6 Polynomials

The possible values of knowledge in terms of properties and constraints could be represented by constants and variables. In theory and practice, there exist relationships among constants and variables. This observation has motivated people to invent entities which are called as polynomials in the technical language. In general, polynomial equations of coordinates in a space represent a large number of geometrical entities such as lines, circles, ellipses, curves, planes, and surfaces, etc.

For example, in Fig. 3.2, if the mobile robot moves along a circular path with center at (x_0, y_0) and radius r_0, the x and y coordinates of the mobile robot's center will be constrained by the following polynomials of order 2:

$$(x - x_0)^2 + (y - y_0)^2 = r_0^2 \tag{3.10}$$

If the mobile robot follows a complex curve on the floor and its motion is subject to time constraint, the time functions of its x and y coordinates could be described by the following equations of polynomials of order 5:

$$x(t) = a_x t^5 + b_x t^4 + c_x t^3 + d_x t^2 + e_x t + f_x \tag{3.11}$$

$$y(t) = a_y t^5 + b_y t^4 + c_y t^3 + d_y t^2 + e_y t + f_y \tag{3.12}$$

where $(a_x, b_x, c_x, d_x, e_x, f_x)$ and $(a_y, b_y, c_y, d_y, e_y, f_y)$ are two sets of coefficients, which could be estimated if a set of positions on the complex curve is given in advance.

3.7 Integration

Summation is an operator which counts the number of instances inside a discrete set. An instance could refer to a presence of entity, a value of property, or a value of constraint, etc. The output from summation is a kind of knowledge. If a discrete set is replaced by a continuous set which is described by an equation, summation becomes integration, which also outputs knowledge.

In geometry, one interesting application of integration is to compute the area between a curve and an axis, or the enclosed area of a shape.

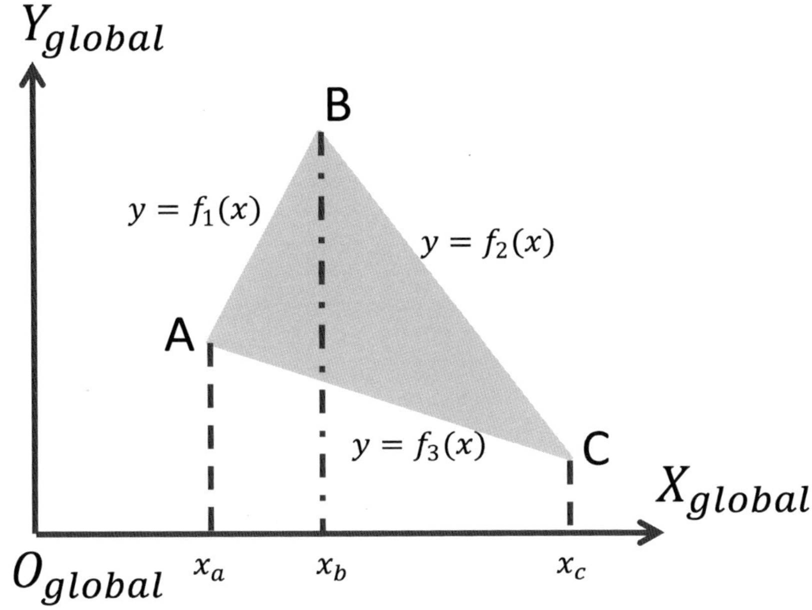

Fig. 3.5 Example of using integration to determine the enclosed area by a triangle ABC in a coordinate system $OXY|_{global}$.

For example, in Fig. 3.5, there is a triangle, in which its three edges are represented by three equations of line: $f_1(x)$, $f_2(x)$, and $f_3(x)$, respectively. Then, the area between line segment AB and the X axis is calculated as follows:

$$A_{AB} = \int_{x_a}^{x_b} f_1(x)dx \tag{3.13}$$

And, the enclosed area of the triangle could be calculated as follows:

$$A_T = \int_{x_a}^{x_b} f_1(x)dx + \int_{x_b}^{x_c} f_2(x)dx - \int_{x_a}^{x_c} f_3(x)dx \tag{3.14}$$

In control theory, the popular PID controller has three terms. One of them is an integrator of input error. If $e(t)$ is input error to a PID controller, control signal $c_i(t)$ from the integrator will be:

$$c_i(t) = \int_0^t e(x)dt \tag{3.15}$$

3.8 Derivative

Refer to Fig. 1.1. The primary knowledge are the properties and constraints of entities in the universe, while the composite knowledge include causalities, interactions, and dynamics. Most importantly, it is the aspect of dynamics which characterizes the fundamental truths of all systems in the universe. Moreover, regardless of the natures of systems in the universe, the dynamics of a system is always described by equations of directives. In mathematics, derivatives are the change rates of systems' properties and/or constraints over time. Therefore, derivatives consist of an important discovery of knowledge undertaken by human beings.

For example, a vehicle of mass m and frontal area A is traveling on a straight road to which axis OX is attached. Assume that the coefficient of air's resistive force is ρ. If the vehicle's engine delivers driving force $f(t)$, the dynamics of the vehicle along axis OX can be described by the following equation of derivatives:

$$m\frac{d^2x}{dt^2} + \frac{1}{2}\rho A\frac{dx}{dt} = f(t) \tag{3.16}$$

In geometry, the first-order derivatives of coordinates of a point on a curve define a vector which is tangent to the curve at the point. If a tangent vector at a point is parallel to one axis, this point is a local (or global), maximum (or minimum) along another axis. For example, Eq. 3.10 can be rewritten as follows:

$$y = y_0 + \sqrt{r^2 - (x - x_0)^2} \tag{3.17}$$

And, one can easily verify that first-order derivative $\frac{dy}{dx}$ is equal to zero at both peak point $(x_0, y_0 + r_0)$ and valley point $(x_0, y_0 - r_0)$.

In control theory, the design of PID controller may make use of a derivative term, which is as follows:

$$c_d(t) = \frac{de(t)}{dt} \tag{3.18}$$

where $e(t)$ is input error to PID controller and $c_d(t)$ is control signal produced by PID controller's derivative term.

3.9 Logic

Our universe contains a huge amount of entities, which act and react. The primary forces behind these interactions are constraints, which create knowledge such as causalities, memberships, or relationships, etc. For example, an entity could be a member of a population, a category, or a family, etc. Therefore, all entities in the universe are not isolated, but inter-related, and form massive networks of memberships, constraints, causalities and interactions as illustrated in Fig. 3.6. Inside these networks, there are not only chaotic orders (i.e. randomness and fuzziness), but also crisp orders (i.e. truthfulness). The existence of relationships among the truths in the universe has motivated human beings to invent the entity called as logic in the technical language.

It is important to take note that when human beings make use of languages to represent knowledge discovered from the universe, the results will be texts in human languages with extensions to both the technical language and programming

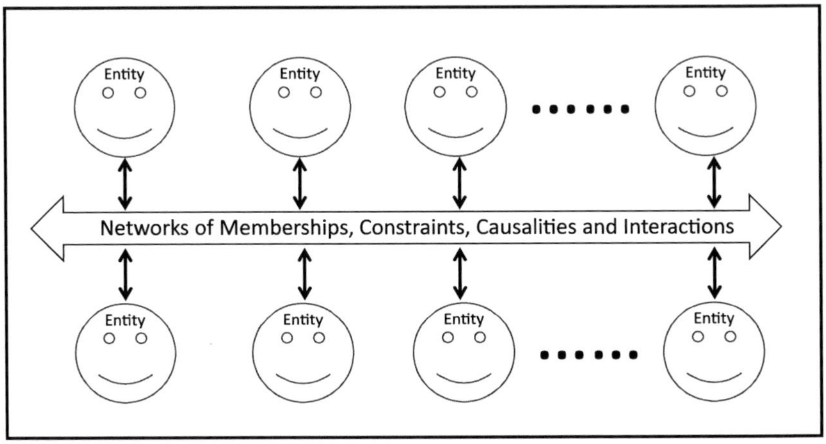

Fig. 3.6 The universe forms massive networks of memberships, constraints, causalities and interactions.

languages. If we analyze published texts found in libraries, we will notice that the knowledge described by all the sentences inside these texts are in logical order. Basically, the logic inside texts follows the logic of the universe. Therefore, texts themselves are standard representations of knowledge and their logic coming from the universe.

However, texts written in human languages are not directly readable by computers. Due to the proliferation of smart devices and smart machines, there is a strong need or demand for producing computer-readable version of representations of knowledge and their logic. In this way, we could achieve the goal of computerizing human intelligence. Therefore, the primary use of logic in the technical language is to translate human-readable versions of texts into computer-readable versions of expressions or equations. Both of them are the representations of a same set of knowledge in the universe.

In mathematics (i.e. the technical language), the term "logic" has many names such as the logic of predicate, the logic of proposition, the logic of relations, the logic of if-then, Boolean logic, etc. Here, we advocate the name of "the logic of truthfulness" for the purpose of unifying the different names attributed to the term "logic". More specifically, we define the logic of truthfulness as follows:

Definition 3.1. The logic of truthfulness is the study of the relational state of being true, or being false, among predicates, propositions, or relations.

In order to illustrate the power of logic, we show an example below, in which a human-readable version of text could be translated into a computer-readable representation with the use of the logic of truthfulness.

Example 3.3. There is a road which connects junction J_1 to junction J_2. The road is composed of two pairs of bridges as shown in Fig. 3.7. Each bridge is in one

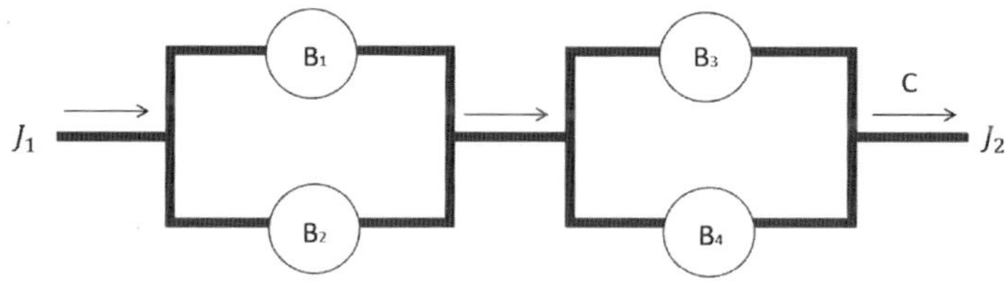

Fig. 3.7 A road connects junction J_1 to junction J_2. Between these two junctions, there are two pairs of bridges which are named as: (B_1, B_2) and (B_3, B_4). Each bridge is in one of these two states: "in use" or "in repair".

of these two states: "in use" or "in repair". If each pair of bridges has at most one being in repair, the road could be open for traffic. Otherwise, the road has to be closed. What will be the computer-readable version of the above description?

Solution: Let B_i denote the state of truth of bridge i where i is equal to 1, 2, 3, or 4. If bridge i is in use, $B_i = 1$. Otherwise, $B_i = 0$. And, let O denote the state of truth underlying the predicate that the road is open to traffic. Then, the equation which determines the logic of truthfulness for O is:

$$O = (B_1 \cup B_2) \cap (B_3 \cup B_4) \tag{3.19}$$

where $\cup$ is the logic operator of disjunction (i.e. "or" operation) and $\cap$ is the logic operator of conjunction (i.e. "and" operation). Now, let C denote the state of truth underlying the predicate that the road has been closed due to bridge's repair work. The equation which determines the logic of truthfulness for C will be:

$$C = \neg O \tag{3.20}$$

where $\neg$ is the logic operator of negation (i.e. "not" operation).

<><><><><>

The above example has introduced three important operators which allow us to formulate expressions or equations in logic. Two additional operators in the logic of truthfulness are:

- Operator of implication denoted by symbol $\rightarrow$, for example:
  If X, Y and Z are three Boolean variables, then:

$$Z = X \rightarrow Y = \neg X \cup Y \tag{3.21}$$

- Operator of exclusion denoted by symbol $\oplus$, for example:
  If X, Y and Z are three Boolean variables, then:

$$Z = X \oplus Y = (X \cup Y) \cap \neg(X \cap Y) \tag{3.22}$$

Assume again that X, Y and Z are three Boolean variables. Then, the following properties in the logic of truthfulness hold:

- Associativity of Conjunction, such as:

$$X \cap (Y \cap Z) = (X \cap Y) \cap Z \tag{3.23}$$

- Associativity of Disjunction, such as:

$$X \cup (Y \cup Z) = (X \cup Y) \cup Z \tag{3.24}$$

- Commutability of Conjunction, such as:

$$X \cap Y = Y \cap X \tag{3.25}$$

- Commutability of Disjunction, such as:

$$X \cup Y = Y \cup X \tag{3.26}$$

- Law of Identity, such as:

$$X = X \cup 0 \tag{3.27}$$

$$X = X \cap 1 \tag{3.28}$$

- Law of Absorption, such as:

$$X = X \cup (X \cap Y) \tag{3.29}$$

$$X = X \cap (X \cup Y) \tag{3.30}$$

- Law of De Morgan, such as:

$$(\neg X) \cup (\neg Y) = \neg(X \cap Y) \tag{3.31}$$

$$(\neg X) \cap (\neg Y) = \neg(X \cup Y) \tag{3.32}$$

In summary, the expressions or equations in logic are the computer-readable translations of truthful knowledge written in texts of human languages. For example, the operator of disjunction $\cup$ is useful for producing computer-readable versions of the truthful statements which are similar to the following ones:

- If either net force or net torque exists, an object will change its state of motion.
- If either force or displacement is zero, there is no work done.
- If either voltage or current is zero, there is no electric power.
- If either sensor or controller is faulty, a closed-loop control system will fail.

Similarly, the operator of conjunction $\cap$ is useful for producing computer-readable versions of the truthful statements which are similar to the following ones:

- If both legs of a humanoid robot are in normal condition, it can undertake biped walking.
- If both John and James have the same parents, they are sibling.
- If both net force and net torque are zero, an object will not change its state of motion.
- If both momentum and kinetic energy are conserved, it is an elastic collision.

3.10 Probability

Causalities produce crisp orders among the entities in the universe. However, competitions among the entities in the universe create chaotic orders such as the ones shown in Fig. 3.8.

Interestingly enough, the universe is full of competitions for success, competitions for failure, competitions for presence, competitions for absence, competitions for having awards, competitions for being famous, etc. The phenomenon of competitions among the entities (or predicates) in the universe has motivated human beings to discover the principle of probability which enriches the technical language.

In short, probability is about the study of competitions of truths (i.e. entities or predicates) in the universe. The outcome of competition is measured by:

- The computed frequency of occurrence of one truth, or a set of truths, in the past.
- The predicted chance of occurrence of one truth, or a set of truths, in the future.

Therefore, the definition of probability could be formulated as follows:

Definition 3.2. The probability of truths is about the study of their frequencies of occurrences in the past, or their chances of occurrences in future.

In terms of the technical language, if predicate (or proposition) E contains a set of multiple truths such as:

$$E = \{e_1, e_2, \ldots, e_n\} \tag{3.33}$$

then probability of E is denoted as $P(E)$ which is equal to the sum of the probabilities of all the individual truths, as follows:

$$P(E) = P(e_1) + P(e_2) + \cdots + P(e_n) \tag{3.34}$$

Refer to Fig. 3.9. The following example shows the way of computing the probability of a predicate:

Fig. 3.8 Competitions for the use of roads in traffic cause chaotic scenes (photo courtesy of free source in Internet with adaptation).

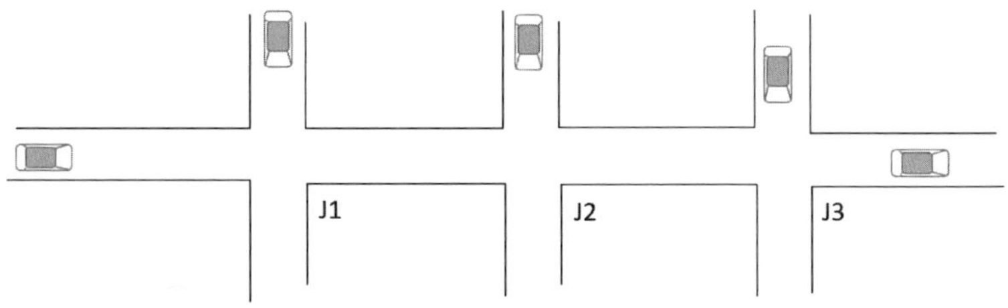

Fig. 3.9 A segment of road consists of three junctions.

Example 3.4. A segment of road which consists of three junctions (J_1, J_2, J_3). The traffic light at each junction could be in either red or green. Let A denote the predicate which states that at least two traffic lights are in green. What is the probability of truth set of predicate A?

Solution: Let R denote red traffic light and G denote green traffic light. Then, the states of lights at these three junctions will form this set of truths: $\{RRR, RRG, RGR, RGG, GRR, GRG, GGR, GGG\}$. And, the probabilities among all these eight truths are all equal. Hence, the probability of each truth is $\frac{1}{8}$. Predicate A states that at least two traffic lights are in green. Hence, predicate A contains the following truths: $\{RGG, GRG, GGR, GGG\}$. As a result, the probability of truth set of predicate A is:

$$P(A) = P(RGG) + P(GRG) + P(GGR) + P(GGG) \tag{3.35}$$

$$= \frac{1}{8} + \frac{1}{8} + \frac{1}{8} + \frac{1}{8} = \frac{1}{2} \tag{3.36}$$

<><><><><>

If there are two related predicates A and B, the following laws of probability hold:

- If truth set of A (or truth set of B) contains all the truths, then:

$$P(A) = 1 \tag{3.37}$$

(or $P(B) = 1$).

- If truth set of A is the complement of truth set of B, then:

$$P(A) + P(B) = 1 \tag{3.38}$$

- If truth set of A and truth set of B are mutually exclusive, then:

$$P(A \cup B) = P(A) + P(B) \tag{3.39}$$

- If truth set of A and truth set of B are not mutually exclusive, then:

$$P(A \cup B) = P(A) + P(B) - P(A \cap B) \tag{3.40}$$

- If truth set of A and truth set of B are mutually exclusive, then:

$$P(A \cap B) = P(A)P(B) \tag{3.41}$$

- If truth set of A and truth set of B are not mutually exclusive, then:

$$P(A \cap B) = P(A)P(B|A) \tag{3.42}$$

where $P(B|A)$ is conditional probability of truth set of B given truth set of A.
- If truth set of A and truth set of B are not mutually exclusive, then:

$$P(B|A) = \frac{P(A \cap B)}{P(A)} \tag{3.43}$$

This is the formulate for calculating conditional probability.

Refer to Fig. 3.9 again. The following example shows the way of computing the conditional probability of truth set of B given truth set of A:

Example 3.5. A segment of road which consists of three junctions (J_1, J_2, J_3). The traffic light at each junction could be in either red or green. Let A denote the predicate which states that at least junction 1's traffic light is in green, and B denote predicate B which states that at least junction 2's traffic light is in green. What is the conditional probability of truth set of B given truth set of A?

Solution: Let R denote red traffic light and G denote green traffic light. Then, the states of lights at these three junctions will form this set of truths: $\{RRR, RRG, RGR, RGG, GRR, GRG, GGR, GGG\}$. And, the probabilities among all these eight truths are all equal. Hence, the probability of each truth is $\frac{1}{8}$. Predicate A states that at least junction 1's traffic light is in green, truth set of A contains $\{GRR, GRG, GGR, GGG\}$. Clearly, the probability of truth set of predicate A is:

$$P(A) = \frac{4}{8} = \frac{1}{2} \tag{3.44}$$

Similarly, predicate B states that at least junction 2's traffic light is in green, truth set of B contains $\{RGR, GGR, RGG, GGG\}$. Truths $\{GRG, GGG\}$ support both predicate A and predicate B. Hence, we have:

$$P(A \cap B) = \frac{2}{8} \tag{3.45}$$

As a result, the conditional probability of truth set of B given truth set of A is:

$$P(B|A) = \frac{P(A \cap B)}{P(A)} = \frac{2}{8} \bullet \frac{2}{1} = \frac{1}{2} \tag{3.46}$$

<><><><><>

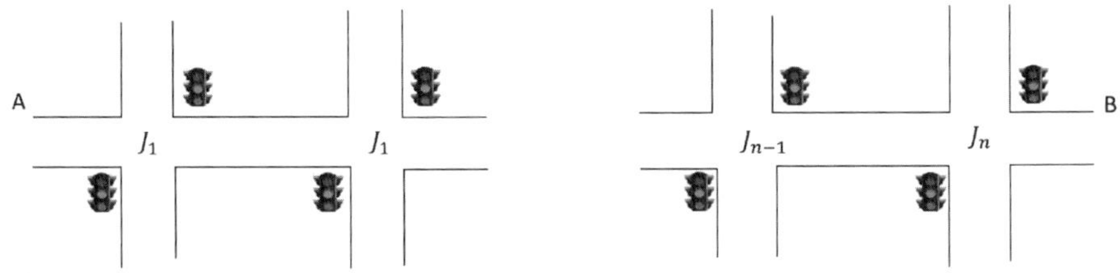

Fig. 3.10 A street from endpoint A to endpoint B has a series of n junctions, each of which has a traffic light system.

It is interesting to take note that with the use of probability in the technical language, knowledge accumulated in the past allow us to predict new knowledge in the future. For example, a team of robots could be deployed to work side by side along a production line of any product. Each robot is either in the state of "being in use" or in the state of "being under repair". And, the states of these robots are continuously recorded into a database over a certain period of years. Assume that the breakdown of one robot will cause the shutdown of the entire production line. Therefore, it is important to predict which robot will most likely cause the next shutdown of the production line. And, the solution for achieving such prediction is the Bayes' theorem or rule. A similar example below shows how the Bayes' theorem works in practice.

Example 3.6. Refer to Fig. 3.10. A street from endpoint A to endpoint B has a series of n junctions which are independent. Each junction has a traffic light system. If none of the traffic light systems is faulty, the street is in normal condition of traffic. Assume that at any one time, only one traffic light system will be faulty. When one traffic light system becomes faulty, the street will be in the status of "under repair". Also, assume that the data about the street's status and all the traffic light systems' working conditions have been recorded and stored into the street's database. Now, let S denote the predicate that the street is in the status of "under repair", and L_i denote the predicate that the traffic light system at junction $i (i = 1, 2, \ldots, n)$ is faulty. How to predict which junction's traffic light system will most likely cause the street to enter the status of "under repair" next time?

Solution: The occurrence of S is caused by $E_i, i = 1, 2, \ldots, n$. Since the junctions are all independent, the probability of S will be:

$$P(S) = P(S \text{ due to } E_1) + P(S \text{ due to } E_2) + \cdots + P(S \text{ due to } E_n)$$

or

$$P(S) = P(S \cap E_1) + P(S \cap E_2) + \cdots + P(S \cap E_n) \tag{3.47}$$

with

$$P(S \cap E_i) = P(S)P(E_i|S) = P(E_i)P(S|E_i) \tag{3.48}$$

where $i = 1, 2, \ldots, n$.

From Eq. 3.48, we have:

$$P(E_i|S) = \frac{P(E_i)P(S|E_i)}{P(S)}) \tag{3.49}$$

Eq. 3.49 is the equation of Bayes' theorem, which computes the probability of E_i causing the occurrence of S. By applying Bayes' theorem, the predicted junction, in which its traffic light system will most likely cause the street to enter the status of "under repair" next time, will be junction j such that:

$$P(E_j|S) = \max_{i=1}^{n} P(E_i|S) = \max_{i=1}^{n} \frac{P(E_i)P(S|E_i)}{P(S)} \tag{3.50}$$

where the terms $P(S)$, $P(E_i)$ and $P(S|E_i)$ could be determined with the use of data from the street's database accumulated over the past years.

<><><><><>

3.11 Possibility

So far, we understand that logic is a tool in the technical language, which was invented for the purpose of representing the truthfulness about causalities, memberships, or relationships. And, probability is also a tool in the technical language, which was invented for the purpose of representing the dynamics of the randomness of competing entities, competing predicates, or competing truths, in the universe, because these entities compete for occurrences. Now, we move on to study another tool in the technical language, which is called as possibility.

As we have already mentioned, among the entities in the universe, there are entities which have self-intelligence. For the case of human beings, there is a very interesting phenomenon in which our sensory systems do not give our brains crisp values, but fuzzy beliefs. For example, our skins do not tell our brains the exact value of temperatures in the surroundings, but fuzzy beliefs such as cold, warm, or hot, etc. Also, our eyes do not tell our brains the exact values of color vectors from a colorful scene or image, but fuzzy beliefs such as red, green, blue, yellow, purple, orange, etc.

This observation actually reveals a discovery of an important principle behind the transformation from quantities (i.e. values) in the physical world into quantities (i.e. words) in a conceptual world. Such principle is a powerful solution to the long-lasting problem of symbol-grounding in artificial intelligence.

For example, a car is moving at the speed of 60.0 km/h on a road. Due to the limitation of human being's vision system, a person's mind will not get the accurate measurement of 60.0 km/h. Instead, the person's mind may produce a set of beliefs such as {very slow speed, slow speed, normal speed, fast speed, very fast speed}, each of which has a certain level of confidence.

Here, a level of confidence in a belief is called as possibility for a belief to be true. And, we could find a function to measure the possibilities of beliefs. In this way, we obtain a set of pairs, each of which has a belief and its value of possibility. Such set is called as belief set, which could be represented as:

$$B = \{(x_i, w_i, \mu_{w_i}(x_i)), i = 1, 2, 3, ..., n\} \tag{3.51}$$

where B is a belief set, x_i is ith value of measurement of a physical quantity (e.g. a car's speed), w_i is the predicate of belief associated with x_i, and $\mu_{w_i}(x_i)$ is the possibility function which measures the level of confidence for value x_i to trigger belief w_i.

In general, one physical value from the universe could trigger multiple beliefs inside the mind of a human being. Inversely, a belief in a human being's mind (i.e. a conceptual world) could be supported by multiple values measured in the physical world. Therefore, the definition of possibility could be formulated as follows:

Definition 3.3. The possibility of beliefs is about the study of the levels of confidences for a quantity in the physical world to trigger a set of related beliefs in a conceptual world, as well as the study of the levels of confidences for a belief in a conceptual world to be supported by a set of related values in the physical world.

In practice, one popular choice for possibility function $\mu_w(x)$, which determines the level of confidence in supporting belief w by value x (i.e. the transformation from value x to belief w), is a Gaussian function as follows:

$$\mu_w(x) = \frac{1}{\sigma\sqrt{2\pi}} e^{-\frac{1}{2}(\frac{x-\bar{x}}{\sigma})^2} \tag{3.52}$$

where coefficients $(\bar{x}, \sigma)$ are the mean value and the standard deviation.

Given a set of training data x, the coefficients $(\bar{x}, \sigma)$ of belief w's possibility function are directly computed. For example, in Fig. 3.11, three circular areas indicate the sub-images in which three sets of image data are taken for the purpose of training a possibility function associated with the belief of "seeing a green color". From these three sub-images, the mean values $\{\bar{x}_{red}, \bar{x}_{green}, \bar{x}_{blue}\}$ and the standard deviations $\{\sigma_{red}, \sigma_{green}, \sigma_{blue}\}$ are directly computed. Then, given a sample pixel with values $\{(R_i, G_i, B_i), i = 1, 2, 3\}$ in the RGB color space, the level of confidence for the belief "seeing a green color" will be:

$$\mu_{green}(R_i, G_i, B_i) = \mu_{green}(R_i)\mu_{green}(G_i)\mu_{green}(B_i) \tag{3.53}$$

with

$$\mu_{green}(R_i) = \frac{1}{\sigma_{red}\sqrt{2\pi}} e^{-\frac{1}{2}\left(\frac{R_i-\bar{x}_{red}}{\sigma_{red}}\right)^2} \tag{3.54}$$

$$\mu_{green}(G_i) = \frac{1}{\sigma_{green}\sqrt{2\pi}} e^{-\frac{1}{2}\left(\frac{G_i-\bar{x}_{green}}{\sigma_{green}}\right)^2} \tag{3.55}$$

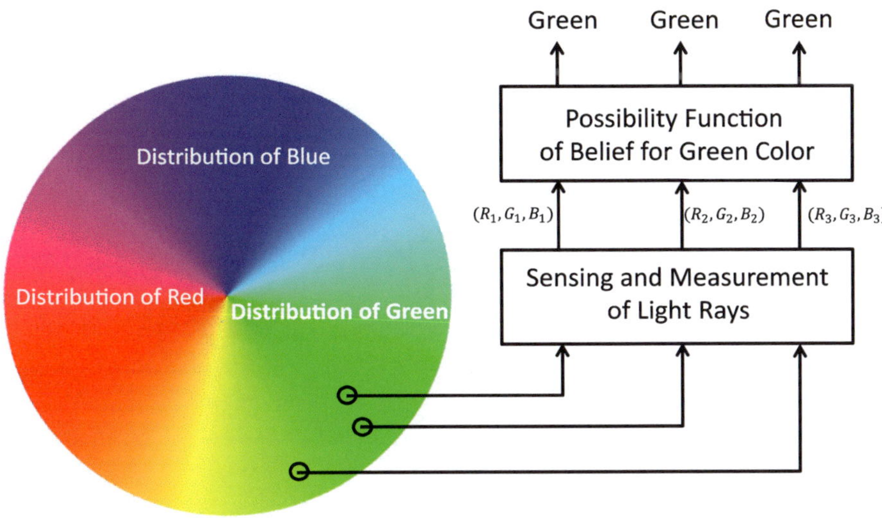

Fig. 3.11 A disk contains the distributions of three major colors, which are red, green and blue. The distributions are not uniform. Inside the distribution of green colors, three sub-images are taken as three sets of training data.

$$\mu_{green}(B_i) = \frac{1}{\sigma_{blue}\sqrt{2\pi}} e^{-\frac{1}{2}\left(\frac{B_i - \bar{x}_{blue}}{\sigma_{blue}}\right)^2} \tag{3.56}$$

In mathematics, $1 - \mu_w(x)$ defines the level of fuzziness of belief w given value x. And, human beings' minds are powerful enough to handle logic with fuzziness. For example, drivers of vehicles could easily learn driving skills such as:

- If a car is moving too close to the left landmark, rotate the steering wheel clockwise. (note: the term "too close" is a fuzzy belief)
- If a car is following a fast-moving car in front, increase the distance to the car in front. (note: the term "fast" is a fuzzy belief)
- If the weather is foggy, reduce a car's speed. (note: the term "foggy" is a fuzzy belief)
- If a car's payload is too heavy, increase the output of the car's engine. (note: the term "too heavy" is a fuzzy belief)

As long as a vehicle could transform sensory data into fuzzy beliefs, it should be able to imitate or learn human beings' actions which involve the use of logic rules with fuzzy beliefs inside. For example, if a car's gearbox has four gear ratios for driving-forward mode (i.e. not reversing mode), it could easily learn the following skills:

- If the car's speed is very slow, change gear ratio to 1.
- If the car's speed is slow, change gear ratio to 2.
- If the car's speed is normal, change gear ratio to 3.
- If the car's speed is fast, change gear ratio to 4.

Hence, possibility is one important tool in the technical language, which contributes to the achievement of artificial self-intelligence on future machines.

3.12 Fourier Transform

Human beings are living inside an ocean of signals. Among the various types of signals, two of them are of particular importance to human beings. One is the type of visual signals, which enable us to see. The other is the type of audio signals, which enable us to hear. In general, the time functions of signals are irregular. However, one of the greatest discoveries in mathematics is about the truth that time functions of any irregular signals are actually the combinations of sinusoids which are described by:

$$h_i(t) = a_i \sin(2\pi f_i t + \phi_i) \tag{3.57}$$

where (a_i, f_i, ϕ_i) are the amplitude, frequency and phase of sinusoid i.

The above truth could be explained in many different ways. From the viewpoint of robotics, one particular explanation is shown in Fig. 3.12. For example, in a coordinate system $OXYZ$, a sum of n sinusoids could be treated as a planar robotic arm with n revolute links in OXY plane, and one prismatic base link which is undergoing linear motion along OZ axis. The linear position of base link is θ, while angular position of each of the revolute link is $k_i\theta$ where $i = 1, 2, \ldots, n$. When θ varies, the end-tip of the robotic arm will draw an irregular curve in the coordinate system $OXYZ$. The projection of this curve onto OXY plane will also be an irregular curve. Such a curve actually resembles a time function of an irregular

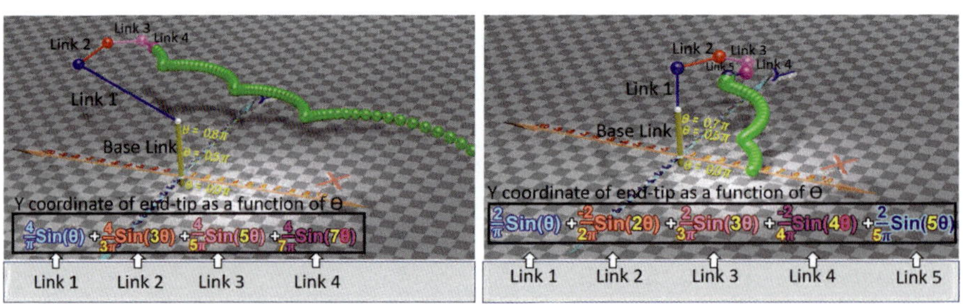

(a) A planar robot with 4+1 links (b) A planar robot with 5+1 links

Fig. 3.12 A combination of n sinusoids could be treated as a planar robotic arm with $n+1$ links, namely: one prismatic base link and n revolute links. The base link's linear position is θ. Each revolute link's length is amplitude a_i of corresponding sin function, where $i = 1, 2, \ldots, n$. Each revolute link's orientation is controlled by $k_i\theta$, where $\{k_i, i = 1, 2, \ldots, n\}$ are coefficients inside sin functions.

signal. In other words, any irregular signal could be decomposed into the sum of a particular set of sinusoids.

Then, what is the solution which allows us to determine the sinusoids, given a time function of any signal? The answer was discovered by Joseph Fourier, a French scientist. And, the solution is Fourier Transform, which was named after his name. A Fourier Transform pair $(G(f), g(t))$ could be written as:

$$G(f) = \int_{-\infty}^{\infty} g(t)e^{-j2\pi ft}dt \tag{3.58}$$

$$g(t) = \int_{-\infty}^{\infty} G(f)e^{j2\pi ft}df \tag{3.59}$$

where j is imaginary number $\sqrt{-1}$ and e is Euler number. Here, it is interesting to mention the great invention of Euler formulate, which is as follows:

$$e^{j\omega t} = \cos(\omega t) + j\sin(\omega t) \tag{3.60}$$

In order to verify the truthfulness of Fourier Transform, let signal $y(t)$ be the sum of the following three sinusoids:

$$y_1(t) = 10.0\sin(20\pi t) \tag{3.61}$$

$$y_2(t) = 30.0\sin(60\pi t) \tag{3.62}$$

$$y_3(t) = 20.0\sin(40\pi t) \tag{3.63}$$

Given time-domain function $y(t)$, the application of Fourier Transform yields $Y(f)$. The time-domain functions $\{y_1(t), y_2(t), y_3(t), y(t)\}$ and the amplitudes of frequency-domain function $Y(f)$ are shown in Fig. 3.13. It is obvious that $Y(f)$ contains three peaks at frequencies 10.0 Hz, 20.0 Hz and 30.0 Hz.

In artificial intelligence, Fourier Transform's most interesting applications are:

- Cognition and Recognition of Speech: We know that speech signals are represented by one-dimensional time functions. Hence, a speech signal is a combination of sinusoids. Some of these sinusoids carry more important features than the rest. By doing Fourier Transform, information about amplitudes, frequencies and phases of those important sinusoids could be analyzed in detail. The results of such study will enable the design of better systems for speech cognition and recognition.
- Synthesis of Speech from Text: People communicate knowledge with intensive use of speech and voice. Hence, text to speech synthesis is an important ability of an entity which aims to become self-intelligent. Fourier Transform allows machines to analyze the composition of sinusoids inside each phoneme of a human language. The outcome of such study could result in a database of artificial phonemes, which in return enables the synthesis of speech from texts.

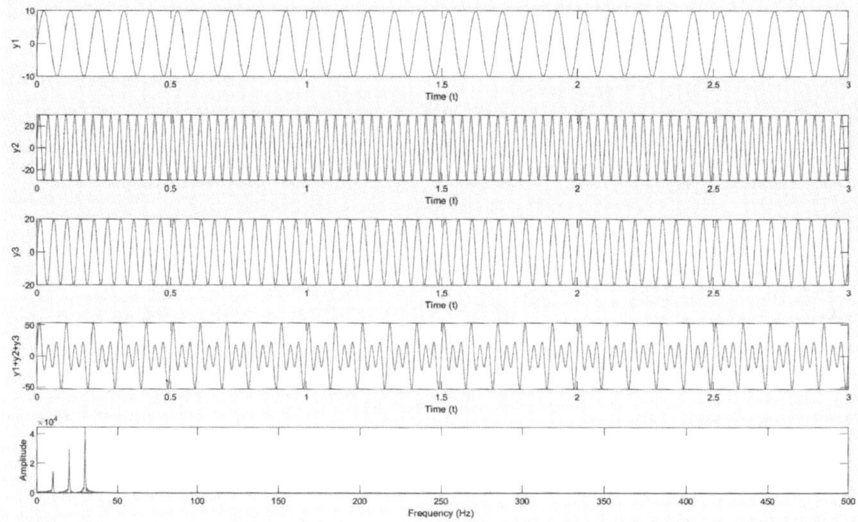

Fig. 3.13 A combination of three sinusoids $\{y_1(t), y_2(t), y_3(t)\}$ produces signal $y(t)$. The forward Fourier Transform yields $Y(f)$ which contains three peaks at the frequencies of these three sinusoids.

3.13 Laplace Transform

In general, an entity with self-intelligence will demonstrate two astonishing capabilities. One is decision-making, while the other is action-taking. Most importantly, the principle behind action-taking is closed-loop control. A closed-loop control system has three basic building blocks, which are systems under control, sensory feedback and controller. To a certain extent, we could say that a closed-loop control system is an entity which has reached the first level of self-intelligence. This is because the properly designed controller inside a closed-loop control system could correctly interpret its input (i.e. error signal) and could meaningfully produce its output (i.e. control signal). For example, a controller basically acts in the following manners:

- If error signal is positive, increase control signal.
- If error signal is negative, reduce control signal.
- If error signal is zero, do not output control signal.

Then, how to design the controller inside a closed-loop control system for it to reach the first level of self-intelligence? Interestingly enough, the indispensable tool to use for the modeling, design and analysis of closed-control systems in engineering is Laplace Transform, which was one of the greatest tools in the technical language and was initially invented by Pierre-Simon Laplace, a French scientist, for the primary purpose of solving differential equations.

In short, given a time-domain function $g(t)$, its Laplace Transform pair will be:

$$G(s) = \mathcal{L}[g(t)] = \int_0^\infty g(t)e^{-st}dt \tag{3.64}$$

$$g(t) = \mathcal{L}^{-1}[G(s)] = \frac{1}{2\pi j}\int_{c-j\infty}^{c+j\infty} G(s)e^{st}ds \tag{3.65}$$

where $j = \sqrt{-1}$, c is a constant, and $s = \sigma + j\omega$, which is called as a complex variable. In practice, it is better to visualize s as a two-dimensional vector in a S-plane, which has horizontal real axis σ and vertical imaginary axis $j\omega$.

Laplace Transform in Eq. 3.64 and Eq. 3.65 may appear to be complicated. However, in practice, Laplace Transform pairs could be calculated in advance and results are put into a look-up table. For example, the following list shows popular pairs of Laplace Transform:

$$\mathcal{L}[k] = k\frac{1}{s} \tag{3.66}$$

$$\mathcal{L}[ke^{-at}] = k\frac{1}{s+a} \tag{3.67}$$

$$\mathcal{L}[kt^n e^{-at}] = k\frac{1}{(s+a)^{n+1}} \tag{3.68}$$

And, two popular properties of Laplace Transform are:

$$\mathcal{L}\left[\frac{d}{dt}g(t)\right] = s\mathcal{L}[g(t)] \tag{3.69}$$

$$\mathcal{L}\left[\int g(t)dt\right] = \frac{\mathcal{L}[g(t)]}{s} \tag{3.70}$$

We believe that the most important application of Laplace Transform is for the modeling, analysis and design of closed-loop control systems. It goes without saying that without Laplace Transform, there will be no modern theory of control, which is so indispensable to the implementation of all intelligent machines.

In control theory, the dynamics of a module inside a closed-loop control system could always be described by a differential equation in time-domain denoted by t, or a polynomial equation in frequency-domain denoted by s. And, the ratio between the Laplace Transform of a module's output and the Laplace Transform of its input is called as a transfer function. In frequency domain denoted by s, a transfer function generally appears to be the ratio between a polynomial equation in numerator and a polynomial equation in denominator. With the use of transfer functions, any closed-loop control system could be depicted by a diagram of interconnected transfer functions, which is also called as a block-diagram in control theory.

In Fig. 3.14, the illustration of S-plane and block-diagram is shown. In general, a block diagram of any complexity could always be simplified into the standard one as shown in Fig. 3.14. The standard block-diagram of a closed-loop control system

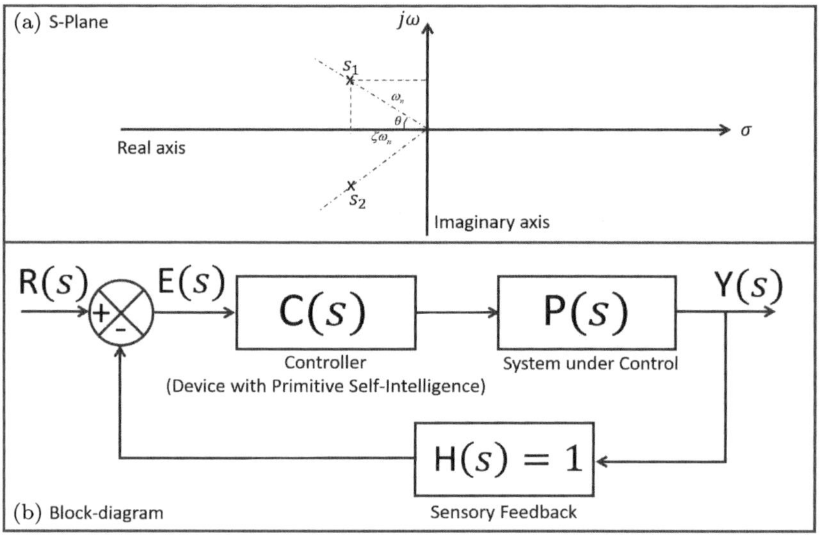

Fig. 3.14 Illustration of: (a) S-plane, and (b) Block diagram of closed-loop control system.

consists of three basic building blocks, which are system under control $P(s)$, sensory feedback $H(s)$ (note: its expected value is 1, which means 100% accuracy in sensing or feedback), and controller $C(s)$.

Refer to Fig. 3.14. The transfer function between output $Y(s)$ and input $R(s)$ is:

$$G_c(s) = \frac{Y(s)}{R(s)} = \frac{C(s)P(s)}{1 + H(s)C(s)P(s)} \tag{3.71}$$

It is important to take note that $R(s)$ is also named as "expected, or reference, output". This means that $Y(s)$ is expected to be equal to $R(s)$ during closed-loop control process. In other words, the objective of closed-loop control is to keep making actual output $Y(s)$ to be equal to expected output $R(s)$ as fast and accurate as possible.

In control theory, the polynomial at denominator in Eq. 3.71 is called as characteristic equation, which is: $1 + H(s)C(s)P(s) = 0$. The roots of the characteristic equation of a closed-loop control system are called as poles in S-plane. A pole could be a real value on σ axis, or a vector (i.e. a complex number). When a pole is a vector, its conjugate is at a location mirrored by σ axis. For example, in Fig. 3.14, poles (s_1, s_2) are a pair of conjugates.

The locations of poles in S-plane allow us to visually judge the behaviors of a closed-loop control system, for example:

- If there are poles in the right-half side of S-plane (i.e. value σ is positive), the closed-loop control system is unstable.

- If all the poles are located in the left-half side of S-plane (i.e. value σ is negative), the closed-loop control system is stable.
- If the pole having the largest value of σ is located on the negative side of σ axis, the closed-loop control system is dominated by the response of a dynamic system of order 1.
- If the conjugate poles having the largest value of σ are located in the left-half side of S-plane, the closed-loop control system is dominated by the response of a dynamic system of order 2.
- If the conjugate poles having the largest value of σ are located on $j\omega$ axis, the closed-loop control system will oscillate.

Moreover, the transfer function $G_c(s)$ of a closed-loop control system could be written into two parts, namely: the dominant transfer function and non-dominant transfer function, such as:

$$G_c(s) = \text{Dominant Transfer Function} + \text{Non-dominant Transfer Function}$$

In control theory, the most popular dominant transfer function is the one as follows:

$$G_d(s) = \frac{\omega_n^2}{s^2 + 2\zeta\omega_n s + \omega_n^2} \tag{3.72}$$

where ζ is called as damping ratio, and ω_n is called as natural frequency of a system under control. For example, in Fig. 3.14, ω_n is the distance from the origin to pole s_1, and $\zeta\omega_n$ is the real part of vector s_1 (i.e. s_1 is a complex number).

In practice, (ζ, ω_n) are two useful specifications for the design of a controller inside a closed-loop control system. And, the best value for ζ is 0.707 in general.

Refer to Fig. 3.14. The designer of a closed-loop control system has the freedom to chooses the controller's transfer function $C(s)$. The popular choices for $C(s)$ include:

- *P* Controller: $C(s) = k_1$, where k_1 is a control gain to be determined by designer.
- *PI* Controller: $C(s) = k_1 + \frac{k_2}{s}$, where (k_1, k_2) are two control gains to be determined by designer.
- *PD* Controller: $C(s) = k_1 + k_3 s$, where (k_1, k_3) are two control gains to be determined by designer.
- *PID* Controller: $C(s) = k_1 + \frac{k_2}{s} + k_3 s$, where (k_1, k_2, k_3) are three control gains to be determined by designer.
- Lead-lag Compensator or Controller: $C(s) = \frac{s+z_0}{s+p_0}$, where (z_0, p_0) are two real values to be determined by designer.

Refer to Fig. 3.14 again. Assume that the transfer function of the system under control is $\frac{1}{s+1}$ (i.e. $P(s) = \frac{1}{s+1}$). A designer chooses *PI* controller. Then, the transfer function of the *PI* controller will be $k_1 + \frac{k_2}{s}$ (i.e. $C(s) = k_1 + \frac{k_2}{s}$). By applying

```
clear all
s1_sigma = -1
s1_omega = 1
omega_n = sqrt(s1_sigma^2+s1_omega^2)
zeta = cos(atan(s1_omega/s1_sigma))
k1 = 2*zeta*omega_n - 1
k2 = omega_n^2
s = tf('s')
TF = (k1*s+k2)/(s^2+(1+k1)*s+k2)
step(TF)
```

(a) MATLAB Program

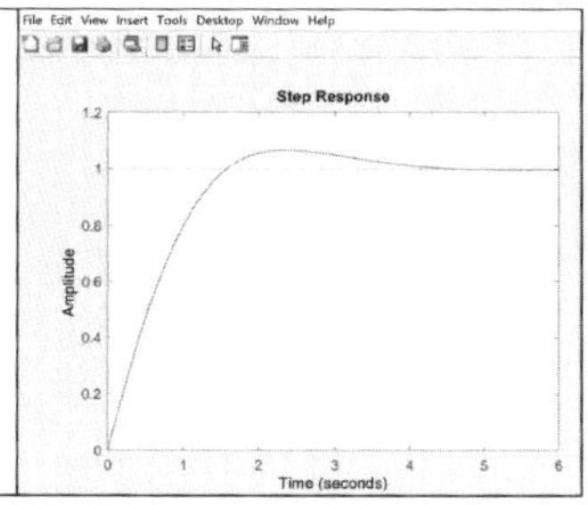

(b) Response to Unit Step Input

Fig. 3.15 Result of Simulation: (a) MATLAB program, and (b) Response to unit step input.

Eq. 3.71, the transfer function of the closed-loop control system will be:

$$G_c(s) = \frac{k_1 s + k_2}{s^2 + (1 + k_1)s + k_2} \tag{3.73}$$

After the examination of Eq. 3.72, the designer wants the polynomial in the denominator to be:

$$s^2 + (1 + k_1)s + k_2 = s^2 + 2\zeta\omega_n s + \omega_n^2 \tag{3.74}$$

and gives the design specification which is to have a pair of conjugate poles at $(1 \pm j)$. This implies that the following two conditions must hold:

$$\omega_n = \sqrt{1^2 + 1^2} \tag{3.75}$$

$$\zeta\omega_n = 1 \tag{3.76}$$

As a result, (ζ, ω_n) will be $(0.7071, 1.4142)$. Finally, from the equalities:

$$1 + k_1 = 2\zeta\omega_n \tag{3.77}$$

$$k_2 = \omega_n^2 \tag{3.78}$$

The two control gains (k_1, k_2) will be $(1.0, 2.0)$. After the above procedure of design, the transfer function of the closed-loop control system becomes:

$$G_c(s) = \frac{s + 2}{s^s + 2s + 2} \tag{3.79}$$

And, the simulated response to the input of unit step function is shown in Fig. 3.15.

3.14 Summary

In this chapter, we have studied the nature of the technical language which is an important extension of human languages. In particular, we have shown some popular examples of using the technical language to represent knowledge such as arithmetic operations, logic operations, probability, possibility, Fourier transform and Laplace transform.

3.15 Exercises

(1) A coordinate system is attached to the floor of a living room. You are at position A, while a service robot is at position B. Translate your command of "approach me by following a straight path" into a description using the technical language.

(2) In a library of robotics software, there are two typical motion functions, which are: MoveTo(arguments) and MoveWith(arguments). What should be the arguments passed to these two functions?

(3) How to represent the orientation of a rigid body in a three-dimensional space?

(4) In a OXY coordinate system, a curve contains a set of known points, such as $(x_i, y_i), i = 1, 2, 3, 4, 5$. What is the equation of polynomial, which represents this curve?

(5) In a OXY coordinate system, a curve starts at point $(0.0, 0.0)$ cm and ends at point $(100.0, 0.0)$ cm. And, the curve passes through point $(30.0, 30.0)$ cm and point $(70.0, 30.0)$ cm. If the curve is represented by an equation of polynomial of order 3. What is the area between the curve and OX axis?

(6) Given an equation such as $y = ax^3 + bx^2 + cx + d$. What are the locally maximum (or locally minimum) values of y?

(7) If you are a goal keeper of a football team, design a set of logic rules for you to follow during a football match.

(8) If you are a college student, design a set of logic rules for you to follow when you study a compulsory course.

(9) If you are a waiter in a restaurant, design a set of logic rules for you to follow during the working hours.

(10) Rains in a city could fall on any day of a week. What is the probability for Wednesday to be a rainy day?

(11) A robot looks at an image and believes that there is a red area inside the image. It then takes a set of five pixels from the red area. The values of the RGB colors of these five pixels are $\{(200, 20, 10), (230, 10, 5), (240, 10, 20), (220, 40, 30), (210, 20, 20)\}$. Design a possibility function which measures the level of confidence associated with the robot's belief.

(12) Given a time domain function such as $f(t) = 5.0 \sin(10.0\pi t + 0.1\pi) + 25.0 \sin(50.0\pi t + 0.2\pi) + 50.0 \sin(100.0\pi t + 0.3\pi)$. What will be the output after the Fourier Transform?

(13) Refer to Fig. 3.14. Assume that transfer function $P(s)$ of the system under control is $\frac{1}{s^2+2}$. Design a controller so that the transfer function of the closed-loop control system will have a pair of conjugate poles at $0.0 \pm j0.7$.

(14) Refer to Fig. 3.14. Assume that transfer function $P(s)$ of the system under control is $\frac{1}{s^2}$. Design a controller so that the transfer function of the closed-loop control system will have a pair of conjugate poles at $1.2 \pm j1.5$.

Chapter 4

Basics of Programming Language

4.1 Introduction

Interactions among human beings for activities, such as knowledge discovery, knowledge representation, knowledge transfer, knowledge communication and knowledge application, heavily depend on the use of human languages. The addition of the technical language to the family of human languages allow us to undertake interactions with deep knowledge, which are better described by equations.

With the ambition of augmenting both the physical and mental capabilities, human beings have invented, and are inventing, various types of machines. In particular, it is one of the greatest hopes in mankind that machines of future could master human languages as well. In this way, interactions among human beings and machines with self-intelligence will become possible. And, the journey toward this goal depends on the achievement of two objectives, namely:

- The interactions between human beings' minds and machines' minds.
- The interactions between machines' minds and machines' bodies.

Interestingly enough, the programming languages invented by human beings could fulfill these two objectives to some extent already. Hence, the purpose of this chapter is to highlight some important features, which are common to all existing high-level programming languages.

4.2 Use of Programming Language

The first functional analogue computer was invented by Sir William Thomson in 1872. The first functional digital computer was invented by Konrad Zuse under the name of $Z3$ in 1941. The first generation of personal computers from IBM and Apple Inc. entered the commercial market in 1975–1976. Today's computers are much more sophisticated than their first generation.

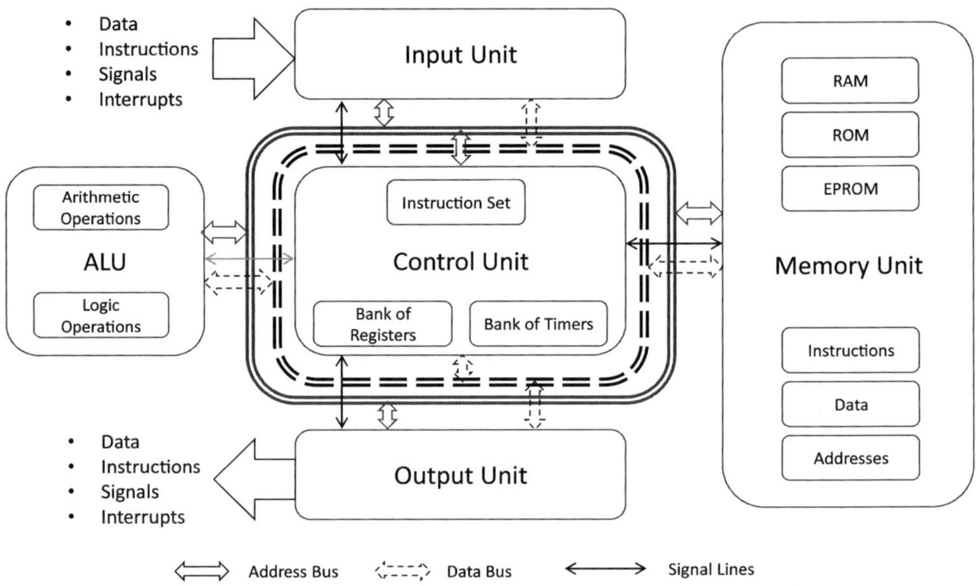

Fig. 4.1 Generic blueprint of digital computers.

A generic blueprint of digital computer is shown in Fig. 4.1, in which the key modules are:

- ALU: It is in charge of arithmetic operations as well as logic operations. This unit plays a similar role of a human brain's cerebrum.
- Control Unit: It is the mini-mind of a digital computer which provides a set of innate instructions (i.e. instruction set which is also called as assembly language) and is in charge of controlling the executions of programs in assembly language. This unit plays a similar role of a human brain's inner mid-brain.
- I/O Unit: It consists of input unit as well as output unit. The input unit is in charge of receiving data from both sensors and communication networks. The output unit is to control the transmission of data from a digital computer to motors, actuators, displays, and communication networks. This unit plays a similar role of a human brain's cerebellum. It is important to take note that a machine with self-intelligence must be equipped with a digital computer as its brain. In this way, the I/O unit of the digital computer will enable the acquisition of sensory data from the machine's body to the machine's mind. The I/O unit of the digital computer will also enable the control of action executions by a machine's body.
- Memory Unit: It provides memory spaces for the storage of data and programs. The memory unit could be split into two separate sub-units such that one unit is for the storage of programs while the other unit is dedicated to the storage of data. This unit (including ROM which stands for read-only memory) plays a similar role of a human brain's long-term memory, while the memory devices

named as registers and random access memory (RAM) in electronics play the role of a human brain's short-term memory.

The emergence of digital computers have propelled two streams of associated developments. One is the invention and development of high-level programming languages, while the other is the invention and development of artificial intelligence. For example, the first high-level programming language with compiler is FORTRAN, which was invented in 1954 and was released in 1956 by John Backus's team at IBM. Since then, many other high-level programming languages have appeared in the commercial market. These include Lisp, COBOL, BASIC, Pascal, Prolog, C, C++, C#, Java, Python, R, and MATLAB.

In 1956, John McCarthy initiated the Dartmouth Summer Workshop of Research Project on Artificial Intelligence. This event has marked the birth of Artificial Intelligence as a new discipline in science and engineering. Also, the terminology "Artificial Intelligence (AI)" was actually due to John McCarthy's pick. It is worth noting that the focus of Dartmouth workshop's discussions was on "thinking machines", which was believed to be able to be developed on top of digital computers and with the help of programming languages. However, this workshop did not realistically spell out the fundamental issues underlying Artificial Intelligence.

In summary, the use of programming languages is versatile. And, two primary objectives are:

- To translate algorithms of data processing and logic inference into programs (i.e. a list of organized instructions) in high-level programming languages, which will further be compiled into programs in assembly languages, so that a digital computer could automatically perform data-centric computing and logic-centric reasoning intended by human beings.
- To implement the physical principles behind human beings' minds into a software system, which will enable machines to demonstrate human-like self-intelligence or artificial intelligence.

4.3 Data Structure

The basic memory unit inside a digital computer is one byte, which has eight bits. Each bit carries the value of either 1 or 0. Therefore, a set of bytes could represent any value or variable in the technical language. For example, a constant or variable of "integer" type could be implemented with one byte (8 bits), two bytes (16 bits), four bytes (32 bits) or eight bytes (64 bits). And, a value or variable of "real" type could be implemented with four bytes (32 bits in single floating-point precision, also called "float" type) or eight bytes (64 bits in double floating-point precision, also called "double" type). As for the representation of characters, each ASCII code requires seven bits. Hence, one byte is good enough to represent one character.

In practice, there may be a huge amount of homogeneous values or variables, which are of the same type. In this case, a programming language generally supports the representation of homogeneous data in the form of vectors and matrices. For example, a vector is a list of values or variables, while a two-dimensional matrix is a list of vectors. Following the same logic of thinking, a three-dimensional matrix is a list of two dimensional matrices, while a four-dimensional matrix is a list of three dimensional matrices, and so on. If $M_{1\times n_1}$ denotes a vector of dimension $1 \times n_1$, $M_{n_1\times n_2}$ denotes a matrix of dimension $n_1 \times n_2$, $M_{n_1\times n_2\times\cdots\times n_k}$ denotes a matrix of dimension $n_1 \times n_2 \times \cdots \times n_k$, then multiple sets of homogeneous values or variables $\{v_i, i = 1, 2, \ldots, n_1\}$ could be represented by one of the following representations:

$$M_{1\times n_1} = \{v_1, v_2, \ldots, v_{n_1}\}|_{1\times n_1} \tag{4.1}$$

$$M_{n_1\times n_2} = \{M_{1\times n_1}, M_{1\times n_1}, \ldots, M_{1\times n_1}\}|_{1\times n_2} \tag{4.2}$$

$$M_{n_1\times\cdots\times n_k} = \{M_{n_1\times\cdots\times n_{k-1}}, \ldots, M_{n_1\times\cdots\times n_{k-1}}\}|_{1\times n_k} \tag{4.3}$$

In the universe, all entities are different. Therefore, a programming language must offer the flexibility of representing a huge amount of heterogeneous values or variables. A common practice is to group heterogeneous data into a single unit which is called as a data structure. And, there should be no limit to the complexity of a data structure which could be created in a programming language.

In C programming language, a data structure could be implemented with the instruction "typedef struct{declarations} name". For example, the following data structure could be easily created:

- Data Structure for Two-Dimensional Point:

```
typedef struct  {
                double  x;
                double  y;
                }       Point2D ;
```

- Data Structure for Three-Dimensional Point:

```
typedef struct  {
                double  x;
                double  y;
                double  z;
                }       Point3D ;
```

- Data Structure for Color:

```
typedef struct  {
                int     red;
                int     green;
                int     blue;
                double  hue
                double  saturation
                double  brightness
                }       Color ;
```

- Data Structure for Color Image:

```
typedef struct  {
                Color   pixmap[HEIGHT][WIDTH];
                int     row;
                int     column;
                }       ColorImage ;
```

where HEIGHT $\times$ WIDTH defines a fixed size of a color image.

4.4 Instruction Set

The central pierce of a digital computer is the control unit which includes instruction set (i.e. assembly language) and an innate mechanism (i.e. a built-in mini-mind) which is in charge of controlling the interpretation and execution of programs in assembly languages. The programs in assembly language could be either the results of the compilation of programs written in high-level programming languages, or the results of direct programming in assembly language by human programmers.

In general, the instruction set of a digital computer's assembly language will include instructions for:

- Data Transfer: The purpose is to store, load, read or write digital numbers into designated locations of memories or registers.
- Arithmetic Operation of Addition: The purpose is to add two operands together and to transfer the result into a designated location of memory or register. The result could trigger the action of setting flags (e.g. zero output, negative output, overflow, etc) in a status register.
- Arithmetic Operation of Multiplication: The purpose is to multiply two operands together and to transfer the result into a designated location of memory or register. The result could trigger the action of setting flags (e.g. zero output, negative output, overflow, etc) in a status register.
- Arithmetic Operation of Division: The purpose is to divide one operand by another operand and to transfer the result into a designated location of memory or register.The result could trigger the action of setting flags (e.g. zero output, negative output, overflow, etc) in a status register.
- Boolean Logic Operations: The purpose is to undertake Boolean operations, such as AND, OR, or XOR (i.e. exclusive OR), between the corresponding bits of two operands.

As for high-level programming languages, instruction sets are also available. And, their syntax look similar to the syntax of the technical language (i.e. mathematics). For example, in C programming language, the following expressions are valid:

- Arithmetic Operations: If (x, y, z) are three variables, then we can do calculations such as: $z = x + y$ (i.e. addition), $z = x * y$ (i.e. multiplication), or $z = x/y$ (i.e. division).
- Logic Operations: If (x, y, z) are three variables, then we can do calculations such as: $z = x\&\&y$ (i.e. x AND y), $z = x\|y$ (i.e. x OR y), or $z = (!x\&\&y)\|(x\&\&!y)$ (i.e. x XOR y).

In addition, high-level programming languages provide instructions which control the flows of computations. For example, computations could be repeated over many times (i.e. multiple loops) with a same set of arithmetic and/or logic operations. In C programming language, instructions "for(condition){*computations*}" and "while(condition){*computations*}" are widely used for implementation of computational loops.

On the other hand, an algorithm behind a program may require multiple paths of computations, which are mutually exclusive. In this situation, a popular solution is to use instruction "switch" as follows:

```
switch(input)    {
case (match 1)   :   {flow 1 of computations}; break;
case (match 2)   :   {flow 2 of computations}; break;
...              :   ...
case (match n)   :   {flow n of computations}; break ;
                 }
```

where flow i of computations will be executed if input is equal to match i, $i = 1, 2, \ldots, n$.

4.5 Function

Programming is a set of mental activities, which translate algorithms represented in the technical language into a list of instructions for digital computers to execute. Many algorithms are generic enough for them to be reusable and shareable. As a result, all high-level programming languages offer the mechanism of creating reusable and shareable functions, which are implementations of reusable and shareable algorithms. For example, functions such as sin(argument), cos(argument), and *sqrt*(argument) are reusable and shareable functions.

In C programming language, the syntax for creating a function is "(output data) function name (input arguments) {instructions;}", which gives programmers all the freedom and flexibility to design functions as standard components inside a software. In this way, a software is considered to be a creative work, which could be treated as a product by design.

For example, the software inside an intelligent vehicle (note: for description purpose, it is named as our vehicle) may include functions such as:

- Functions for Vehicle Control:
  - IsEngineOn(): It outputs the status of our vehicle's engine. The status is a Boolean value (i.e. true or false).
  - TurnOnEngine(): It takes action of switching on our vehicle's engine. It may return a Boolean value to indicate whether the action is successful or fails.
  - ShutdownEngine(): It takes action of turning off our vehicle's engine. It may return a Boolean value to indicate whether the action is successful or fails.
  - SpeedUpWith(): It increases our vehicle's speed by an amount of value given by input argument of the function.
  - SlowDownWith(): It decreases our vehicle's speed by an amount of value given by input argument of the function.
  - SteerClockwiseWith(): It increases our vehicle's steering wheel clockwise by an amount of value given by input argument of the function.
  - SteerCounterClockwiseWith(): It increases our vehicle's steering wheel counterclockwise by an amount of value given by input argument of the function.
  - OpenDoor(): It opens one of our vehicle's doors. The identity of the door to be opened is given by the function's input argument.
  - LockDoor(): It locks one of our vehicle's doors. The identity of the door to be locked is given by the function's input argument.
  - CloseDoor(): It closes one of our vehicle's doors. The identity of the door to be closed is given by the function's input argument.
  - OpenDoorWindow(): It opens the window of a door of our vehicle. The identity of the door is given by the function's input argument.
  - CloseDoorWindow(): It closes the window of a door of our vehicle. The identity of the door is given by the function's input argument.
  - OnWindScreenWiper(): It turns on the wiper of a windscreen of our vehicle. The identity of the windscreen (either front one or back one) is given by the function's input argument.
  - OffWindScreenWiper(): It turns off the wiper of a windscreen of our vehicle. The identity of the windscreen (either front one or back one) is given by the function's input argument.
  - OnAirConditioner(): It turns on our vehicle's air conditioner. This function may take an input argument which indicates the desired temperature inside the vehicle.
  - OffAirConditioner(): It turns off our vehicle's air conditioner.
  - OnHeadLight(): It switches on a head light of our vehicle. The identity of the head light is given by the function's input argument.
  - OffHeadLight(): It switches off a head light of our vehicle. The identity of the head light is given by the function's input argument.

- OnInteriorLight(): It switches on an interior light of our vehicle. The identity of the interior light is given by the function's input argument.
- OffInteriorLight(): It switches off an interior light of our vehicle. The identity of the interior light is given by the function's input argument.

• Functions for Vehicle Diagnosis:

- CheckEngineOil(): If the engine of our vehicle is equipped with a sensor which measures the engine's oil level, this function reads the sensor's output and compares the reading with reference value (e.g. the lower limit). This function returns one value which indicates that the oil level is too low, low, or normal.
- CheckVehicleFuelTank(): It reads the output from the sensor which measures the fuel level inside our vehicle's fuel tank. This function returns the reading.
- CheckEngineCoolant(): It reads the output from the sensor which measures the coolant liquid's level inside our vehicle's coolant top-up tank. This function returns the reading.
- CheckTyrePressure(): It reads the output from the four sensors which measure the pressures of our vehicle's four tyres. This function returns all the readings.
- CheckTyreTreadDepth(): If our vehicle is equipped with a set of four sensors, each of which measures the tread depth of a wheel's tyre on our vehicle, this function reads the output from these four sensors. This function returns all the readings.
- CheckBatteryStatus(): If our vehicle is equipped with a sensor which measures the output voltages of in-vehicle batteries, this function reads the output from the sensor, and compares the readings with reference values in order to determine whether a battery is to be replaced or not. This function returns the status of in-vehicle batteries. And, the status of a battery could be either "in good condition" or "in weak condition".

• Functions for Vehicle's Vision System:

- IsFrontViewCameraReady(): It verifies whether the camera system for our vehicle's frontal view is in working condition or not. This functions returns a Boolean value.
- IsRearViewCameraReady(): It verifies whether the camera system for our vehicle's rear view is in working condition or not. This functions returns a Boolean value.
- IsLeftSideViewCameraReady(): It verifies whether the camera system for our vehicle's left-side view is in working condition or not. This functions returns a Boolean value.
- IsRightSideViewCameraReady(): It verifies whether the camera system for our vehicle's right-side view is in working condition or not. This functions returns a Boolean value.

- DistanceToLeftSideLandmark(): It detects the landmark at the left-side of our vehicle, and measures the distance from our vehicle to the detected landmark. This function returns the value of the measured distance.
- DistanceToRightSideLandmark(): It detects the landmark at the right-side of our vehicle, and measures the distance from our vehicle to the detected landmark. This function returns the value of the measured distance.
- DistanceToFrontOccupiedSpace(): It identifies the empty space or free-moving space in front of our vehicle, and measures the longitudinal depth of the empty or free-moving space. Beyond this distance, it is occupied space. This function returns the value of the measured distance.
- DistanceToRearOccupiedSpace(): It identifies the empty space behind a vehicle, and measures the longitudinal depth of the empty space. Beyond this distance, it is occupied space. This function returns the value of the measured distance.
- DistanceToTrafficJunction(): It identifies stop line in front of our vehicle, and measures the distance between our vehicle and the identified stop line. This function returns the value of the measured distance.
- GetDirectionToFollow(): It detects the landmarks (i.e. left-side landmark and/or right-side landmark) in front of our vehicle, and estimates the parameters (i.e. radius and center) of a circular path which approximates the detected landmarks. This function outputs the value of tangent angle to the estimated circular path.
- GetSpeedToChange(): Based on the available information about the distance to frontal occupied space, it determines the amount of speed value to increase or reduce by our vehicle. This function returns the determined value.
- GetDistanceToStopLine(): This function returns the value of the distance between our vehicle and a frontal stop line.
- GetTrafficLightColor(): It detects the presence of a traffic light system in front of our vehicle, and identifies the actual color of the light which is turned on. This function returns a string of characters, which represents the name of identified color.
- StopAtStopLine(): It implements vision-guided motion control for the scenario of driving our vehicle forward until it stops in front of the stop-line of a junction. This function returns a Boolean value, which indicates whether the motion control is successful or fails.
- IsItSafeToChangeLane(): It analyzes the traffic situation on the lane which is next to either the left-side or the right-side of our vehicle. The function returns a value which indicates one of these three situations: safe to change to the lane at the left-side, safe to change to the lane at the right-side, or not safe to do lane-changing.
- DoChangeLane(): It implements vision-guided motion control for the maneuver of lane-changing by our vehicle. This function returns a Boolean value, which indicates whether the motion control is successful or fails.

- AbortChangeLane(): It implements vision-guided motion control for the maneuver of aborting the attempt of doing lane-changing by our vehicle, and controlling our vehicle back to its initial lane. This function returns a Boolean value, which indicates whether the motion control is successful or fails.
- IsItSafeToOvertakeFrontVehicle(): It analyzes the driving behaviors of the frontal vehicle, which could be "consistently slow", "consistently fast enough", or "consistently faster". If the frontal vehicle is consistently slow, this function returns the Boolean value "true". Otherwise, it returns the Boolean value "false".
- DoOvertakeFrontVehicle(): It implements vision-guided motion control for the maneuver of overtaking a frontal vehicle. This function returns a Boolean value, which indicates whether the motion control is successful or fails.
- AbortOvertakeFrontVehicle(): It implements vision-guided motion control for the maneuver of aborting the attempt of overtaking the frontal vehicle, and controlling our vehicle back to its initial lane. This function returns a Boolean value, which indicates whether the motion control is successful or fails.
- ReadRoadSignBoardText(): It implements image understanding algorithms and extracts texts which are relevant to the travel by our vehicle. The texts may include information such as speed limit, road condition, next destination in front, next carpark, next petrol stations, etc. This function returns a data structure which contains a list of tagged texts.
- GetSpeedLimitFromRoad(): It verifies the latest information from road's signboard, and returns the value of speed limit, if any.
- IsCarparkNearby(): It verifies the latest information from road's signboard, and returns the Boolean value which indicates the presence or absence of nearby carpark.
- IsParkingLotAvailable(): When our vehicles enters a carpark, it could obtain the information about empty parking lots in two ways. One way is to get such information from radio communication with the parking management system. The second way is to reply on our vehicle's vision systems to detect the presence and location of empty parking lots. This function returns a Boolean value which indicates the presence or absence of an empty parking lot.
- DoVerticalParking(): It implements vision-guided motion control for the maneuver of driving our vehicle into an empty parking lot which is perpendicular to the road next to it.
- DoParallelParking(): It implements vision-guided motion control for the maneuver of driving our vehicle into an empty parking lot which is parallel to the road next to it.
- TravelToDestination(): It implements the highest level of vision-guided autonomous driving from current location to the location of an intended destination which is given by the input argument to the function. The locations are represented in the format of GPS data (note: GPS stands for Global

Positioning System). This function returns a Boolean value which indicates whether the destination has been reached or not.

4.6 Library

Usually, a high-level programming language offers an IDE, which stands for Integrated Development Environment. Refer to Fig. 4.2. An IDE includes graphic-user interfaces and a list of menus. With an IDE, a programmer will be at ease in composing functions, programs, and layouts of user interfaces. Also, an IDE allows the direct launch of a compiler, which translates a programmer-composed texts of functions, programs and layouts of user interfaces, into computer-readable codes. If the use of shared functions is invoked, an IDE will link the computer-readable codes of shared functions with the programmer's codes obtained after the compilation.

With the advance of software development, there are millions and millions of shared functions in the market and public domain. Therefore, it will be more efficient to classify and assembly shared functions into individual entities which are called as libraries, instead of using a bundle of individually shared functions separately.

In short, a library in a high-level programming language is a collection of shared functions in one specific domain of the technical language. For example, a high-level programming language may include libraries such as: library of basic mathematics functions, library of graphic functions, library of image processing functions, library of vision computing functions, library of speech processing functions, library of statistics functions, library of control theory functions, library of matrix operation functions, library of logic operation functions, library of dynamics functions, library of kinematics functions, library of motion planning functions, etc.

In practice, a library includes the compiled codes of shared functions, instead of their human-readable versions written in a high-level programming language. This is to protect the intellectual right of the owner(s) of these shared functions.

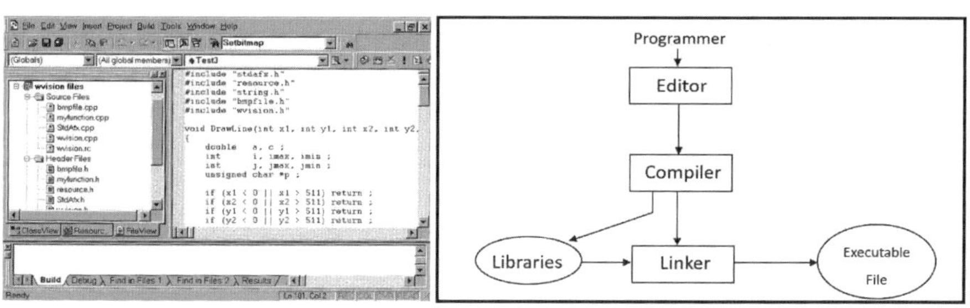

Example of an IDE (a) Workflow of an IDE (b)

Fig. 4.2 Integrated Development Environment (IDE) includes graphic-user interface, a list of menu, built-in editor, access to non-proprietary compiler, and access to non-proprietary linker.

4.7 Dynamic Class

The physical world consists of a huge amount of entities in the universe. Among these entities, some of them are very proactive. The actions and reactions among these proactive entities produce many meaningful scenes on earth, which are full of interactions, events, episodes and stories. Most importantly, the descriptions of the knowledge underlying the dynamics of the physical world result in the so called conceptual worlds.

It is interesting to take note that a conceptual world represented by a human language with extensions to both the technical language and programming languages also contains proactive entities which are called as dynamic classes written in object-oriented programming languages such as C++, Java, or C#, etc. In general, a dynamic class consists of both a set of static properties and constraints, and a set of dynamic functions which are members of the dynamic class. Since a class could be a container of a set of reusable and shareable functions, classes are alternatives of libraries in an object-oriented programming language.

For example, in C++ programming language, an intelligent vehicle's software system may include the following types of classes (note: members inside each class are illustrative entries, which are not exhaustive):

- Class of Vehicle's Driving Engine:

```
Class Engine  {
public:                  // accessible by users
              double     power;
              double     speed;
              char       status;
private:                 // accessible by owner only
              char       *manufacturer ;
              char       *serial_number ;
public:                  // accessible by users
              char       EstablishEngineConnection();
              char       TurnOn();
              char       TurnOff();
              char       SetPowerTo(double power) ;
              char       SetSpeedTo(double speed) ;
              char       GetCurrentStatus() ;
              }          ;
```

- Class of Vehicle's Power-Steering System:

```
Class Steering  {
public:                   // accessible by users
                double    steering_angle;
                char      status ;
private:                  // accessible by owner only
                char      *manufacturer ;
                char      *serial_number ;
public:                   // accessible by users
                char      EstablishSteeringConnection() ;
                char      SetSteeringAngleTo() ;
                char      GetCurrentStatus() ;
                }         ;
```

- Class of In-vchicle Camera:

```
Class Camera    {
public:                   // accessible by users
                short     image_width;
                short     image_height;
                byte      *image_matrix;
                char      status ;
private:                  // accessible by owner only
                char      *manufacturer ;
                char      *serial_number ;
public:                   // accessible by users
                char      EstablishCameraConnection() ;
                char      TurnOn();
                char      TurnOff();
                char      AllocateImageMemory() ;
                char      FreeImageMemory() ;
                char      CaptureOneImage() ;
                byte      *GetOneImage() ;
                char      GetCurrentStatus() ;
                }         ;
```

- Class of In-vehicle Vision:

```
Class Vision  {
public:                 // accessible by users
              Camera    *camera_list;
              short     number_camera;
              char      status;
private:                // accessible by owner only
              char      *chief_developer;
              char      *owner_name ;
public:                 // accessible by users
              char      AllocateCamera() ;
              char      TurnOn();
              char      TurnOff();
              char      LandmarkDetection() ;
              char      RoadSignBoardDetection() ;
              char      ParkingLotDetection() ;
              char      ObstacleDetection() ;
              char      GetCurrentStatus() ;
              }         ;
```

- Class of Vehicle's External Lighting System:

```
Class ExternalLight  {
public:                        // accessible by users
                     double    light_intensity;
                     double    light_orientation[2];
                     char      status ;
private:                       // accessible by owner only
                     char      *manufacturer ;
                     char      *serial_number ;
public:                        // accessible by users
                     char      EstablishLightConnection() ;
                     char      TurnOn();
                     char      TurnOff();
                     char      SetIntensityTo() ;
                     char      SetOrientationTo() ;
                     char      GetCurrentStatus() ;
                     }         ;
```

- Class of Vehicle's Interior Lighting System:

```
Class InteriorLight   {
public:                         // accessible by users
                      double    light_intensity;
                      char      status ;
private:                        // accessible by owner only
                      char      *manufacturer ;
                      char      *serial_number ;
public:                         // accessible by users
                      char      EstablishLightConnection() ;
                      char      TurnOn();
                      char      TurnOff();
                      char      SetIntensityTo() ;
                      char      GetCurrentStatus() ;
                      }         ;
```

- Class of In-vehicle Wheel:

```
Class Wheel   {
public:                 // accessible by users
              double    diameter;
              double    width ;
              double    tread_depth ;
              double    pressure ;
              char      status ;
private:                // accessible by owner only
              char      *manufacturer ;
              char      *serial_number ;
public:                 // accessible by users
              char      EstablishWheelConnection() ;
              char      TurnOnBreak();
              char      TurnOffBreak();
              char      UpdatePressureReading() ;
              char      GetTyrePressure() ;
              char      UpdateTreadDepthReading() ;
              char      GetTreadDepth() ;
              char      GetCurrentStatus() ;
              }         ;
```

- Class of In-vehicle Door:

```
Class Door    {
public:                 // accessible by users
              char      window_status ;
              char      door_status ;
private:                // accessible by owner only
              char      *manufacturer ;
              char      *serial_number ;
public:                 // accessible by users
              char      EstablishDoorConnection() ;
              char      OpenDoor();
              char      CloseDoor();
              char      LockDoor() ;
              char      OpenWindow() ;
              char      CloseWindow() ;
              char      LockWindow() ;
              char      GetCurrentDoorStatus() ;
              char      GetCurrentWindowStatus() ;
              }         ;
```

- Class of Vehicle:

```
Class Vehicle   {
public:                               // accessible by users
                Engine                engine ;
                Steering              steering ;
                Vision                vision ;
                ExternalLight         *external_light ;
                InteriorLight         *interior_light ;
                Wheel                 wheels[4] ;
                Door                  doors[5] ;
private:                              // accessible by owner only
                char                  *manufacturer ;
                char                  *serial_number ;
public:                               // accessible by users
                char                  AllocateExternalLight() ;
                char                  AllocateInteriorLight() ;
                char                  TurnOnEngine();
                char                  TurnOffEngine() ;
                char                  SpeedUp() ;
                char                  SlowDown() ;
```

```
char    FollowRoad() ;
char    FollowFrontalVehicle() ;
char    OvertakeFrontalVehicle() ;
char    DriveThroughJunction() ;
char    ParkVehicle() ;
char    ReadRoadSignBoard() ;
char    TravelToDestination() ;
}       ;
```

4.8 Programmable Script

Human languages with extensions to both the technical language as well as programming languages allow the translation of knowledge in the minds of human beings into computer-readable codes. In this way, machines become fully controllable by human beings. Also, such translation from human ways of thinking in terms of knowledge into computer-readable codes makes it possible for us to achieve the objective of computerizing human intelligence.

However, the topmost objective in artificial intelligence is to design machines with self-intelligence. And, the first step toward this direction is to make machines to master human languages. In this way, human languages will enable interactions not only among human beings, but also between human beings and machines at a cognitive level, which is in contrast to today's interactions with machines at a signal level.

Interestingly enough, the invention of programmable scripting languages such as Javascript, MATLAB, Python, etc, helps us to narrow down the gap between a human language and a programming language. Actually, a script is a list of phrase-like commands, which could be interpreted and executed by a digital computer.

For example, a robotic arm equipped with a gripper could perform "pick-up" action at location A and "place-down" action at location B, by interpreting and executing the following script:

	Command Set	Interpretation and Execution
1	Define Point H	Assign coordinates to H
2	Define Point A	Assign coordinates to A
3	Define Point B	Assign coordinates to B
4	Define Point C	Assign coordinates to C
5	Define Point D	Assign coordinates to D
6	Open Gripper	Open the gripper
7	Move To C	Move robotic arm to location C
8	Move To A	Move robotic arm to location A
9	Close Gripper	Grasp an object at location A

10	Move To C	Move the object to location C
11	Move To D	Move the object to location D
12	Move To B	Move the object to location B
13	Open Gripper	Place the object at location B
14	Move To D	Lift up the robotic arm to location D
15	Move To H	Move robotic arm to its home position
15	Close Gripper	Put robotic arm into standby state

4.9 Summary

In this chapter, we have studied the basics of programming languages which are also important extensions of human languages. In particular, we have explained the use of data structures, functions, and libraries to represent prepositional knowledge as well as procedural knowledge. In particular, we have outlined the important concept of dynamic class, which is very useful for representation of dynamic entities in the physical world.

4.10 Exercises

(1) Discuss about the difference between computerizing human intelligence and designing machine's self-intelligence.
(2) Design a data structure for books in a library.
(3) Design a data structure for items in a shop.
(4) Design a data structure for clients of a bank.
(5) Design a list of functions which process a list of matrices.
(6) Design a list of functions which process a string of characters.
(7) Design a set of classes for the purpose of recording and managing information about students in an university.
(8) Design a set of classes for the purpose of programming and controlling a robotic arm in industry.
(9) Design a set of scripting commands for programming an intelligent vehicle.

Part 2
Knowledge Acquisition

Knowledge acquisition is one of the most difficult tasks on earth. On average, a human being spends more than ten years in attending schools, colleges, and universities, for the sole purpose of learning knowledge and skills. In general, there are three ways for a human being to acquire knowledge. The first way is supervised-learning, in which interactions between teachers and learners play a major role. The second way is self-study, in which learners could gain knowledge from reading texts and doing experiments without too much interventions of teachers. The third way is autonomous perception or discovery, in which learners' minds transform received signals into cognitive states of knowing the meanings (i.e. knowledge) behind these signals. Actually, it is the mental capability behind the transformations from signals to meanings that lays the foundation of self-intelligence. Without such capability, an entity will not be able to acquire knowledge at all. Hence, this part of the book focuses on the study of the principles behind the transformation from signals to meanings. Due to the huge amount of knowledge in the universe, it is not possible to cover the entire spectrum of knowledge acquisition. Instead, the chapters in this part of the book limit the scope to two important scenarios of knowledge acquisition. The first one is the acquisition of knowledge from speech signals and texts. The second one is the acquisition of knowledge from visual signals and images. These two mental capabilities are relevant to the design of self-intelligence for future vehicles or drivers in the form humanoid robots.

Chapter 5

Knowledge Acquisition from Text

5.1 Introduction

There are two important scenarios of knowledge acquisition. One scenario is to acquire new knowledge from visual signals coming from the physical world. In this case, the transformation from visual signals to the cognitive states of knowing the meanings (i.e. knowledge) behind these visual signals plays a vital role. The second scenario is to acquire knowledge which have already been discovered or invented by human beings. In this case, these knowledge have already been represented in the form of texts, or inside the minds of human experts, which could be transformed into sound signals or speeches. Therefore, the acquisition of knowledge from speeches or printed versions of texts plays a vital role.

In addition, speeches are important sound signals which enable interactions among human beings. And, such interactions primarily occur at cognitive level in the form of voices, spoken languages, and written languages. However, today's interactions between human beings and machines are still occurring at signal level with the use of signal emitters such as lights, acoustic vibrations, mouse clicks, keyboards, keypads, touch screens, buttons, switches, and other types of haptic devices, etc. Therefore, it is a challenge to make machines of future to master human languages. In this way, interactions between human beings and machines could also occur at cognitive level.

5.2 Representation of Human Language

The invention of human languages has been motivated by the need or demand of representing knowledge related to entities in the physical world. The outcome of knowledge representation is a collection of documents in the form of texts. Interestingly enough, a text includes words, phrases and sentences, etc. These entities form a conceptual world, and could also interact according to their grammatical properties and constraints. Therefore, there is a correspondence between the interactions among the entities in the physical world and the interactions among the entities in a conceptual world (i.e. a specific human language).

Although any specific human language could be considered as a combination of vocabulary and grammar, it is necessary to study language representation from the perspective of knowledge or meanings encoded by human languages.

5.2.1 *Representation of Physical Meanings*

In the physical world, the meanings or knowledge about the physical entities (e.g. objects) in a scene are organized at two levels, namely: (a) the elementary meanings and (b) the composite meanings.

5.2.1.1 *Elementary Meanings in Physical World*

Given an object in the physical world, this object should have its own properties in various domains such as physics, chemistry, and geometry. In addition, an object could exert constraints onto other objects. Also, an object may be subject to constraints imposed by other objects.

Therefore, the simplest way of organizing the elementary meanings in the physical world is to construct a list of objects, each of which possesses: (a) properties and (b) constraints.

In practice, an object in the physical world could be treated as a dynamic class in a high-level programming language. For example, in C++ programming language, the dynamic class of an object could include:

- Properties:
  - List of properties of material, which could be another class.
  - List of properties of mechanics, which could be another class.
  - List of properties of electronics, which could be another class.
  - List of properties of biology, which could be another class.
  - List of properties of geometry, which could be another class.
  - List of properties of appearance, which could be another class.
  - List of properties of time, which could be another class.
  - List of properties of space, which could be another class.
- Constraints:
  - List of constraints of material, which could be another class.
  - List of constraints of mechanics, which could be another class.
  - List of constraints of electronics, which could be another class.
  - List of constraints of biology, which could be another class.
  - List of constraints of geometry, which could be another class.
  - List of constraints of appearance, which could be another class.
  - List of constraints of time, which could be another class.
  - List of constraints of space, which could be another class.

- Motions:
  - List of motions related to perception, which could be another class.
  - List of motions related to locomotion, which could be another class.
  - List of motions related to manipulation, which could be another class.
  - List of motions related to grasping, which could be another class.
- A pointer to a class in a conceptual world, which represents the object.

5.2.1.2 *Composite Meanings in Physical World*

In the physical world, composite meanings refer to the properties and constraints of a group of objects, which act, interact, and evolve over time as well as in space.

Refer to Fig. 5.1. The composite meanings in the physical world could be organized into a hierarchically-layered networks such as:

- Interaction Network:
  When two objects could exert valid constraints upon one another, these two objects are engaged into an interaction (e.g. a person is riding a bicycle). If a

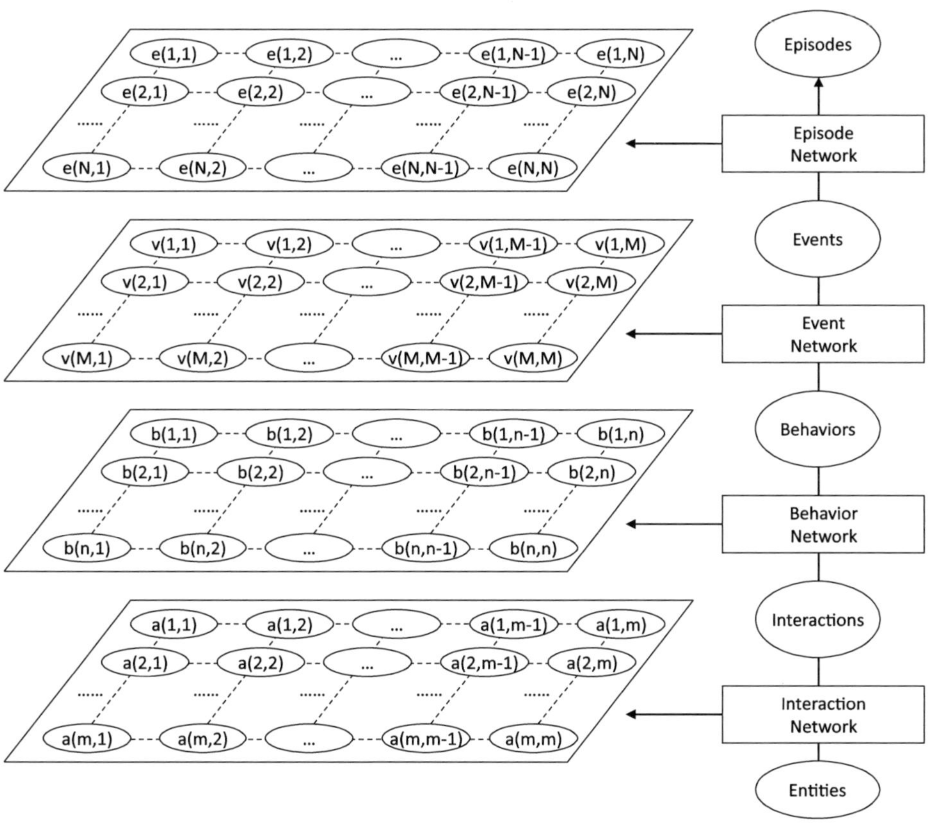

Fig. 5.1 Hierarchically-layered networks for the representation of the physical composite meanings, in which the input at the bottom are objects, and the final output are episodes (i.e. scenarios).

vehicle or robot has cognised a list of objects, which are denoted by

$$\{O_i,\ i = 1, 2, \ldots, m\},$$

then all the possible pairings of interacting objects will form a network, which is called as being an *interaction network.* In an interaction network, the input are objects and the output are interactions, as shown in Fig. 5.1. Because of the variety of constraints, an interaction network will consist of a multi-layered sub-networks, each of which can be described by a connection matrix such as

$$\{a_{j,k},\ j = 1, 2, \ldots, m;\ k = 1, 2, \ldots, m\}$$

where $a_{j,k}$ takes a value of either 1 or 0. For instance, if object O_j could interact with object O_k under a particular constraint, $a_{j,k}$ equals to 1. Otherwise, it equals to zero.

- Behavior Network:
When an interaction evolves in time for the purpose of achieving a specific outcome (e.g. a person is riding a bicycle from point A to point B), the result will be called as a *behavior.* Alternatively, when a set of interactions occur in space and in time according to a specific sequence, the result is also called as a *behavior.* For instance, a sequence of actions that a vehicle or robot executes in order to open a door is a typical behavior. Therefore, if a vehicle or robot has cognized a set of interactions, which are denoted by

$$\{A_i,\ i = 1, 2, \ldots, n\},$$

then all the possible evolutions and orderly grouping of these interactions in time and in space will create what is called as a *behavior network,* in which the input are interactions and the output are behaviors, as shown in Fig. 5.1. Mathematically, a behavior network could be described by a connection matrix such as

$$\{b_{j,k},\ j = 1, 2, \ldots, n;\ k = 1, 2, \ldots, n\}$$

where $b_{j,k}$ takes a value of either 1 or 0. For instance, if interactions A_j and A_k could form two adjacent interactions within a sequence, $b_{j,k}$ equals to 1. Otherwise, it equals to zero.

- Event Network:
When at least two behaviors cooperatively evolve in time and in space for the purpose of achieving a specific outcome (e.g., two persons playing a tennis match), the result will be what is called as an *event.* If a vehicle or robot has cognized a set of behaviors, which are denoted by

$$\{B_i,\ i = 1, 2, \ldots, M\},$$

then all the possible cooperations among these behaviors will result in what is called as an *event network,* in which the input are behaviors and the output are events, as shown in Fig. 5.1. Mathematically, an event network could be described by a connection matrix such as

$$\{v_{j,k},\ j = 1, 2, \ldots, M;\ k = 1, 2, \ldots, M\}$$

where $v_{j,k}$ takes a value of either 1 or 0. For instance, if behaviors B_j and B_k could cooperate within an event, $v_{j,k}$ equals to 1. Otherwise, it equals to zero.

- Episode Network:
  By definition, an episode refers to a sequence of events occurring in time as well as in space (e.g., two teams playing a football match). Therefore, if a vehicle or robot has cognised a set of events, which are denoted by

$$\{V_i,\ i = 1, 2, \ldots, N\}$$

then all the possible arrangements of these events in time and in space will result in what is called as an *episode network*, in which the input are events and the output are episodes, as shown in Fig. 5.1. Mathematically, an episode network could be described by a connection matrix such as

$$\{e_{j,k},\ j = 1, 2, \ldots, N;\ k = 1, 2, \ldots, N\}$$

where $e_{j,k}$ takes a value of either 1 or 0. For instance, if events V_j and V_k could occur sequentially, or in parallel, within an episode, $e_{j,k}$ equals to 1. Otherwise, it equals to zero.

5.2.2 *Representation of Conceptual Meanings*

In a conceptual world, the meanings about the conceptual entities (e.g. words) are also organized at two levels, namely: (a) the elementary meanings, and (b) the composite meanings.

5.2.2.1 *Elementary Meanings in Conceptual World*

The basic constituent entities in the conceptual world are the words in natural languages. Interestingly, each word in a natural language has its own properties and constraints. In English, the lexical properties include noun, pronoun, verb, adjective, adverb, preposition, and auxiliary verb, etc. And, the syntactical constraints include noun-verb, adjective-noun, verb-preposition, and preposition-noun, etc.

In addition, a word may have a reference to a physical entity such as an object, a behavior, an event, or an episode. For instance, the word "car" refers to a physical car. And, the word "fly" refers to a physical behavior of an object evolving in the sky. Therefore, when a vehicle or robot has cognized a set of words, the simplest way is to organize the meanings of these words into a list.

Similarly, a word in a conceptual world could also be treated as a dynamic class in a high-level programming language. For example, in C++ programming language, the dynamic class of a word could include:

- Properties:
  - List of graphic models for writings in different languages, which could be another class.
  - List of acoustic models for utterances in different languages, which could be another class.
  - List of variations due to tense, which could be another class.
  - List of lexical types, which could be another class.
- Constraints:
  - List of pre-determiners (e.g. many cars), which could be another class.
  - List of pre-modifiers (e.g. sport car), which could be another class.
  - List of post-determiners (e.g. left door of car), which could be another class.
  - List of post-modifiers (e.g. rules of traffic), which could be another class.
- A pointer to a class in the physical world, to which the word refers.

5.2.2.2 *Composite Meanings in Conceptual World*

In a conceptual world, composite meanings refer to the properties and constraints of a group of words, which act and interact according to syntactical constraints, or *logics*.

Refer to Fig. 5.2. The composite meanings in a conceptual world could be organized into a hierarchically-layered networks such as:

- Phrase Network:
  When two words could be put together according to a valid syntactical constraint, the result will be called as a *phrase*. For instance, we can form "noun-verb" phrases in English (e.g. "She smiles"). Therefore, if a vehicle or robot has cognized a set of words, which are denoted by

  $$\{W_i,\ i = 1, 2, \ldots, u\},$$

  then all the possible pairings of these words will form what is called as a *phrase network*, in which the input are words and the output are phrases, as shown in Fig. 5.2. Because a word may impose, or be subject to, multiple constraints, a phrase network will consist of a multi-layered sub-networks, each of which could be described by a connection matrix such as

  $$\{p_{j,k},\ j = 1, 2, \ldots, u;\ k = 1, 2, \ldots, u\}$$

  where $p_{j,k}$ takes a value of either 1 or 0. For instance, if words W_j and W_k could be put together to form a phrase, $p_{j,k}$ equals to 1. Otherwise, it equals to zero. Because a word may correspond to a physical entity, a phrase may describe an interaction in the physical world.
- Sentence Network:
  At a higher level, when a set of phrases could be put together according to valid logics (e.g. the if-then rule), the result will be what is called as a *sentence* in a

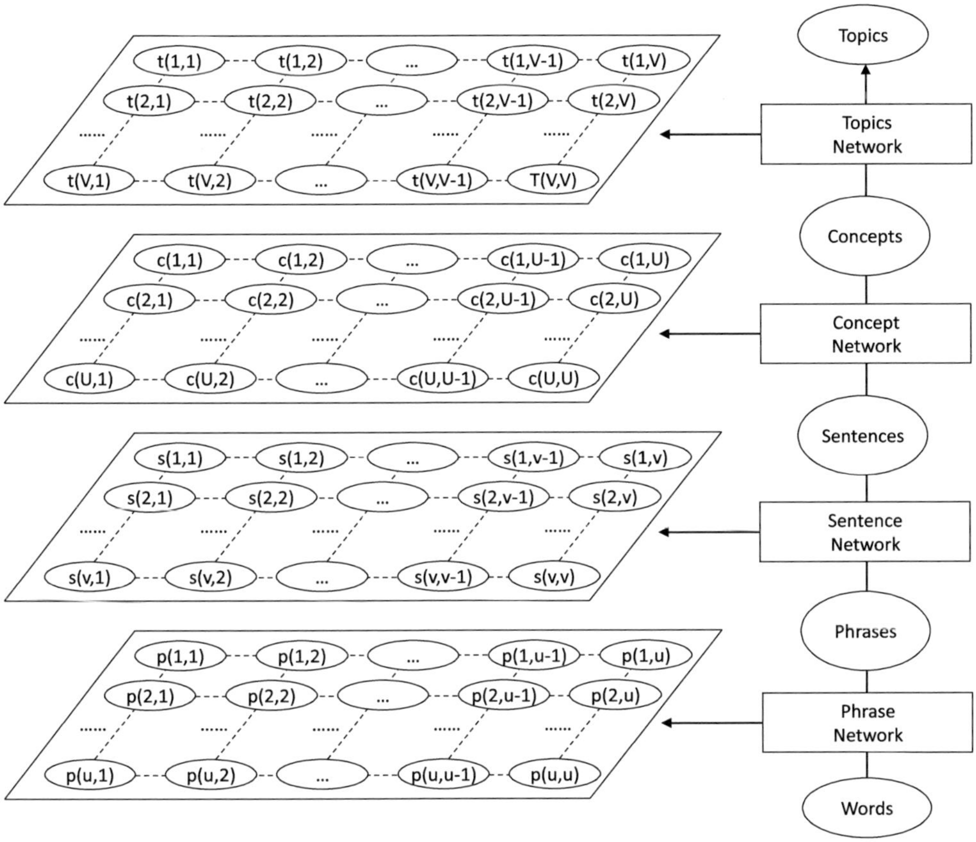

Fig. 5.2 Hierarchically-layered networks for the representation of the conceptual composite meanings, in which the input at the bottom are words, and the final output are topics.

conceptual world. If a vehicle or robot has cognized a set of phrases, which are denoted by

$$\{P_i, \; i = 1, 2, \ldots, v\},$$

then all the meaningful groupings among these phrases will result in what is called a *sentence network*, in which the input are phrases and the output are sentences, as shown in Fig. 5.2. Mathematically, a sentence network could be described by a connection matrix such as

$$\{s_{j,k}, \; j = 1, 2, \ldots, v; \; k = 1, 2, \ldots, v\}$$

where $s_{j,k}$ takes a value of either 1 or 0. For instance, if phrases P_j and P_k could be put together to be part of a sentence, $s_{j,k}$ equals to 1. Otherwise, it equals to zero. Because a phrase may correspond to an interaction in the physical world, a sentence normally describes a behavior in the physical world.

- Concept Network:
  In the conceptual world, a concept is a conceptual entity, which describes an event (including a situation, or a phenomenon) in the physical world. Therefore, an event in the physical world may create the occurrences of many sentences, each of which describes a permissible behavior occurring within an event. As a result, when such sentences are grouped together, the result will be what is called as a *concept* in a conceptual world. And, if a vehicle or robot is able to cognize a set of sentences, which are denoted by

$$\{S_i,\ i = 1, 2, \ldots, U\},$$

  then all the meaningful groupings among these sentences will form what is called as a *concept network*, in which the input are sentences and the output are concepts, as shown in Fig. 5.2. Mathematically, a concept network could be described by a connection matrix such as

$$\{c_{j,k},\ j = 1, 2, \ldots, U;\ k = 1, 2, \ldots, U\}$$

  where $c_{j,k}$ takes a value of either 1 or 0. For instance, if sentences S_j and S_k describe two behaviors occurring within a same event, $c_{j,k}$ equals to 1. Otherwise, it equals to zero.
- Topic Network:
  In a conceptual world, a topic is a conceptual entity, which describes an episode (including a process) in the physical world. And, an episode in the physical world may create the occurrences of many concepts, each of which describes a permissible event occurring within the episode. Therefore, when the concepts related to a same episode are grouped together, the result will be called as a *topic* in a conceptual world. And, if a vehicle or robot has cognized a set of concepts, which are denoted by

$$\{C_i,\ i = 1, 2, \ldots, V\},$$

  then all the meaningful groupings among these concepts will form a network, which is called as a *topic network*. In a topic network, the input are concepts and the output are topics, as shown in Fig. 5.2. Mathematically, a topic network could be described by a connection matrix such as

$$\{t_{j,k},\ j = 1, 2, \ldots, V;\ k = 1, 2, \ldots, V\}$$

  where $t_{j,k}$ takes a value of either 1 or 0. For instance, if concepts C_j and C_k describe two events within a same episode, $t_{j,k}$ equals to 1. Otherwise, it equals to zero.

5.2.3 *Representation of Concept-physical Meanings*

Clearly, any entity in the physical world has both the physical meanings as well as the conceptual meanings. As a result, a human language must possess the power of

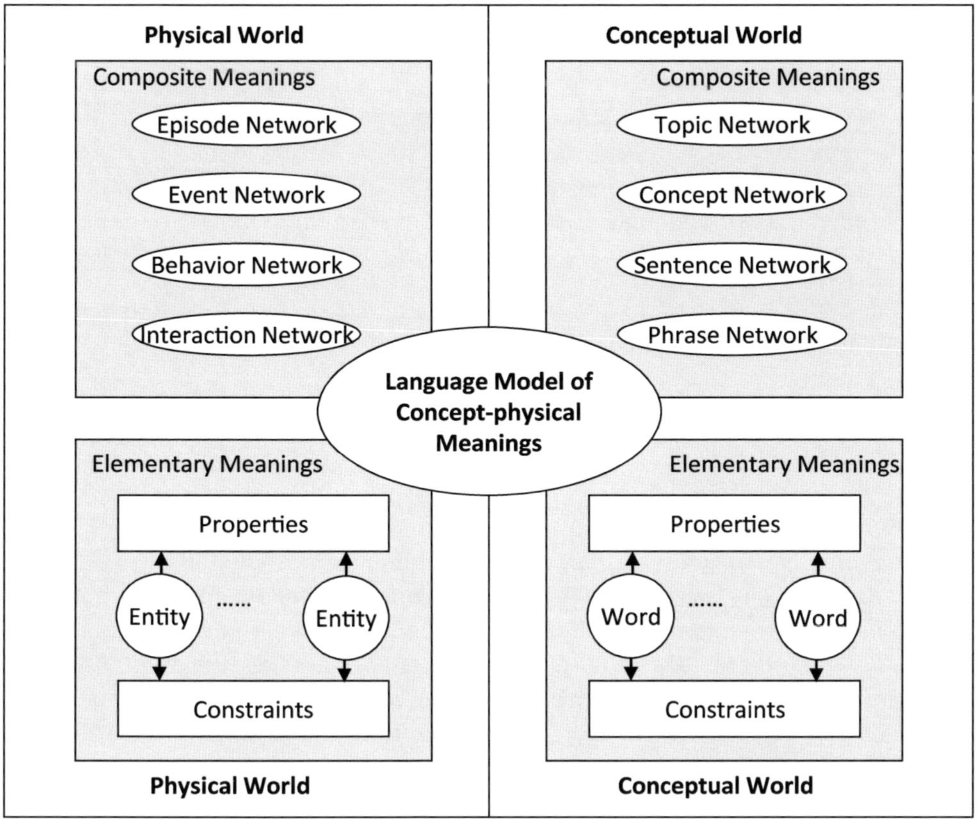

Fig. 5.3 Language model of concept-physical meanings, in which the elementary meanings are organized as both a list of objects in the physical world and a list of words in the conceptual world. And, the composite meanings are organized in the form of hierarchically-layered networks in both the physical world and a conceptual world.

representing such hybrid knowledge or meanings. Therefore, it is better to represent a human language with a model which encapsulates the concept-physical meanings as illustrated by Fig. 5.3.

As shown in Fig. 5.3, the properties and constraints are actually organized at two levels, namely: (a) the elementary meanings, and (b) the composite meanings. In particular, the elementary meanings are organized as both a list of (physical) objects in the physical world, and a list of (conceptual) words in a conceptual world. And, the composite meanings are organized in the form of hierarchically-layered networks. In each network, a node represents an entity, which can be either physical or conceptual. But, a link in a network indicates a constraint, which makes two adjacent nodes to act and react for the purpose of achieving a valid outcome.

For instance, when a cup is placed on top of a table, the kinematic constraint of zero degree of freedom (i.e. zero-DOF) will make the table and the cup to achieve a valid outcome, which is to say that a cup is at rest on top of a table. However, the kinematic constraint of six degrees of freedom will not make the table and the cup to achieve a valid outcome. This is because a cup cannot fly, by itself, on top of a table.

With the properly designed representation of a human language, its implementation with the use of a programming language will enable a vehicle or robot to enter the practice of supervised learning of a human language, in which a human teacher could interactively do knowledge transfer. However, the success of supervised learning of a human language still requires these additional mental capabilities such as

- Mental ability of transforming text into speech.
- Mental ability of transforming speech into text.
- Mental ability of transforming image into text.
- Mental ability of transforming text into meanings.

5.3 Transformation From Text to Speech

Speaking and listening are a pair of astonishing mental capabilities of human beings. Actually, the invention of a human language are for two primary purposes. One objective is to support speaking, while the other objective is to enable writing. In general, the lexicon of a spoken language is also the lexicon of a written language. But, there are exceptions. For example, in Chinese, the lexicon of spoken language (e.g. the system of pinyings) is totally different from the lexicon of written language.

From the perspective of a spoken language, a text is a collection of sentences, while a sentence is a collection of words. With some exceptions (e.g. Chinese spoken language in which one word corresponds to one pair of phonemes), a word is a collection of phonemes (i.e. basic units of sounds). And, phonemes are divided into two major types, which are phonemes of vowels and phonemes of consonants. For human languages invented in western countries, there are over 20 vowels and 100 consonants. Therefore, a word in a human language (excluding exceptions) plays two roles. One role is to guide its writing, while the other role is to guide its utterance.

We all know that a human being must spend many years to learn (or to imitate) the use of a human language before he or she could reach a certain level of proficiency in both spoken language and written language. Most importantly, supervised learning is the common method which is widely adopted in education over a long period of history. This implies that teachers and learners, who are involved in the process of language learning, must have the mental abilities of speaking as well as the mental abilities of listening.

It is a matter of fact that any human being could learn to speak any human language. This truth implies that all human beings have the same production system of speech, which is called as mouth in layman term. The mouth of a human being contains the organs of speech, which play the roles of (active or passive) articulators

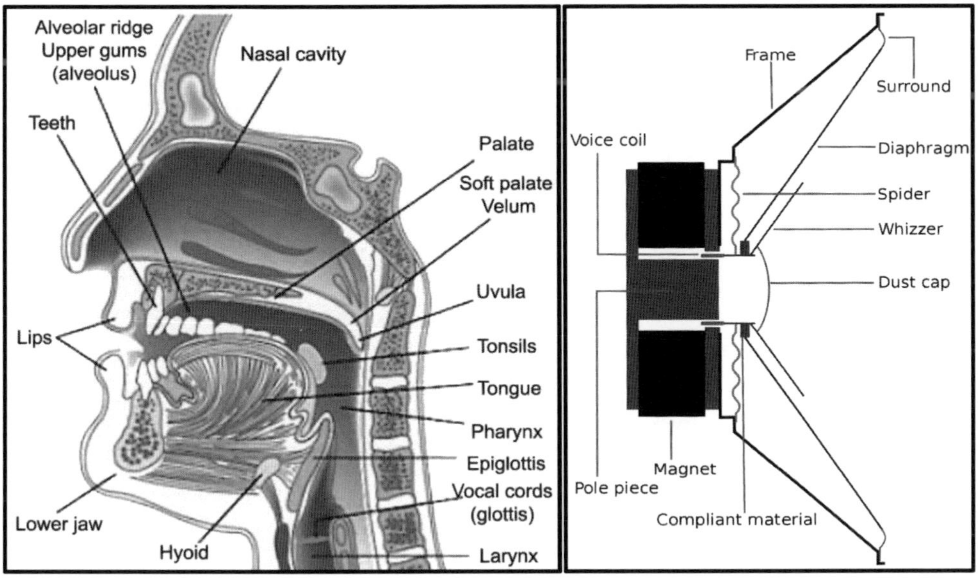

(a) (b)

Fig. 5.4 Comparison of (a) speech production system in human being and (b) speech production system in machine.

during the process of producing sound waves which are intended by phonemes inside words.

Refer to Fig. 5.4. The major organs of speech in a human being's mouth include lips, teeth, tongue, alveolar ridge, hard palate, soft palate, uvula, pharynx, glottis (i.e. vocal cords), and larynx. The placements of these articulators and manners of their articulations are under the control of human brain. In this way, sound waves corresponding to different phonemes in a human language could be produced. Such process could be considered as a hybrid modulation, in which the magnitude and frequency of a periodic air pressure wave inside a human being's mouth are modulated simultaneously in real-time before it exits the mouth.

In communication, the purpose of modulation is to encode information or knowledge from one signal (i.e. information signal) to another signal (i.e. carrier signal). For example, the two typical modulations are

- Modulation of magnitude: In this case, a carrier signal's frequency remains unchanged, while its magnitude is modified by an information signal. Then, the modulated carrier signal will be transmitted out.
- Modulation of frequency: In this case, a carrier signal's magnitude remains unchanged, while its frequency is modified by an information signal. Similarly, it is the modulated carrier signal which will be transmitted out.

We know that a human being could learn to speak any human language. Surprisingly enough, today's machines could also speak any human language without too much difficulty. This is due to the invention of electronic speaker or machine mouth as shown in Fig. 5.4(b), in which the key elements are

- A permanent magnet: It produces a constant magnetic field.
- A voice coil: It is also called as an electronic magnet, which receives electronic signals in order to produce variable magnetic field. This variable magnetic field interacts with the magnetic field of the permanent magnet. Such interaction causes the voice coil to vibrate. The magnitude and frequency of voice coil's vibration depend on the time signals applied to the voice coil.
- A diaphragm: It is a plate fixed onto the voice coil. In this way, the vibration of the voice coil will move the diaphragm forward and backward rapidly. Such movement from the diaphragm will produce a pressure wave in air, which is actually a sound wave.
- A spider: It is an elastic ring. Its outer edge is fixed on the frame of electronic speaker, while its inner edge is connected to the outer edge of the diaphragm. Its function is to supply elastic force which tends to restore the diaphragm back to its neutral position (i.e. the diaphragm's position when the voice coil does not vibrate).

Now, the next question will be: how to make a machine to apparently speak like a human being? In other words, given a text in a human language, will a machine be able to produce a sound wave which is very close to the one produced by a human being when he or she reads and utters the same text? A general solution is shown in Fig. 5.5, in which the key modules include

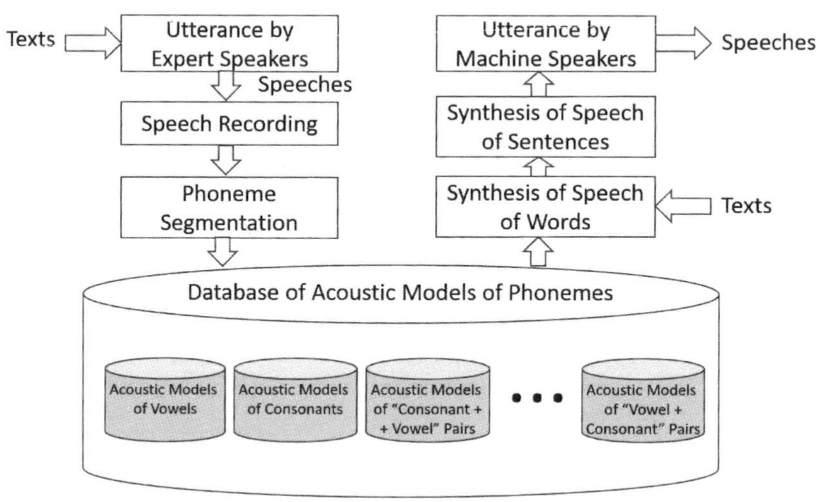

Fig. 5.5 A text to speech synthesis system.

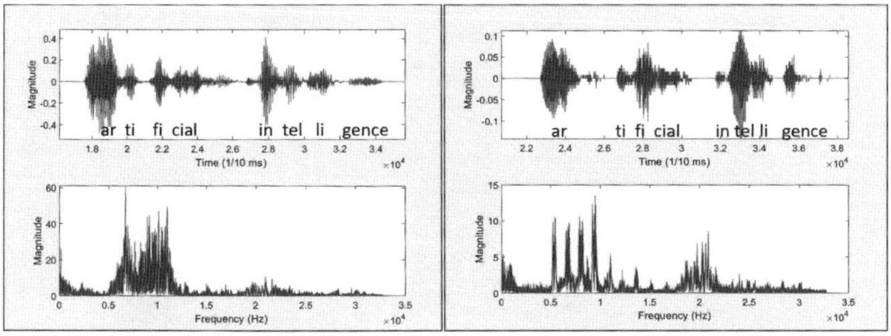

Text to Speech by Male Voice (a) Text to Speech by Female Voice (b)

Fig. 5.6 Example of text to speech synthesis. In this case, the text is "artificial intelligence". Two speeches are synthesized: (a) one by a male voice and (b) the other by a female voice. For each case, both the speech signal in time domain and its Fourier transform are shown.

- Construction of databases of phonemes: The databases contain acoustic models of phonemes in a spoken language. This task involves the participation of expert speakers (e.g. TV presenters, Radio presenters, film actors, etc) and the selection of sample texts which are comprehensive enough to cover all the phonemes and their possible combinations. The invitation of expert speakers may consider the criteria of gender, age and accent, etc. When expert speakers and texts are ready, the process underlying this task consists of speech recording and phoneme segmentation. The results of phoneme segmentation are finally classified into acoustic models of vowels, acoustic models of consonants, acoustic models of "consonants+vowels" pairs, acoustic models of "vowel+consonant" pairs, etc. All these constructed models are stored inside the database of phonemes.
- Synthesis of speech of words: Given a text as input to a machine, this module segments the text into a list of sentences, and further segments each sentence into a list of words. For each word, this module will identify the phonemes inside the word and will put the corresponding acoustic models of identified phonemes together so as to form the acoustic model of the word. Therefore, the outcome from this module will be the speech of each individual word.
- Synthesis of speech of sentences: For each sentence inside the input text, this module extracts the words which belong to the sentence, and assemblies the speeches of these words together. In this way, the outcome from this module will be the speech of each individual sentence.
- Utterance by machine speakers: This module plays back sequentially the speech of each individual sentence inside an input text.

In Fig. 5.6, we show an example in which the input text is "artificial intelligence" and the output are two speeches. One speech is synthesized from a database of phonemes of a male voice, while the other speech is synthesized from a database

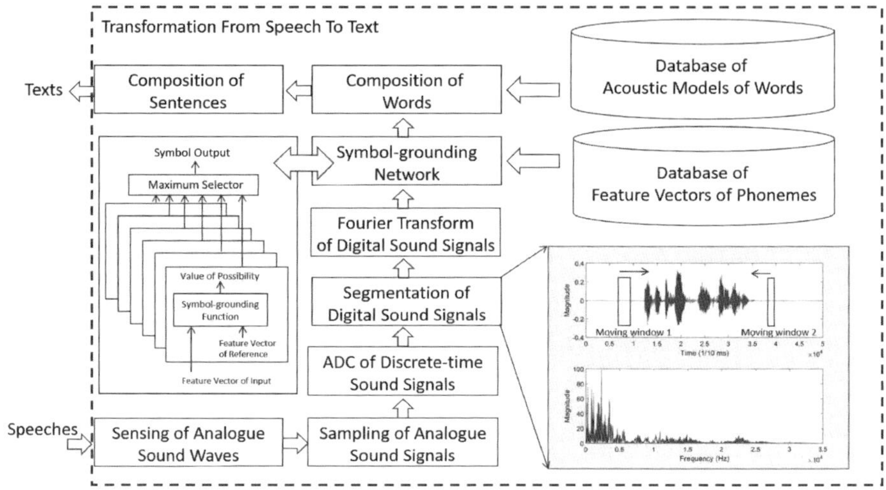

Fig. 5.7 A blueprint of transformation from speech to text.

of a female voice. In each case, the speed signal in time domain and its Fourier transform are shown.

5.4 Transformation From Speech to Text

When a teacher speaks, a learner must have the mental capabilities of transforming a teacher's speech into its corresponding text. Such mental capabilities must be equipped by future machines which intend to have self-intelligence.

Technically speaking, transformation from speech to text is the inverse process of converting text to speech. However, a human being's brain could not always control perfectly the placements of his or her mouth's articulators and the manners of his or her mouth's articulations. As a result, due to the variation of speech production by human beings, a same text may have slightly different versions of speeches. Therefore, it makes the transformation from human-produced speech to text a challenging task for today's machines to perform.

Refer to Fig. 5.7. The process of transforming speech to text will involve the following key modules:

5.4.1 *Sensing of Analogue Sound Waves*

The purpose of this module is to sense incoming sound waves and to convert the pressure waves of air into electronic signals in time domain. This task could be accomplished by devices called as microphones.

The structure of a microphone is quite similar to the one of an electronic speaker. A microphone has a diaphragm which receives the pressure waves of sounds. The pressure waves of sounds make the diaphragm to vibrate, which in turn drives a voice coil to vibrate. The vibration of the voice coil will produce time-variant electronic signals, which are called as analogue sound signals.

5.4.2 *Sampling of Analogue Sound Signals*

Analogue sound signals are continuous both in time and in amplitude. The purpose of this module is to convert continuous-time signals into discrete-time signals. This task could be accomplished by any general-purpose micro-controller which has a device called as analogue-to-digital converter.

An analogue-to-digital converter has built-in sampling devices, which take samples of incoming signals at regular time interval. The time interval for taking samples is programmable. For example, some applications could take millions of samples per second. The output from this module will be sound signals which are discrete in time, but still analogue (i.e. continuous) in amplitude.

5.4.3 *ADC of Discrete-time Sound Signals*

Here, ADC stands for analogue-to-digital conversion. The purpose of this module is to convert continuously variant amplitudes into digital numbers. This task could be accomplished by analogue-to-digital converter as well. Hence, the output from a micro-controller's ADC are digital sound signals, which are discrete in time and digital in amplitude.

5.4.4 *Segmentation of Digital Sound Signals*

Sound signals of speech are series of sound units, each of which corresponds to a phonemes. Most importantly, when the time interval of taking speech's sound samples is chosen to be very short, there will be clear pauses between two adjacent sound units. Therefore, the purpose of this module is to segment incoming digital sound signals into series of digital sound units in time domain.

Since digital sound signals are one-dimensional time signals, the process of segmenting digital sound signals into digital sound units could be operated in three phases as follows:

- Localization of Peaks: In this phase, the module uses moving window 1 of certain time-width (e.g. between 0.1 s to 0.5 s which depends on the speed at which the speech is produced) to scan the incoming digital sound signals. At each location during the scan along the time axis, it calculates the sum of squared amplitudes of signals inside the moving window 1. Such sum is a measurement of signal's energy within the moving window 1. Then, a peak will be identified at the location in which the signal's energy within the moving window 1 is locally maximum.
- Localization of Valleys: In this phase, the module uses moving window 2 of certain time-width (e.g. 10% of window 1's width) to scan the incoming digital sound signals between each pair of two adjacent peaks. At each location during the

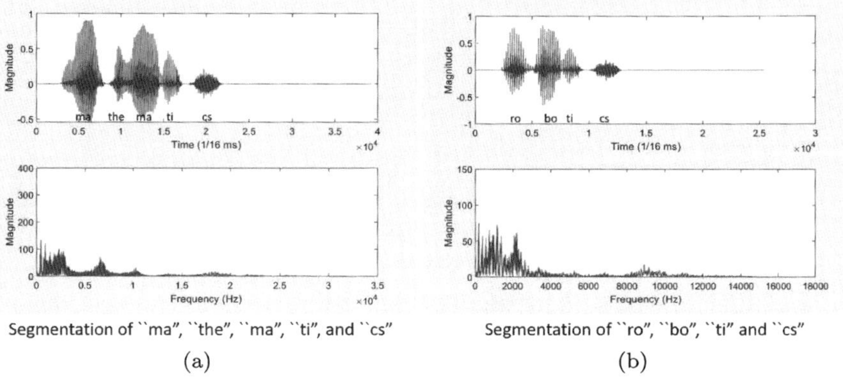

Fig. 5.8 Example of (a) segmenting incoming digital sound signal into phonemes {"ma", "the", "ma", "ti", "cs"}, and (b) segmenting incoming digital signal into phonemes {"ro", "bo", "ti", "cs"}.

scan, it calculates the sum of squared amplitudes of signals inside the moving window 2. A valley will be identified at the location in which the signal's energy within the moving window 2 is locally minimum.

- Removal of Basins: Prolonged pauses between the pronunciation of two adjacent phonemes will produce basins in the digital sound signals. The basins of digital sound signals do not carry useful information. They have to be removed. In this phase, the module uses moving window 1 again to scan each segmented digital sound unit which is between two adjacent valleys. During the process of scan, it removes time interval in which the signal's energy is negligible.

In Fig. 5.8, two examples of segmenting digital sound signals into digital sound units are shown. It is clear that at this step, each digital sound unit has a time information. If the number of digital sound units from an incoming digital sound signal is denoted by Q, the input of speed data could be described by the time information of a set of digital sound units as follows:

$$D_s = \{t_j, j = 1, 2, \ldots, Q\} \tag{5.1}$$

where only time information are available. Hence, we must continue to extract other types of useful information.

5.4.5 *Fourier Transform of Digital Sound Signals*

The input to this module is a series of digital sound units. Hence, the purpose of this module is to apply Fourier transform on each digital sound unit. The result will be two series of digital numbers. One series of digital numbers corresponds to the magnitudes in frequency domain, while the other series of digital numbers are the phase angles in frequency domain.

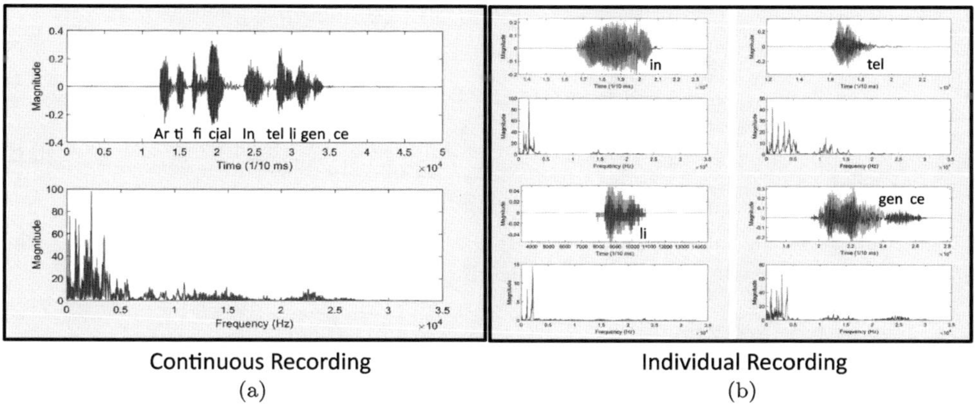

Fig. 5.9 Comparison between (a) results of Fourier transform applied on a digital sound signal of the word "intelligence" and (b) results of Fourier transform applied to digital sound units corresponding to phonemes of "in", "tel", "li", and "gence".

For example, in Fig. 5.9, a digital sound signal is obtained from the utterance of the phrase "artificial intelligence". We can apply Fourier transform to this whole digital sound signal or to its individual digital sound units. Clearly, it is better to do Fourier transform on individual sound units. This is because a simple series of digital values (e.g. magnitudes of frequencies) in frequency domain has lesser number of alternative combinations than a crowded series of digital values in the same domain. Lesser number of alternative combinations means lesser chance of ambiguities.

We know that the frequencies of audible sounds fall into the interval between 20 Hz and 20k Hz. We could choose the peak frequencies within interval [20 Hz, 20k Hz] to form the feature vector attributed to a digital sound unit. In frequency domain, a peak frequency is a frequency at which the magnitude is locally maximum.

Assume that a list of P peak frequencies is retain to form a feature vector of a digital sound unit. At each peak frequency, there are three metrics which are {the frequency, the amplitude, the phase angle}. Therefore, the feature vector for digital sound unit j will be as follows:

$$v_j = (f_j, m_j, a_j) \tag{5.2}$$

with

$$f_j = (f_{j,1}, f_{j,2}, \ldots, f_{j,P}) \tag{5.3}$$

$$m_j = (m_{j,1}, m_{j,2}, \ldots, m_{j,P}) \tag{5.4}$$

$$a_j = (a_{j,1}, a_{j,2}, \ldots, a_{j,P}) \tag{5.5}$$

where

- $f_{j,k}$ $(k = 1, 2, \ldots, P)$ is the kth peak frequency,

- $m_{j,k}$ $(k = 1, 2, \ldots, P)$ is the magnitude at the kth peak frequency,
- $a_{j,k}$ $(k = 1, 2, \ldots, P)$ is the phase angle at the kth peak frequency.

Therefore, the digital speech data from input could be described by

$$D_s = \{[v_j = (f_j, m_j, a_j), t_j], j = 1, 2, \ldots, Q\} \tag{5.6}$$

Now, assume that a human language has N reference phonemes (including vowels, consonants and their possible pairings). Then, database D_p of feature vectors of reference phonemes will have the following entries

$$D_p = \{[v_i = (f_i, m_i, a_i), s_i], i = 1, 2, \ldots, N\} \tag{5.7}$$

with

$$f_i = (f_{i,1}, f_{i,2}, \ldots, f_{i,P}) \tag{5.8}$$

$$m_i = (m_{i,1}, m_{i,2}, \ldots, m_{i,P}) \tag{5.9}$$

$$a_i = (a_{i,1}, a_{i,2}, \ldots, a_{i,P}) \tag{5.10}$$

where

- s_i represents the symbol of ith reference phoneme,
- $f_{i,k}$ is the kth peak frequency,
- $m_{i,k}$ $(k = 1, 2, \ldots, P)$ is the magnitude at the kth peak frequency,
- $a_{i,k}$ $(k = 1, 2, \ldots, P)$ is the phase angle at the kth peak frequency.

In a database of reference phonemes, feature vector v_i represents a reference measurement of the digital sound unit corresponding to reference phoneme s_i. The storage of feature vector v_i in a database of reference phonemes is called as a feature vector of reference phoneme s_i.

In practice, feature vector v_i of reference phoneme s_i will be the result of training in which a set of digital sound units corresponding to reference phoneme s_i will be provided. In this way, feature vector v_i of reference phoneme s_i actually is the mean vector, which is the computed average of the feature vectors from the training set (i.e. each digital sound unit in the training set has its own feature vector).

In addition, we could also compute the distance from feature vector v_i of reference phoneme s_i to each of the feature vectors in the training set. These values of distance determine another metric σ_i which is a standard deviation of feature vector v_i of reference phoneme s_i. Hence, database D_p of feature vectors of reference phonemes is finally described by

$$D_p = \{(v_i, \sigma_i, s_i), i = 1, 2, \ldots, N\} \tag{5.11}$$

5.4.6 *Symbol Grounding of Digital Sound Unit*

Given a feature vector v_x as input, this module compares it with all the feature vectors of reference phonemes, and outputs the symbol of phoneme, of which the feature vector is the closest one to v_x. This is a typical problem of symbol-grounding in artificial intelligence. Interestingly enough, such problem could be tackled by using possibility functions.

For example, if v_i is a feature vector of reference phoneme s_i, the possibility for v_i to be the closest one to v_x is determined by the following symbol-grounding function

$$p(v_x, v_i) = \frac{1}{\sigma_i \sqrt{2\pi}} e^{-\frac{1}{2\sigma_i^2}\|v_x - v_i\|^2} \tag{5.12}$$

where $\|v_x - v_i\|$ is the Euclidean distance between feature vector v_x and feature vector v_i.

In a database of feature vectors of phonemes, each reference phoneme has its own symbol-grounding function. Therefore, the symbol-grounding functions from a set of reference phonemes actually form a network which has three layers as follows:

- Input layer: It takes feature vector v_x as input. The size of input layer is fixed.
- Middle layer: It contains all the symbol-grounding functions. The number of these functions could dynamically grow when more training or experience will be available. Hence, its size is not fixed.
- Output layer: It is a selector which outputs the symbol of phoneme corresponding to the symbol-grounding function which has the maximum value of possibility (i.e. $\max p(v_x, v_i), i = 1, 2, \ldots$). The size of output layer is fixed.

Such network could be called as a symbol-grounding network.

5.4.7 *Composition of Words*

The input to this module is a time sequence of identified phonemes, while the output is a time sequence of words. This module will also use the database of acoustic models of words. Inside the database, the acoustic model of a word is actually a sequence of phonemes. The length of a word's sequence of phonemes could be from 1 up to K. In Chinese language, K is always equal to 1. This is because a single Chinese word corresponds to a single phoneme.

Let $\{s_i, i = 1, 2, \ldots L\}$ be the input sequence of identified phonemes. The process of word composition will be as follows:

- Prediction of possible words: If $K < L$, it predicts the following possible words $\{(s_1, s_1 s_2, \ldots, s_1 s_2 \ldots s_K)\}$. If $K > L$, it predicts the following possible words $\{(s_1, s_1 s_2, \ldots, s_1 s_2 \ldots s_L)\}$.
- Verification of predicted words: For each predicted word, this module compares it with the acoustic models of words in a database. The metric of comparison is percentage of match between the phonemes of a predicted word and the phoneme of a reference word from the database.

- Output and re-iteration: It outputs the reference word, which is the longest and best match to a predicted word. Then, it removes the phonemes of the matched word from the input sequence of identified phonemes. After removal, it updates the value of L. Then, it repeats the above process of prediction and verification.

5.4.8 *Composition of Sentences*

This module takes a sequence of words as input, and outputs groups of words, each of which corresponds to a sentence (i.e. a list of related words).

Usually, the duration of pause between two adjacent sentences will be much longer than the duration of pause between two adjacent words. This information is useful to help do simple segmentation of a sequence of words into clusters of words, each of which corresponds to a list of words in a sentence.

The pause between two adjacent words is the time difference between these two words. And, the time information of a word is the average of the times $\{t_j\}$ of the word's phonemes, which are available in D_s (i.e. input of speed data).

5.5 Transformation From Image to Text

The mental capabilities of speaking and listening with the use of a human language enable human-machine interactions to occur at a cognitive level. On top of these mental capabilities, reading and writing is another pair of mental capabilities which help a lot supervised learning of human languages. In particular, reading is another way of acquiring knowledge which are written in the form of printed versions of texts. With today's advance in image processing and computer vision, it is an achievable task to transform images of printed versions of texts into digital files of texts, which are computer-readable.

Refer to Fig. 5.10. The process of transforming printed versions of texts into digital files of texts will involve the following key steps:

5.5.1 *Digitization of Input Image*

The purpose of this module is to convert light rays coming from printed versions of texts into digital color images. This task could be accomplished by a machine's vision system. We will study the process of machine vision in the next chapter.

At this step, the input to this module are illuminated texts printed on media such as papers, whiteboard, blackboard, walls, road surfaces, etc. The output from this module are digital color images, which are two-dimensional matrices of color vectors.

In general, a color vector has three elements, or three coordinates in a color space. One of the popular color spaces is called RGB color space, in which the three principal axes indicate the values of the three component colors {red, green, blue}. Hence, a color image in RGB color space could be described by

$$I_{h\times w} = \{c_{i,j}, i = 1, 2, \ldots, h; j = 1, 2, \ldots, w\} \tag{5.13}$$

with

$$c_{i,j} = (r_{i,j}, g_{i,j}, b_{i,j}) \tag{5.14}$$

where

- h is the height of the color image in terms of the number of rows,
- w is the width of color image in terms of the number of columns,
- $r_{i,j}$ is the value of red component color at image location (i, j),
- $g_{i,j}$ is the value of green component color at image location (i, j),
- $b_{i,j}$ is the value of blue component color at image location (i, j).

5.5.2 *Identification of Background Color*

The purpose of this module is to identify the background color which will be used to subsequently segment the input of a color image into the output of two parts. One part of output only contains the text image, while the other part of output contains the background image.

We know that reading texts from books, whiteboards or blackboards is different from seeing a general scene. This is because the background of texts is generally uniform and unique in color. On the other hand, the color of texts themselves is also generally uniform and unique in color.

As a result, during the process of reading texts, the background in the input of a color image will occupy more area that the one occupied by text. Hence, a simple

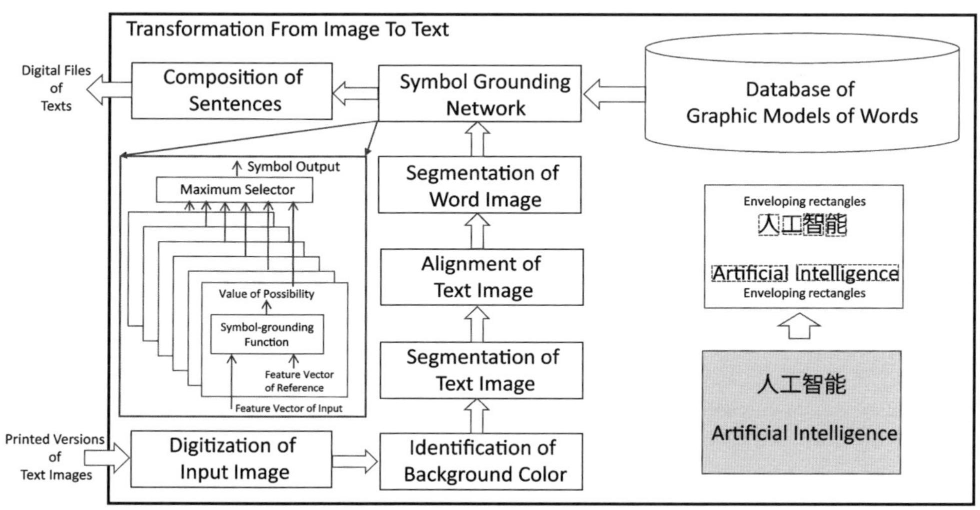

Fig. 5.10 Blueprint behind the process of transforming printed versions of texts into digital files of texts, which are computer-readable.

algorithm is to count the numbers of occurrences of major colors in the input of a color image. The color with the maximum number of occurrences will be the color of the background.

5.5.3 *Segmentation of Text Image*

During the process of reading texts, if the background color has been identified by the previous module, color image $I_{h\times w}$ could be divided into background image $I_{h\times w}(B)$ and text image $I_{h\times w}(T)$. And, text image $I_{h\times w}(T)$ without background will be the output from this module.

5.5.4 *Alignment of Text Image*

In general, it is a difficult task to obtain scale-invariant and rotation-invariant features from images. This is because the change of scale or rotation will systematically change the appearance of an image. Therefore, it is important to assume that all text images are horizontally aligned. In this way, we bypass the issue of considering rotation-invariant features of text images.

The process of aligning a text image horizontally could be accomplished as follows:

- Rotate text image $I_{h\times w}(T)$ within the angle range of $[-90.0^0, +90.0^0]$ with a small incremental angle $\triangle\theta$.
- At each orientation, it counts each row's number of pixels (i.e. picture element) which belong to text. The value of such counting is called as the pixel count of a row.
- Identify the orientation at which the maximum pixel count of a row occurs. Align text image $I_{h\times w}(T)$ at this orientation.

5.5.5 *Segmentation of Word Image*

The purpose of this module is to segment the input of aligned text image into a set of word images. Each word image only contains the image of one word.

This task of segmenting aligned text image into word images could be accomplished in two steps as follows:

- Horizontal segmentation: It divides an aligned text image into horizontally-separated sub-images. Each horizontally-separated sub-image contains a list of word images which are horizontally aligned. This segmentation is achievable with the use of the values of pixel counts of rows. For example, when an text image is horizontally aligned, the rows which do not contain texts will have zero pixel count. Therefore, the rows with zero pixel counts actually separate an aligned text image into horizontally-separated sub-images.

- Vertical segmentation: For each horizontally-separated sub-image, it counts each column's number of pixels which belong to text. The value of such counting is called as the pixel count of a column. Clearly, words are separated by multiple columns with zero pixel count.

For each individually segmented word image, its enveloping rectangle could be determined (i.e. the upper limit, the lower limit, the left limit, the right limit). And, the dimension of its enveloping rectangle could be further scaled up (or down) to a standard size. In this way, images of reference words could be learnt and be stored into a database of graphic models of words.

Assume that the standard size of the graphic model of a word is $a \times b$ (i.e. a rows by b columns). If a human language has W words, database D_g of graphic models of words will contain the following entries:

$$D_g = \{g_k = [I_{a \times b,k}, S_k], k = 1, 2, \ldots, W\} \tag{5.15}$$

where

- g_k denotes the graphic model of kth word in a human language,
- $I_{a \times b,k}$ denotes graphic model g_k's reference image,
- S_k denotes the string of characters (or a sequence of strokes for Chinese language) associated with graphic model g_k.

In Chinese language, word is the smallest unit. But, for western languages, the smallest unit is character. If a word consists of a string of characters, a word image could be further segmented into a sequence of character images (e.g. the width of characters in a text is almost a constant). And, the above procedure is still applicable to character images.

5.5.6 *Symbol Grounding of Word Image*

If word image $I_{a \times b,x}$ is given as input to this module, the output will be the string of characters (or a sequence of strokes for Chinese language) from graphic model g_k ($k = 1, 2, \ldots, W$), in which $I_{a \times b,k}$ is the best match to $I_{a \times b,x}$. Clearly, this is also a typical problem of symbol grounding in artificial intelligence.

Although there is no concern of obtaining scale-invariant features and rotation-invariant features at this step, there is another issue which is the variation of the thickness of characters (or strokes) for any given word image. As a result, the popular technique of template matching in computer vision may not be a good candidate to serve as the symbol-grounding function in this module.

Refer to Fig. 5.11. Within a word image, two adjacent pixels form an on-off transition if one pixel belongs to the word's character (or stroke in Chinese language) while the other pixel belongs to the background (or empty) area. Then, a better solution is to estimate a feature vector from word image $I_{a \times b,x}$ or $I_{a \times b,k}$ as follows:

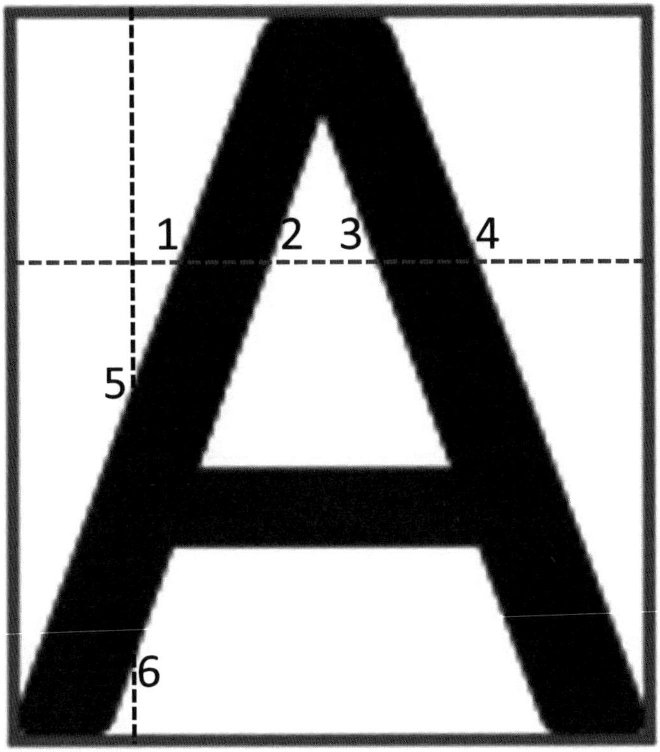

Fig. 5.11 Example of showing the way to determine both the row count of on-off transition and the column count of on-off transition. In this case, locations 1, 2, 3, 4 are on-off transitions in a row, while locations 5 and 6 are on-off transitions in a column.

- For each row of word image, compute the number of on-off transitions and the result will be a vector of size $a \times 1$, which contains the row counts. Such a vector is called as a row count vector.
- For each column of word image, compute the number of on-off transitions and the result will be a vector of size $b \times 1$, which contains the column counts. Such a vector is called as a column count vector.
- Put the row count vector and the column count vector together so as to form the feature vector of size $(a + b) \times 1$, which is a measurement of word image $I_{a\times b,x}$ or $I_{a\times b,k}$. For simplicity, let e_k denote such feature vector of kth word image $I_{a\times b,k}$ in a database.

In practice, each graphic model in a database is obtained from at least one set of training data. And, the number of training data could dynamically grow over time when more training opportunities arise. Since each entry (i.e. a word image of reference) in a training set has its feature vector, feature vector e_k of kth word image $I_{a\times b,k}$ in a database will be the mean vector, which is the computed average of the training set's feature vectors. In addition, from the Euclidean distances between feature vector e_k and the training set's feature vectors, the metric of standard deviation of feature vector e_k could be calculated.

As a result, database D_g of graphic models of words is expanded into the following representation:

$$D_g = \{g_k = [I_{a\times b,k}, S_k, e_k, \sigma_k], k = 1, 2, \ldots, W\} \tag{5.16}$$

where e_k is the feature vector of graphic model g_k, and σ_k is the standard deviation of e_k.

Now, if word image $I_{a\times b,x}$ is given as input to this module, the possibility for word image $I_{a\times b,x}$ to match with word image $I_{a\times b,k}$ in the database could be calculated as follows:

$$p(e_x, e_k) = \frac{1}{\sigma_k\sqrt{2\pi}} e^{\frac{-1}{2\sigma_k^2}\|e_x - e_k\|^2} \tag{5.17}$$

where e_x is the feature vector of word image $I_{a\times b,x}$. And, this equation is called as a symbol-grounding function in this module.

Similarly, the symbol-grounding functions from a set of reference word images also form a network which has three layers as follows:

- Input layer: It takes feature vector e_x as input. The size of input layer is fixed.
- Middle layer: It contains all the symbol-grounding functions. The number of these functions could dynamically grow when more training or experience will be available. Hence, its size is not fixed.
- Output layer: It is a selector which outputs the string of characters of the word corresponding to the symbol-grounding function which has the maximum value of possibility (i.e. $\max p(e_x, e_k), k = 1, 2, \ldots$). The size of output layer is fixed.

Such network could also be called as a symbol-grounding network.

5.5.6.1 *Composition of Sentences*

The purpose of this module is to group incoming words together so as to form sentences. These incoming words have their own spatial orders imposed by the input of word images. Therefore, the focus of this module is to identify the locations of symbol ".", which are separators of sentences in a text.

In general, a short range of locations after symbol "." are empty space. Hence, template image which consists of "." and empty space could be created. The use of such template image is to scan horizontally-separated sub-images of texts, and to mark the locations of symbol ".", at which there are perfect matches with the template image. Figure 5.12 shows an example which illustrates the process of using template image to localize symbol "." in horizontally-separated sub-images of text.

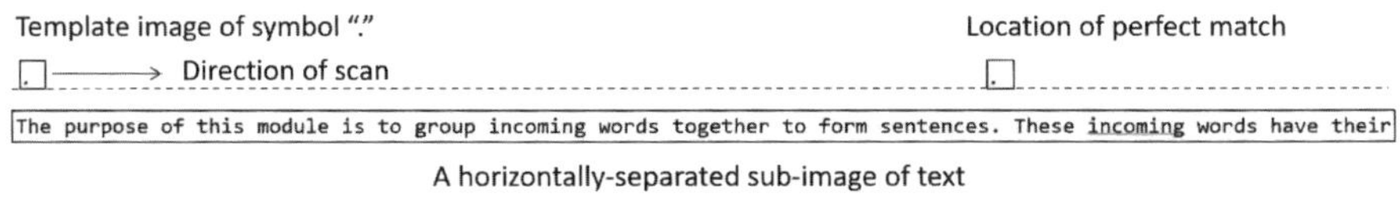

Fig. 5.12 Localization of symbol "." in a horizontally-separated sub-image of text.

5.6 Transformation From Text to Meanings

In general, any human being could learn any human language. The purpose of such learning is to gain the mental capability of transforming texts into meanings behind the texts. Moreover, such learning could be either institutional learning guided by teachers or social learning assisted by people engaged into daily interactions. Most importantly, the outcome of such learning is for learners to enrich the meaning networks as well as knowledge databases inside their minds. For example, the meaning networks include

- Organized networks of the meanings in the physical world (as shown in Fig. 5.1), and
- Organized networks of the meanings in conceptual worlds (as shown in Fig. 5.2).

In addition, the knowledge behind the meaning networks could be re-organized around the following keywords

- Who: The keyword "who" refers to the category of entities with self-intelligence, which include human beings, animals, and machines with self-intelligence, etc.
- What: The keyword "what" refers to the category of results such as properties, things, objects, works, missions, jobs, tasks, actions, activities, etc. These results could be past results, present results, or future results.
- Where: The keyword "where" refers to the category of spatial constraints or places such as planets, continents, countries, cities, villages, towns, districts, factories, schools, hospitals, buildings, shops, stations, roads, bridges, etc.
- When: The keyword "when" refers to the category of time constraints which could be measured in terms of centuries, decades, years, months, weeks, hours, minutes, seconds, etc.
- Why: The keyword "why" refers to the category of pairs of causes and effects (i.e. causalities) which are represented by predicates in the form of "if conditions, then consequences".
- How: The keyword "how" refers to the category of constraints under the names of strategies, methodologies, decision rules (e.g. "if situations, take actions"), and mission plans, etc.

The results of such re-organization are to be stored into individual databases such as the database of "who", the database of "what", the database of "where", the database of "when", the database of "why", and the database of "how". In this

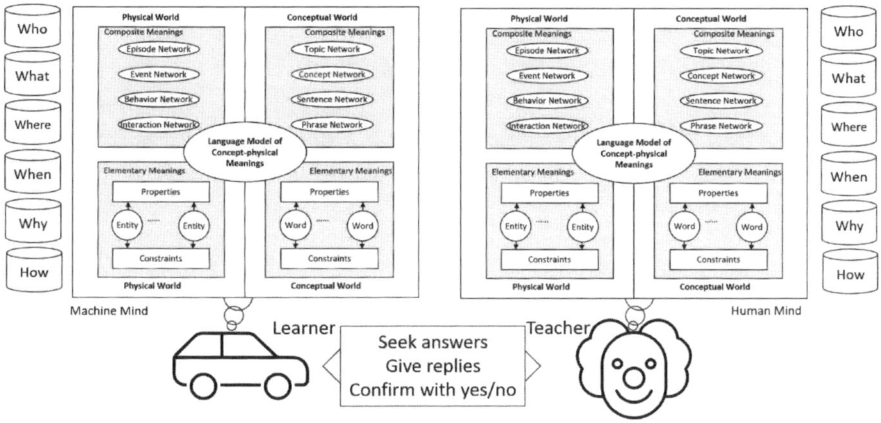

Fig. 5.13 Scenario of supervised learning by machine learners guided by human teachers.

way, the process of transforming texts into meanings will consist of two major steps such as:

- Extract words from texts, and categorize them into "who", "what", "where", "when", "why", and "how".
- Match the categorized words with the corresponding ones inside the learnt databases of "who", "what", "where", "when", "why", and "how".

In artificial intelligence, one of the primary goals is to design machines with self-intelligence so that they could participate in the process of supervised learning of both human languages and knowledge as outlined in Fig. 5.13. This task will become possible if machines of future could have their built-in minds which are similar to human beings' mind. In this way, machine learners could demonstrate their mental capabilities of seeking answers, giving replies and confirming yes or no. These mental capabilities are the necessary and sufficient conditions for machines to interact with human teachers with the use of human languages.

The breakthrough in this domain will have tremendous impact on human societies. This is because machines' minds could be easily copied from one machine to another machine. Such phenomenon will greatly accelerate the proliferation of machines with self-intelligence in future.

5.7 Summary

In this chapter, we have studied in detail the scenario of acquiring knowledge from speeches or printed versions of texts. In particular, we have explained the processes behind the transformations such as from text to speech, from speech to text, from image to text, and from text into meanings. These mental capabilities are very important for machines of future to acquires knowledge from speeches and printed versions of texts.

5.8 Exercises

(1) Deign a set of classes in C++, which represent the networks as outlined in Fig. 5.1.
(2) Deign a set of classes in C++, which represent the networks as outlined in Fig. 5.2.
(3) Deign a set of classes in C++, which represent the networks as outlined in Fig. 5.3.
(4) If a time domain function is $10\sin(20\pi t + \pi/10) + 20\sin(40\pi t + \pi/20) + 40\sin(80\pi t + \pi/40)$, what will be the result of its Fourier transform?
(5) Explain the working principle modulating a signal's magnitude by another signal.
(6) Explain the working principle of modulating a signal's frequency by another signal.
(7) Given a set of phonemes $\{ar, ti, ci, al, in, tel, li, gence\}$, how many utterances of correct words could you generate?
(8) Write a function in C programming language, which segments the input of digital sound signals into the output of digital sound units.
(9) Write a class in C++, which implements a symbol-grounding function which includes the training of the parameters of reference feature vector and its standard deviation.
(10) Write a class in C++, which implements a network of symbol-grounding functions, and offers the methods of deleting, creating and training symbol-grounding functions.
(11) Write a class in C++, which segments the input of text image into the output of a list of word images.

Chapter 6

Basics of Visual Perception

Human beings are living inside an ocean of signals, in which speech sounds and visual signals are extremely important to knowledge acquisition. We also know that knowledge come from the physical world and are represented with human languages with extensions to the technical language as well as programming languages. However, before knowledge are being written into the form of texts in human languages, the discovery and invention of knowledge heavily depend on the perception of visual signals. It is interesting to take note that without learning, a human being could not reach a professional level of proficiency in the use of a human language. But, it is not the case for visual perception because a human being has a perfect vision system since the first day of his or her birth.

Due to the wide spectrum of possible applications in industry and society, machine vision is an exciting field in engineering and science. When we study machine vision (i.e. vision by machine such as robot or vehicle), it is inevitable for us to have the attempt to compare it with human vision (i.e. vision by human being). In other words, it will not be a vain attempt to draw some useful conclusions out of some inspiring properties of human vision. A simple reason is that human vision is robust, and that it works in both real-time and real-environment.

Nevertheless, machine vision is an invention by human beings, but not by the nature. Therefore, human designers do have the total freedom to design the blueprint underlying machine vision. This means that the development of machine vision does not necessarily be restricted by the properties of human vision.

In this chapter, we will focus on the study of some interesting phenomena of visual perception, which may certainly inspire human designers to achieve both better understanding of human vision and better design of machine vision.

6.1 Nature of Human Vision

Human vision is the result of the evolution, or natural selection. And, the evolution is primarily characterized by the process of optimizing the inner properties and constraints of a living creature so as to make the final functions or behaviors of the living creature globally optimal. Interestingly enough, design in engineering embraces a similar process of optimization. Hence, from an engineering point of view, evolution can be considered as a kind of design.

It is a matter of fact that despite decades of efforts devoted to the study of human vision, there is no clear answer to the question of what the working principle behind human vision is. As a result, the study of human vision still remains at the physiological and psychological levels. Because of the lack of scientific proofs about the inner processes behind human vision, the study of human vision is best qualified to be a phenomenology, but not a stand-alone science. This section shares some common findings with regard to human vision.

6.1.1 *Human Vision Is Composite*

Human vision is composite. This claim is supported by the following three interesting facts:

- First of all, the retina in a human eye contain two kinds of photo-receptors, namely: (a) *cone* and (b) *rod*. In addition, there are three types of cone cells. For instance, the cone cells of one type are particularly sensitive to the wavelengths around 420 nanometers (i.e. violet, or blue). The cone cells of another type are very sensitive to the wavelengths around 530 nanometers (i.e. green). And, the cone cells of the third type are especially sensitive to the wavelengths around 560 nanometers (i.e. red). As a result, the cone cells make human vision capable of discerning the wavelengths of incoming lights, within the range from 390 to 780 nanometers. Most importantly, human vision will trigger the psychological sensation of colors when it perceives a scene illuminated by sun-light. This fact indicates that human vision is color-based.
- Second, the rod cells in a human eye have a peak response around the wavelength of 500 nanometers. But, they are much more sensitive to the energy of a light than to its wavelength. Hence, the rod cells make human vision capable of perceiving a poorly-illuminated scene (e.g. a moon-lit scene). Therefore, human vision works fine in monochrome.
- Third, the cone and rod cells are not evenly distributed across the inner surface in a human eye's retina. In particular, a small area called *fovea* has a very high density of cone cells. And, this area serves the so-called *central vision*, while the remaining area serves the so-called *peripheral vision*. As a result, human vision is a combination of both central and peripheral visions.

6.1.2 *Human Vision Is Parallel*

Human vision is amazingly fast enough for us to perceive a real scene in real-time. It seems that the latency of visual perception in human vision is negligible. This astonishing power in processing the huge amount of visual signals coming into the eyes should be attributed to the innate neural-processing architecture, which is highly parallel in nature. Therefore, human vision is running in parallel, when it processes visual data in real-time.

From an engineering point of view, parallel vision also means that the innate metal processes underlying human vision should be simple enough in order to be effectively implemented in a parallel architecture for data processing in real-time. Therefore, the development of machine vision should preferably focus on the design of innate solutions, or algorithms, which could be effectively executed by a parallel computer.

6.1.3 *Human Vision is Quantitative*

Human vision could perceive depth. This ability is called the *stereopsis*. Especially, our visual perception system do give us the sensation of being in a three-dimensional space. Then, this question arises: What is the principle, which enables human vision to perceive a three-dimensional scene?

Now, it is a common consensus that binocular vision permits stereopsis. For instance, the process of triangulation in binocular vision is an effective way to determine the pose (i.e. position and/or orientation) of objects in a scene. But, the principle of triangulation will not work, if one of the following two conditions does not fulfill:

- Condition 1:
  The determination of pose by triangulation in binocular vision requires that both eyes in human vision are functioning at the same time.
- Condition 2:
  The determination of pose by triangulation in binocular vision is only valid within a sub-volume of a scene, which is visible to both the eyes in human vision.

By introspection, we know that human vision will not suffer any disruption of visual output if we intentionally close, or obstruct, one eye. On the other hand, we know that the final output from human vision is the sum of the contents perceived by both the eyes, instead of the commonly-perceived areas. In other words, human vision is immune to occlusion.

These observations strongly suggest that human vision is also able to derive the pose of objects in a scene from the visual signals captured by one eye only. In other words, human vision could work perfectly well as a monocular vision. Because there are two eyes in human vision, it is easy to claim that human vision should be treated as a combination of two cooperative monocular visions, instead of a single binocular vision, as illustrated in Fig. 6.1.

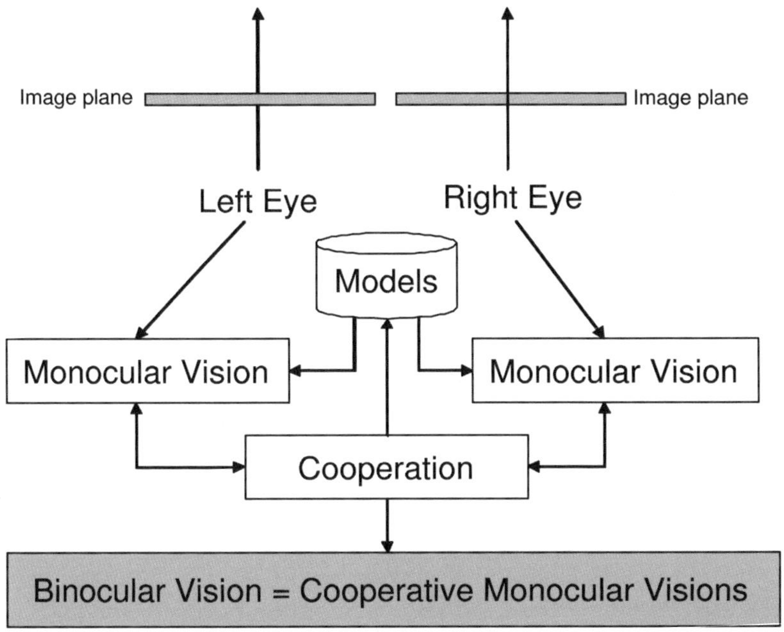

Fig. 6.1 Binocular vision should be treated as a combination of two cooperative monocular visions, in which the pose of objects in a scene could be determined by either a bottom-up approach, or top-down approach.

6.1.4 *Human Vision Is Qualitative*

Our visual experience shows that no matter how long we observe a scene, we are not able to quantitatively tell the exact dimension of the scene at a reasonable precision, or accuracy. However, we are able to qualitatively describe a scene at a glance of it. This fact may hint to us that human vision is qualitative.

Also, our visual experience tells us that we could easily name colors and shapes, but not the numerical parameters in their physical representations. This fact indicates that visual signals entering the eyes of human vision will undergo a series of massively-parallel transformations inside the brain until the level of cognition. As a result, the final outcome of visual perception by human vision is conceptual, but not numerical.

Interestingly, the qualitative nature of human vision helps explain why human vision is robust. It is possible that visual signals in human vision are first assigned with conceptual labels, which could be externally learnt, or internally computed. Subsequently, human vision manipulates labelled visual signals in order to derive the meanings of a scene in both the physical world and the conceptual worlds. In this way, human vision will not be sensitive to numerical variations of visual signals. For instance, a same color will trigger a relatively stable response by human vision (i.e. color sensation), even if the ambient lighting varies within a reasonable range.

In contrast to the qualitative nature of human vision, today's machine, or computer, vision achieves very limited success. Partly, it is due to the fact that in general a process will not be robust in the presence of noises, if such a process purely manipulates numerical values of signals. Hence, today's machine, or computer, vision systems are not robust enough, because their computations are purely based on the manipulation of numerical values of visual signals.

6.1.5 *Human Vision Is Situated*

Stereopsis indicates that human vision is able to perceive the pose of the objects in a three-dimensional scene. Also, stereopsis implies that human vision could generate a three-dimensional representation of a scene.

By introspection, there is no doubt that the visual impression of a scene produced by human vision is three-dimensional. For instance, the mental processes for our visual perception are running inside our brain. But, our visual perception makes us feel that we are inside the world. In other words, we do not have the impression that the world is inside our brain.

Although it is clear that the geometrical representation of a scene produced by human vision is three-dimensional, one may still raise the following question: Is this three-dimensional representation referenced to a coordinate system (i.e. frame) assigned to a scene, or to a coordinate system associated to a viewer? In other words, the question is how human vision is situated with respect to a scene, and a viewer (i.e. the body of a human being).

From an engineering point of view, the geometrical representation of a scene produced by human vision should be viewer-centric. And, the evidence to this claim could be apprehended from the existence of view-frame and body-frame, both of which are centered at a viewer.

6.1.5.1 *View Reference*

Refer to Fig. 6.2. First of all, the two geometrical representations as the result of the outputs from both the left eye and the right eye in human vision must be fused, or combined, together within a common coordinate system (or frame), which is called a *view-frame*. For instance, although each eye in human vision could produce its own geometrical representation of a scene referenced to its own frame (e.g. the left-eye frame, or the right-eye frame), we do have the impression that the two representations from both of our eyes are perfectly aligned together, and that the final result is referenced to a single coordinate system (i.e. a view-frame).

Second, a view-frame should maintain a stable (or fixed) pose with respect to both the left-eye frame and the right-eye frame. In this way, any rapid change in the motion of a viewer's body will have no effect on the perception of the geometry of a scene. Now, it is clear that the three-dimensional representation of a scene produced by human vision should be referenced to a view-frame. But, one may further raise this question: Where is a view-frame located with respect to a viewer's body?

By introspection, it is easy for us to verify that the close of one eye in human vision will not cause any sensation of disruption to visual perception. This evidence

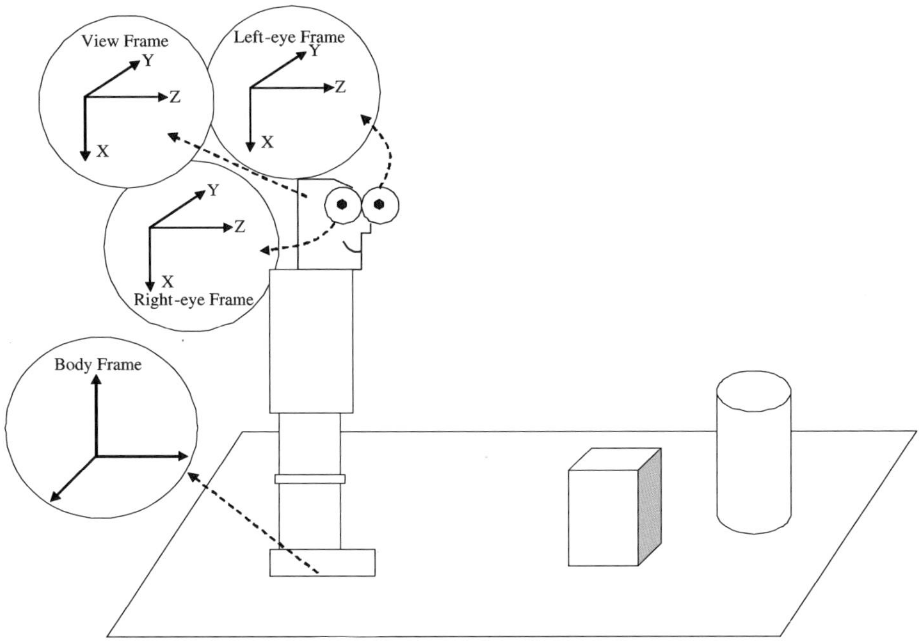

Fig. 6.2 The geometrical representation of a scene produced by human vision should be viewer-centric. And, a view-frame must keep a stable (or fixed) pose with respect to both the left-eye frame and the right-eye frame. Also, a view-frame must be related to a body-frame.

suggests that a view-frame should not be the left-eye's frame nor the right-eye's frame. Otherwise, if we assume that a view-frame coincides with the left-eye's frame, then the close of the left-eye will make human vision fail. Similarly, if we assume that a view-frame coincides with the right-eye's frame, the close of the right-eye will also make human vision fail.

On the other hand, we have said that a view-frame must maintain a stable (or fixed) pose with respect to the eyes. It means that a view-frame should not be fixed in a scene. Instead, a view-frame should be physically, or virtually, attached to a viewer's body. And, it is likely that a view-frame should be located at an equal distance from both the left eye and the right eye. Interestingly enough, a particular location to place a view-frame is the mid-point between the left eye and the right eye.

6.1.5.2 *Body Reference*

Visual perception is an effective way for us to understand the meanings of a scene in both the physical world and the conceptual world. And, an important purpose to understand the meanings of a scene is to plan and guide our actions to be performed by our body. Therefore, it is necessary to define a coordinate system assigned to a viewer's body, to which the geometry of a scene perceived by a viewer will be transformed. In other words, a view-frame must be related to a body-frame in order to link visual perception to various actions being undertaken by a viewer's body.

In engineering terms, a body-frame is a coordinate system assigned to a viewer's body, which represents the body's pose with respect to a scene. In order to suit a specific behavior, a body will require an appropriate body-frame. For instance, when a body is supported by one of its legs during an action such as running, a body-frame should be placed at a contact point between the supporting leg's foot and a scene. However, when a body is supported by one of its hand-arms during an acrobatic movement, a body-frame should be placed at a contact point between the supporting hand-arm and a scene.

Clearly, a body-frame should not be fixed in the physical world. Instead, it should be virtually attached to a viewer. Because the assignment of a body-frame is viewer-centric, the geometrical representation of a scene produced by human vision should also be viewer-centric.

6.1.6 *Human Vision Is Developmental*

Vision enables a human being to first attain the visual awareness about the external world, and then to plan and guide his, or her, behaviors for various purposes such as avoiding dangers, exploring an unknown environment, and undertaking things for specific gains. Human vision is admirably powerful. However, the prowess of human vision is not gained once at the time of birth, but is gradually developed especially at the early stage of one's childhood.

From an engineering point of view, there are at least two good reasons for human vision to be developmental.

6.1.6.1 *Autonomous Learning*

First of all, the perception of a conceptual world by human vision (e.g. reading a textbook) has to be achieved through incremental learning (i.e. a developmental process) after the birth, but not be pre-programmed or designed in advance (i.e. before the birth). The reason is as follows:

We know that in the history of mankind, human intelligence has grown exponentially since the gradual emergence of human languages many thousand years ago. And, the emergence of human languages allows us to conceptually represent the physical world. As a result, the representation of the physical world in the form of human languages defines what we call *the conceptual worlds*. Therefore, we can say that since the early period of the civilization in human society, there are two worlds in the mind of human beings, namely: (a) the physical world, and (b) the conceptual worlds. In the physical world, all natural objects and creatures are constituent entities. And, all the man-made creatures also belong to the physical world. However, the constituent entities in the conceptual worlds include things such as the words in the human languages as well as the technical languages dedicated to science and engineering. For instance, a library is a typical manifestation of the conceptual worlds.

Because human vision is the result of the natural evolution (but not human creation), it is likely that human vision is genetically designed to be able to perceive the physical world (i.e. to transform visual signals into the cognitive state of knowing the physical properties and constraints of a scene). For instance, a new-born baby can see colors, shapes and forms, etc. And, there is no re-programming needed in order to assist, or reinforce, a baby's visual perception of the physical meanings about a scene.

However, a conceptual world is not the creature of any natural evolutionary process, but the mind of human beings. It is likely that human vision is genetically designed in such a way that the conceptual meanings of a scene will be incrementally learnt after the birth. And, it is easy to verify that without learning (e.g. linguistic programming), a person will never be able to perceive the conceptual meanings of a scene. For instance, a person will not be able to describe a simple scene in Chinese language if he or she does not have any learnt-knowledge about Chinese.

The above discussions should be convincing enough for us to believe that human vision is developmental. But, a developmental vision is primarily characterized by autonomous learning. In engineering terms, learning generally involves three successive steps, namely: (a) modeling, (b) optimization, and (c) representation. However, for human vision, the autonomous learning specifically consists of:

- modeling visual properties and constraints,
- optimizing the mental processes which transform visual signals into visual properties and constraints,
- representing visual properties and constraints in both the physical world and the conceptual worlds.

6.1.6.2 *Sensory-motor Mapping*

In addition, the skill of visual guidance by human vision has to be incrementally learnt, instead of being designed in advance.

For instance, let us consider the case of vision-guided manipulation (including grasping), which all of us are familiar with. In general, the skill of visual guidance could not be computed by human brain under a similar framework of planning (i.e. using inverse kinematics) and control as in robotics. Otherwise, by modeling (i.e. kinematics) and computation (i.e. planning), everyone of us can be a master of any sportive game without pre-training. But, the reality is not so. Most importantly, all skills underlying a vision-guided manipulation do require repeated practices or training.

This fact indicates that the mental process behind a vision-guided manipulation relies on some sorts of transformations that a nervous system can easily learn, but not on complex computations that a nervous system has trouble to cope with. For instance, the nervous systems in human brain are not like a modern computer, which is remarkable for complex computations. Although a human brain has about 100 billions of neural cells, it is not good at all in doing computations (e.g. can we rapidly tell the solutions to the equation $0.9x^2+1.234x+5.678=0$?). However, the flexible connectivity and configurability of the nervous systems in human brain make it astonishingly powerful enough to undertake massively-parallel transformations from signal to cognition.

6.1.7 *Human Vision Is Cognitive*

Human vision is developmental because the perception of the meanings in the conceptual worlds has to be learnt by an autonomous learning process. Interestingly enough, this fact also reveals another important aspect of human vision, which is cognitive vision. The cognitive aspect of human vision states that the output from human vision are not signals, but knowledge about a scene, or an image.

By introspection, it is easy to convince ourselves that our vision is cognitive. First of all, it is easy to verify that human vision does not output any signal. For instance, could we tell the values of signals inside our brain, which represent a particular color? Or, could we tell the numerical values inside our brain, which represent the physical aspect of a shape or form? Obviously, the answer is negative.

Second, the output from human vision is conceptual and descriptive. Although the input stimuli to human vision are signals, the final output from human vision is the knowledge about the meanings of a visual scene, or an image, in both the physical world and the conceptual worlds. This is because human vision internally undertakes successive transformations from visual signals to cognition via features.

6.1.7.1 *Physical Meanings*

From an engineering point of view, all entities in the physical world possess their own properties, and are constraining one another in many different ways. Because of the constraints exerted mutually by the physical entities, it is easy to understand that interactions, behaviors, events and episodes will occur in the physical world. Therefore, the physical world exhibits its meanings at two levels, namely: (a) the elementary meanings, and (b) the composite meanings.

It goes without saying that there will be nothing in the physical world, if there is no entity at all. If there exist properties and constraints in the physical world, there must be at least one entity. Hence, the properties and constraints of an entity in the physical world, and their evolution in space and time, define what are called as the *elementary meanings* in the physical world.

On the other hand, the physical world consists of entities, which act and interact. As a result, the composite meanings in the physical world refer to the properties and constraints of interactions in the physical world, and their evolution in space and time, which define what are called as the *composite meanings* in the physical world.

6.1.7.2 *Conceptual Meanings*

In the physical world, an interaction between two (or more) entities, which evolves in time and space, will give rise to the notion of *behavior.* When a behavior in turn evolves in time and space, such a situation is called an *event.* And, the evolution of one, or multiple related, event in space and time is called an *episode.*

Interestingly, the emergence of human languages together with the technical languages dedicated to science and engineering allow us to create the abstracted version of the physical world. This kind of abstraction gives rise to what is called the *conceptual world.*

In the conceptual worlds, the basic entities are *words.* And, the interaction between two (or more) words according to admissible grammatical rules (i.e. a kind of logic) will engender the notion of *phrase.* When phrases are put together according to admissible grammatical rules (i.e. also a kind of logic), the result is called a *sentence.*

In a broad sense, the constraints in the conceptual worlds can be called the *logics.* For instance, the if-then rule is a kind of logic in English. And, behind every sentence in the conceptual world, there must be physical meanings. Most importantly, when a set of physical meanings evolves in space and time, a series of sentences could be generated. And, according to specific rules which are called *logic,* such sentences could be put together to form a *paragraph,* which describes what is called a *concept.*

Interestingly enough, concepts could also be engendered from the process of abstraction with the entities in the physical world. In other words, the process of abstraction manipulates the properties of a class of similar physical entities, and transforms these properties into a hierarchical representation of categories. Because of the result of such a categorization, new concepts will emerge. For instance, the concept of "vehicle" refers to the category of "vehicle". And, the concept of "vegetable" refers to the category of "vegetable". It is important to note that the mental power of human beings in undertaking the process of categorization propels the rapid expansion of the conceptual world.

Finally, when related paragraphes are logically composed together, the result will be a *text,* which describes what is called a *topic.* Therefore, the conceptual worlds also exhibit the meanings at two levels, namely: (a) the elementary meanings, and (b) the composite meanings.

Because the basic entities in the conceptual worlds are words, the properties and constraints of a lexical entity in a natural or technical language define what are called as *elementary meanings* in the conceptual worlds.

Similarly, as the conceptual worlds consist of words which act and interact, the properties and constraints of sentences, and their logical compositions, in the form

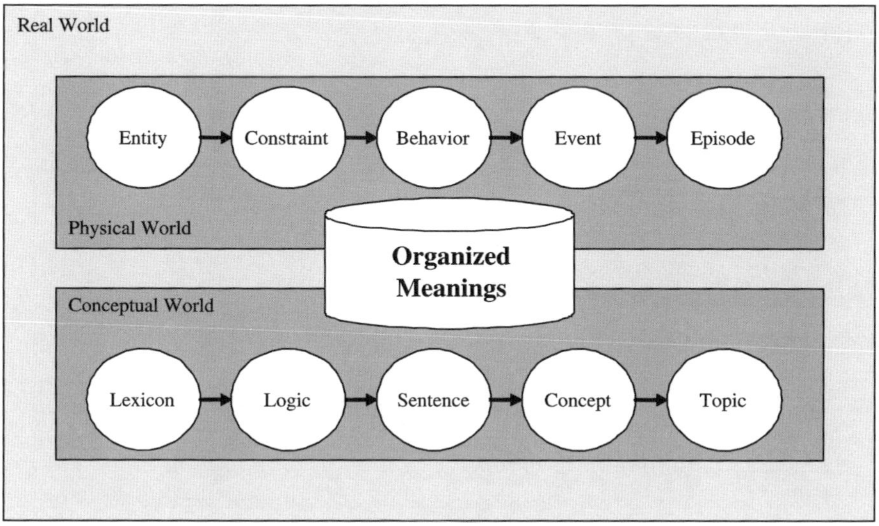

Fig. 6.3 The real world consists of: (a) the physical world, and (b) the conceptual worlds. In the physical world, every entity has its own properties and constraints. Because of constraints, two, or more, entities could act and interact. Such a phenomenon will engender behaviors, events and episodes. In a conceptual world which is represented by a human (or natural) language with extension to the technical language and programming languages, every lexical entity has its own properties and constraints (i.e. logics). And, the interaction of lexical entities according to logic constraints will create sentences and descriptions for concepts and topics.

of a natural or technical language, define what are called as *composite meanings* in the conceptual world.

In summary, the cognitive nature of human vision tells us that the real world can be considered as a combination of two worlds, namely: (a) the physical world and (b) the conceptual worlds, as shown in Fig. 6.3. And, the essences of both the physical world and the conceptual worlds are the meanings, which consist of properties and constraints.

6.1.7.3 *Visual Intelligence*

Intelligence is an attribute engendered by a brain. In other words, intelligence is manifested in the form of the dynamic response to meaningful signals by a brain. But, most importantly, intelligence is characterized by a brain's ability to acquire and manipulate meanings. Generally speaking, the mental behaviors, which manipulate the organized meanings inside a memory, include learning, analysis, synthesis, reasoning, decision-making, and result-making, etc.

In the physical world, we know that there are two main sources of signals, namely: (a) the visual signals, and (b) the audio signals. Therefore, the ability for a brain to transform visual signals into the cognitive state of knowing the properties and constraints behind visual signals should be considered as part of intelligence.

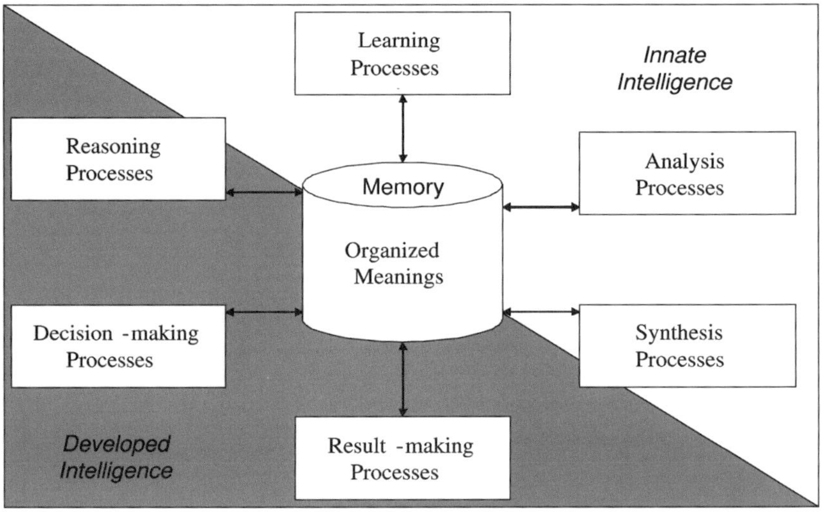

Fig. 6.4 A mind system, in which the mental behaviors of manipulating the organized meanings will give rise to the sensation of intelligence. And, the typical mental behaviors of manipulating the organized meanings include learning, analysis, synthesis, reasoning, decision-making, and result-making. As a human being has innate mental processes for intelligence at the time of birth, intelligence could be divided into two parts, namely: (a) the innate intelligence before the birth, and (b) the developed intelligence after the birth.

Refer to Fig. 6.4. We believe that all the mental processes in a living brain are centered around the *memory*, which internally stores the *organized meanings* perceived from the external world. And, the visual perception of the physical meanings of a scene should be the innate ability of a new-born baby. This is because there is no alternation of a new-born baby's brain after the birth.

Also, we believe that intelligence will consist of two parts. The first part is the so-called *innate intelligence*, which a person possesses at the time of birth. And, the second part is the so-called *developed intelligence*, which a person acquires after the birth. As intelligence is manifested in the form of the dynamic response to meaningful signals by a brain, now the question will be: What is a brain from an engineering point of view?

In Fig. 6.5, we depict a diagram, which outlines the inner modules at the lowest level of a so-called *visual intelligence system* (i.e. a system which outputs *intelligence* as a result of the dynamic response to visual signals by a brain). Then, from an engineering point of view, a brain can be considered as a combination of: (a) the memory, in which the internalized meanings are incrementally acquired, and (b) the mind, which heavily depends on four basic mental processes such as

- Meanings-Controller:
  When the information-flows coming from visual signals feed into a brain, the first step is to prioritize the information-flows according to the degree of relevance.

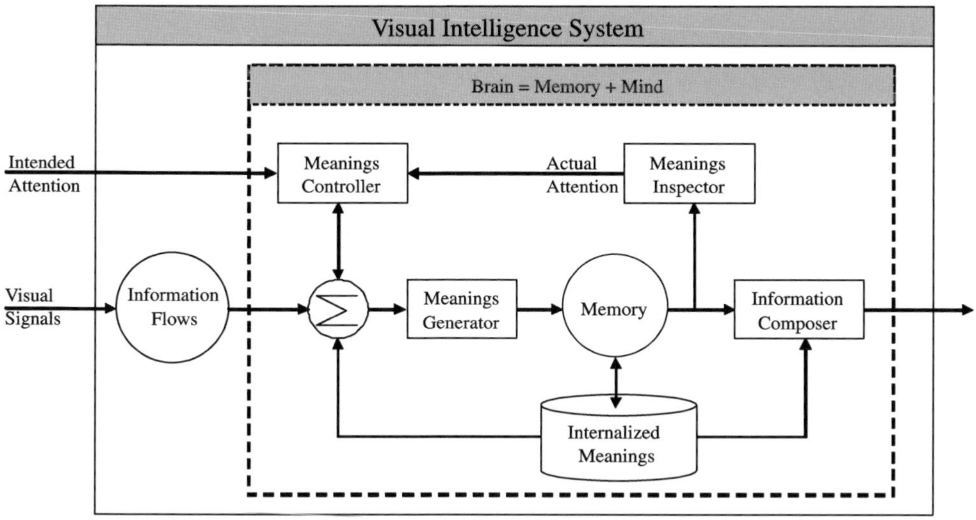

Fig. 6.5 A visual intelligence system, in which the input of information-flows as well as the feedback of internalized meanings are transformed into newly-learnt meanings by a closed-feedback loop, which consists of three basic mental processes such as: (a) meanings-generator, (b) meanings-controller, and (c) meanings-inspector. And, the mental process of information-composer will output the understood, or intended, meanings in the form of composed texts, synthesized images, or control-signals.

For instance, an information-flow, which makes the actual visual attention converge to the intended visual attention, will be marked with a higher priority. And, such a function of regulating the priority of information-flows will be the task of the mental process called the *meanings-controller*. On the other hand, for the purpose of extrapolating, or synthesizing, learnt meanings into new meanings, it is necessary to feed learnt meanings back to the input of the meanings-controller. Therefore, the meanings-controller will also regulate the priority of learnt meanings at the input.

- Meanings-Generator:
  Here, the meanings-generator could play three roles, namely: (a) meanings-extractor, (b) meanings-predictor, and (c) meanings-synthesizer. For instance, when an information-flow arrives as input, the meanings-generator will extract the concept-physical meanings behind the information-flow. Alternatively, with the feedback of learnt meanings, it is possible to extrapolate learnt meanings into new meanings. In this case, the meanings-generator plays the role of meanings predictor. Also, it is possible to combine a certain number of small pieces of learnt meanings into a larger piece of meanings. In this case, the meanings-generator actually undertakes the task of meanings synthesis.
- Meanings-Inspector:
  In order to focus the attention on intended meanings (or outcome), it is necessary to provide a feedback mechanism, which could measure the actual visual

attention. And, the mental process, which scans and detects the focused activities inside the memory, is called the *meanings-inspector.*

- Information-Composer:
  The dynamic response to meaningful signals by a brain could manifest in various ways such as action-takings, or decision-makings. This implies that a brain is able to transform the internalized meanings back to various forms of output, such as control-signals, texts, or mental-images. Here, the mental process, which translate understood, or intended meanings, into different types of output, is called the *information-composer.*

Based on Fig. 6.5, we believe that the interactions among the above-mentioned four mental processes will give rise to various mental behaviors such as learning, analysis, synthesis, reasoning, decision-making, and result-making, etc. For instance, the transformation of visual signals into the cognitive state of knowing the meanings of the objects in a scene depends on the closed-feedback loop, which will consist of the meanings-controller (i.e. to focus on intended visual attention), the meanings-generator (i.e. visual cognition and recognition), and the meanings-inspector (i.e. to measure actual visual attention).

6.1.7.4 *Visual Awareness*

In the literature of cognitive science, many discussions were devoted to visual awareness, and consciousness. Unfortunately, there is no commonly-accepted explanation about what the consciousness is in general, and what the visual awareness is in particular. For example, one may state that the consciousness is something close to what is described in the following hypothetical definition:

Definition 6.1. Consciousness is a state of certitude, in which one is aware of the sensory input, the effector output, the functioning of one's body and mind, and most importantly one's existence in harmony with one's beliefs, and the external world.

As the ultimate goal of vision is to enable a living (or artificial) creature to attain the state of visual awareness, it is important to reach a consistent answer to the question of what the visual awareness is.

As we discussed earlier, human vision not only outputs the meanings of a scene, or an image, in the physical world, but also the meanings in the conceptual worlds. But, one may say that human vision can only output the meanings of a scene, or an image, in the physical world in the ancient time when there was no natural language, or sign. However, it was probably the desire to communicate the perceived meanings in the physical world, that has motivated our ancestors to create human languages in the first instance, and then the conceptual worlds hereafter. With this consensus in mind, it will be possible to define what the visual awareness should be. In particular, we advocate the following definition:

Definition 6.2. Visual awareness is the cognitive state of knowing the meanings of a scene, or an image, in both the physical world and the conceptual worlds (i.e. representations by human languages).

As the consciousness is strongly correlated with the cognitive state of awareness, we can even postulate the following hypothetical definition of consciousness:

Definition 6.3. Consciousness is the dynamic response of a creature, as a result of interactions among the mental processes which manipulate the organized meanings derived from sensory inputs.

It is worth noting that the complete answer to the question of what the consciousness should be will be reached, only if we will be able to verify any theory of consciousness by building a conscious creature such as a conscious vehicle, robot or computer.

6.1.7.5 *Cognitive Output*

As we mentioned earlier, human vision is cognitive in nature. The output from human vision includes not only the physical meanings of a scene, but also its conceptual meanings. Since the conceptual world is the abstracted version of the physical world in the form of human languages as well as technical languages dedicated to science and engineering, this observation suggests that human vision adopts a hybrid representation in order to store the meanings of a scene in both the physical world and the conceptual world.

In engineering terms, it is to say that the representation of the physical meanings of a scene must embrace the representation of the conceptual meanings of the same scene, or vice versa. Interestingly, this inspiring property of human vision indicates that machine vision should also adopt a hybrid approach in order to adequately represent the meanings of a scene.

6.1.8 *Human Vision Is Attentive*

In the literature, visual attention is one of the often highlighted properties about human vision. And, there are many explanations with regard to visual attention. Most of them are based on the psychological experiments, or introspection. Here, we propose to examine visual attention from an engineering point of view.

We know that the physiology of human vision is very complex. However, analogy do exist between human vision and machine vision. For instance, image formation in both human vision and machine vision is realized in two steps, namely: (a) optical imaging (i.e. from light beams to the formation of optical images), and (b) digital imaging (i.e. from optical images to the generation of discrete visual signals or digital images). Interestingly, the optical imaging systems in both human vision and machine vision are quite similar.

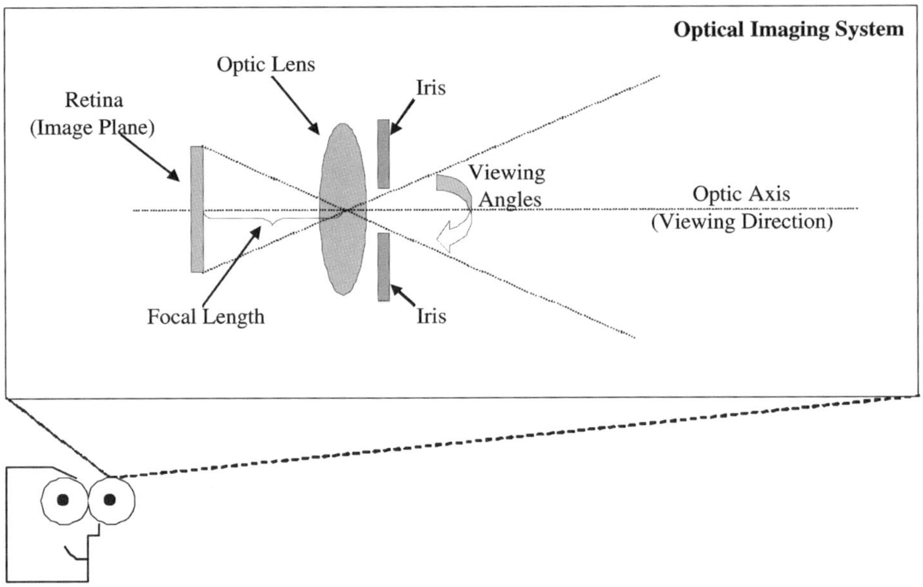

Fig. 6.6 The optical imaging system inside an eye (or camera) generally consists of: (a) an iris, (b) an optic lens, and (b) a retina (with or without fovea).

As shown in Fig. 6.6, the optical imaging system of an eye in human vision (or a camera in machine vision) generally consists of: (a) an iris, (b) an optic lens, and (c) a retina (with or without fovea). Inside an eye, the iris controls the important parameters, which are called the *viewing angles*. And, the lens controls another two important parameters, namely: (a) the optic axis (i.e. the viewing direction), and (b) the focal length. Also, the retina determines two parameters, which are: (a) the image resolution, and (b) the distribution of photo-receptors.

Clearly, the optical imaging system of an eye depends on a set of parameters. In theory, all these parameters could be time-varying. However, in human vision, only viewing angles, viewing direction, and focus lengths are time-varying parameters. Most importantly, any change of these time-varying parameters will cause the sensation of the change of *visual focus*, or *visual attention*.

As a visual attention corresponds to a set of specific values assigned to the parameters governing the optical imaging system of an eye, we can postulate the following definition of visual attention:

Definition 6.4. The sensation caused by a set of specific values assigned to the parameters in the optical imaging system of an eye is called as a *visual attention*.

Now, one may ask this question of what the advantages of having changeable visual attentions are. We believe that there are at least five good reasons for human vision to actively alter its visual attention.

6.1.8.1 *Resolution-driven Change of Visual Attention*

The image formation in human vision follows a geometrical principle, which is commonly known as the *perspective projection.* And, in a perspective projection, an infinitely large scene could be projected onto an infinitely small retina.

However, the retina in human vision has a fixed size. Thus, if the visual attention could not be changed, scenes of different dimensions will be perceived at different resolutions. Obviously, it would be advantageous, if scenes of different dimensions could be perceived at an equally-good resolution. In order to achieve this result, the visual attention should not be fixed.

6.1.8.2 *Distribution-driven Change of Visual Attention*

As we discussed earlier, human vision is composite, because of: (a) the high-resolution central vision, and (b) the low-resolution peripheral vision. Moreover, the distribution of photosensitive cells in the retina is not uniform. For instance, the highest density of photosensitive cells is found in the fovea of an eye.

This property implies that a same scene will not be perceived at an uniform resolution. For instance, when we look at a photo on a wall, the photo is not perceived at an uniform resolution inside an eye. Therefore, in order to achieve the visual perception at an uniform resolution (e.g. the best resolution possible), it is necessary to rapidly alter the viewing direction in order to randomly scan a scene, or a photo.

6.1.8.3 *Curiosity-driven Change of Visual Attention*

When a person enters an unknown environment, the first thought in his or her mind should be to gain the visual awareness of an unknown environment as quick as possible. In order to achieve such a result, it is necessary to alter the visual attention so as to rapidly build a mental "picture" about a new scene.

Similarly, when we read a new book (i.e. a new "scene" in a conceptual world), the early stage of our visual attention may be on the table of contents, and also graphic illustrations, in order for us to quickly assess whether it is worth reading it or not.

6.1.8.4 *Expectation-driven Change of Visual Attention*

When events, or behaviors, occur and evolve in space and time, it may be necessary to attend the visual focus to an immediately-expected event or behavior. For instance, when we hear some expected auditory signals, we may orientate our visual attention toward the source of these auditory signals.

Another example is that when we suddenly remember to look for a car-key, it is likely that we expect it to be on a desk, or inside a drawer. This expectation will drive our visual attention to specific lieu, or direction.

6.1.8.5 *Priority-driven Change of Visual Attention*

A complex scene usually contains multiple objects. But, not all of them are equally important, or relevant, to a particular task of visual perception. For instance, when we drive on a road, the visual attention is more on the road than into the sky. As a result, it is necessary to selectively attend the visual focus to one prioritized object, or lieu.

In summary, the inspiring property of visual attention by human vision suggests that human vision is active. Hence, machine vision should also be active.

6.2 Nature of Machine Vision

All creatures on earth are the results of evolution, or natural selection, which can be considered as a kind of design process from an engineering point of view. It is a matter of fact that most living creatures on earth do have vision. Then, why is it advantageous for them to possess vision? A possible answer is that vision allows them to be aware of the external world. And, the ability to gain such an awareness is a necessary condition for them to survive, or to achieve intended outcomes. In other words, the ability to see is a sufficient condition for a living creature to attain a certain level of autonomy and intelligence.

Machine such as vehicle or robot is a creature of human being. If a machine has to see and understand the external world, then it is necessary for a human designer to equip such a machine with vision. Also, the ability for a machine to see and understand a scene will be a sufficient condition for it to attain a certain level of autonomy and intelligence.

We know that human vision is astonishingly powerful. We are able to visually perceive and understand a complex scene instantaneously and effortlessly. However, despite the study on human vision for decades, not much is understood about the inner processes of human vision. This implies that it is not possible yet for us to follow the blueprint of human vision in order to design and implement machine vision. Nevertheless, we can still make our own effort to design a workable blueprint of machine vision. Hopefully, such an engineering attempt of developing machine vision will eventually pave the way to better understand human vision in future.

6.2.1 *What Is Machine Vision*

Vision appears so natural to human beings. Then, it seems difficult to give a durable answer to the question of what machine vision is all about. For instance, it is easy to say that humans, or animals, possess vision because of the needs for survival, such as finding foods, avoiding dangers, adapting to changing environment, and interacting in groups for intellectual, social and economic gains, etc.

Indeed, vision do provide a necessary condition for a living-creature's survival. However, the actual question is: What is the unique cause which makes vision a

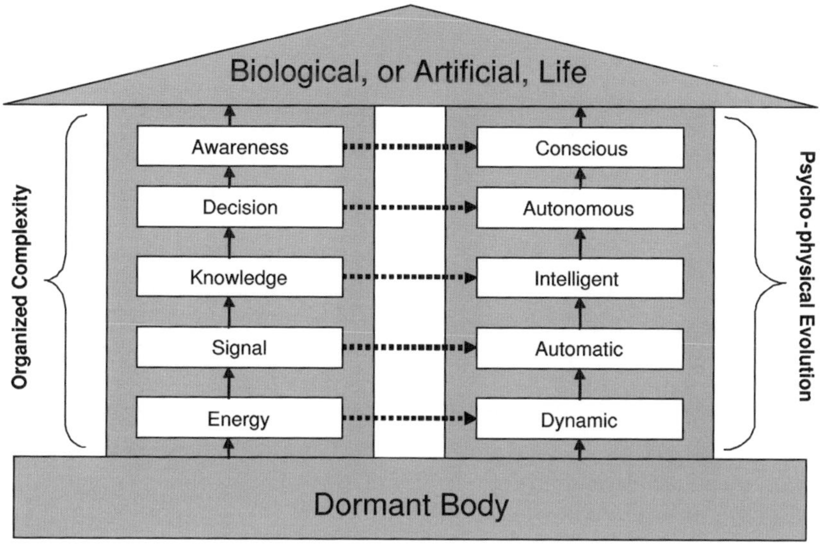

Fig. 6.7 A biological, or artificial, life is an organized complexity, which is the result of the psycho-physical development. For instance, if a dormant body can respond to energy, it becomes a dynamic entity. If a dynamic body can respond to signals, then it becomes an automated entity which can achieve intended behaviors. If a dynamic body can respond to not only signals but also the meanings behind signals, it will become an intelligent entity. If an intelligent entity has the ability of making its own decisions and acts according to its own decisions, it will become an autonomous entity. Finally, if a dynamic body can transform input signals into the cognitive state of knowing its inner and outer worlds, such a body will become a conscious creature.

necessity? In other words, what makes vision a necessary condition which a biological, or artificial, creature must satisfy?

In order to reach a convincing answer to the above question, it is necessary to examine the answer to the question of what a biological, or artificial, life is from an engineering point of view.

6.2.1.1 *Artificial Life*

As shown in Fig. 6.7, no one will claim that a dormant body, such as a rock or stone, has life according to our common sense. However, a dormant body will become a dynamic entity if it can respond to energy. For instance, a car could respond to energy, and accelerates when the input energy increases.

Then, at the next level of psycho-physical development, a dynamic body will become an automated entity if it is able to respond to signals in order to achieve intended behaviors, or motions. For instance, all automated machines in a production line could perform pre-programmed motions, actions, or even tasks.

Subsequently, a dynamic body will become an intelligent entity if it can respond to not only signals, but also to the meanings behind signals (i.e. knowledge). In fact, intelligence, or the ability to understand the meanings behind signals, is a necessary step for a dynamic body to reach the level of psycho-physical development, which is called *autonomy*. This is because such a mental ability will enable a dynamic body to make its own decisions. And, a dynamic body will become an autonomous entity if it can act according to the decisions made by itself.

Finally, if a dynamic body possesses sophisticated sensory systems which allow it to gain the awareness of its inner and outer worlds, and also to know the consequences of its own decisions and actions, such a dynamic body should be qualified to be a conscious entity. And, the consciousness will be the highest level of psycho-physical development, which a biological, or artificial, life could attain.

Therefore, from an engineering point of view, a biological, or artificial, life could be defined as follows:

Definition 6.5. A biological, or artificial, life is an organized complexity, which is the result of the psycho-physical development from the state of being a dormant body to the ultimate state of being a conscious entity.

6.2.1.2 *Artificial Vision*

Today, it is relatively easy for us to design a machine's body, and also to incorporate all kinds of sensors to a machine's body. With the help of sensory systems, two important sources of external signals that a machine's body could receive will be: (a) the visual signals, and (b) the audio signals.

Then, one may raise these two interesting questions: (a) How to transform visual signals into the cognitive state of knowing the meanings behind visual signals? and (b) How to transform audio signals into the cognitive state of knowing the meanings behind the audio signals? The former is the fundamental question, which cognitive vision must answer. And, the latter is the fundamental question, which cognitive linguistics should answer. In this book, we are concerned about the first question.

Based on the first question, it is easy for us to state that the ultimate purpose of artificial vision is for a machine to acquire the meanings of a scene from visual signals. And, it is necessary for a physical agent, such as a humanoid robot, to possess vision in order to gain the visual awareness about the external world for various purposes related to decision-making and action-taking, etc.

We know that a scene usually consists of entities called *objects*. The interaction between two (or more) objects will create what we call a *behavior*. When a behavior evolves in time and space, the result will be what we call an *event*. And, when multiple events act, and evolve, together in time and space, this will give rise to the notion of *episode*. Therefore, there will be no meanings if a scene has no object inside. And, the meanings of a scene generally refer to the properties and constraints of objects, behaviors, events and episodes in a scene. In engineering terms, vision could be simply defined as follows:

Definition 6.6. Vision is a process, which cognises and recognizes the meanings of a scene from visual signals.

6.2.2 *Machine Vision Is Computational*

Today's vision theory is dominated by the reconstructionism (or constructionism), or its variations. The doctrine of reconstructionism was advocated by David Marr in the early 1980s.

6.2.2.1 *Reconstructionism*

The main idea behind David Marr's theory is that vision should be concerned about the computational methods. The ultimate goal of these computational methods is to derive the representations of the geometrical properties behind the input visual signals. And, these representations are hierarchical at three levels, namely: (a) the primal sketch, which represents intensity changes in terms of features, (b) the 2.5D sketch, which represents the positions and orientations of surfaces or contours, and (c) the 3D sketch, which represents the three-dimensional models of objects.

In line with David Marr's theory, the major focus underlying computational vision in general, and machine vision in particular, is on the R&D of computational solutions, or algorithms, which could eventually lead to the success of: (a) image segmentation, (b) feature extraction, (c) feature grouping, (d) 3D reconstruction, and (e) object recognition.

Refer to Fig. 6.8. Reconstructionists believe that knowledge, or physical representations, are to be extracted from visual signals via features by computational solutions, or algorithms. As a result, the reconstructionists are in quest of answers to these two fundamental questions:

- What to compute? and
- How to compute?

However, despite many decades of scientific investigations in pursuit of David Marr's theory, the crucial obstacles in validating the doctrine of reconstructionism still remain.

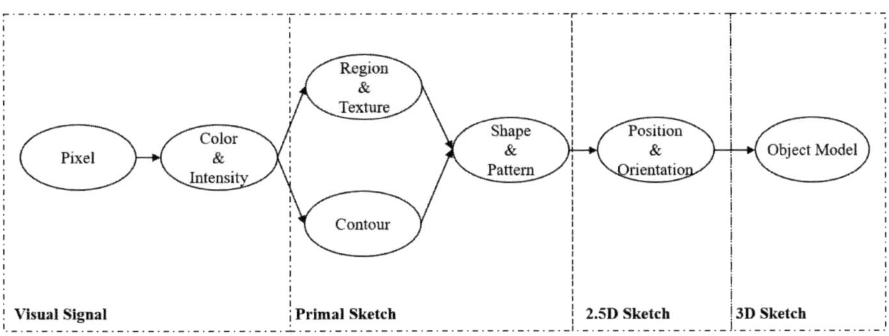

Fig. 6.8 Computational framework of the reconstructionism inspired by David Marr's vision theory, in which the knowledge, or the physical representation, of the objects in a scene are to be extracted from visual signals via features by computational solutions or algorithms.

6.2.2.2 *Problem of Robustness*

For instance, vision algorithms within the computational framework are primarily focused on the transformation of visual signals from one type of physical representation into another type of physical representation. As a result, these algorithms heavily depend on the manipulation of numeric values of input visual signals.

Unfortunately, the computations with numeric values are sensitive to noises, and variations. Thus, any extensive manipulation of noisy, or incomplete, numeric values will only make the final outcome worse.

6.2.2.3 *Problem of Adaptation*

On the other hand, the doctrine of computational vision does not emphasize enough on the importance of visual learning. In particular, it does not consider the importance to explicitly represent the meanings of a scene in a conceptual world.

For instance, reconstructionists think that the visual task of object recognition is a step to be undertaken after the success of 3D reconstruction. Unfortunately, it is still problematic with 3D reconstruction within the framework of reconstructionism, which heavily depends on the computations with input visual signals.

Moreover, we know that the visual task of recognition depends on the success of the visual task of cognition. In other words, recognition must be model-based. And, the models for recognition should be first designed, and then be autonomously built by a machine (or computer) through the process of cognition. However, within the framework of computational vision, it is not clear whether the models are autonomously learnt by a machine, or pre-programmed by a human being. But, it is clear to us that without the innate ability of cognition, the vision algorithms behind recognition will systematically suffer from the problem of adaptation, if there will be any change of the context in a scene.

6.2.3 *Machine Vision Is Cognitive*

In the previous discussions, we have shown that the doctrine of reconstructionism aims at computing the geometrical representation of a scene in the physical world. It advocates the computational approaches, which heavily depend on the manipulation of complex, and numerical, values. Therefore, the main concern of the reconstructionism is about what to compute and how to compute.

In contrast, cognitive vision stresses that both design and computation are equally important. In particular, cognitive vision pays equal attention to the questions of

- What are the visually-perceivable meanings in a scene?
- How to represent the visually-perceivable meanings in a scene?

- How to cognise the visually-perceivable meanings in a scene?
- How to recognize the visually-perceivable meanings in a scene?

6.2.3.1 *Meaning-centric Framework*

As we have discussed in previous chapter, there are two worlds in the mind of contemporary human beings, namely: (a) the physical world, and (b) the conceptual worlds. This observation indicates that the design of machine vision must consider the representation of the meanings of a scene in both the physical world and the conceptual world. In particular, it is important that the meanings of a scene in both the physical world and the conceptual worlds should be represented within an unified framework.

As it is clear to us now, the ultimate purpose of (cognitive) vision is to transform the input visual signals into the cognitive state of knowing the color, shape, pose, identity, category and behavior of the objects in a scene. Therefore, in order to achieve such a transformation, we must have clear answers to the following crucial questions:

- How to represent the meanings of an image's visual features in both the physical world and the conceptual worlds?
- How to represent the meanings of a scene's objects in both the physical world and the conceptual worlds?
- How to transform the input visual signals into the meanings of both the visual features in an image and the objects in a scene?

In Fig. 6.9, a meaning-centric framework is depicted, which aims at devising workable solutions to the design and implementation of cognitive machine vision. And, the basic principle behind such a meaning-centric framework consists of

- Top-down design in order to model the visual properties and constraints in both an image and a scene.
- Bottom-up transformation in order to cognise and recognize the visual properties and constraints in both an image and a scene.
- Hybrid representation of the meanings of the objects in a scene, in terms of color, shape, pose, identity, category, and behavior.

6.2.3.2 *Top-down Design*

We know that human vision can perceive the physical meanings of the objects in a scene. Also, human vision can learn the conceptual meanings of the objects in a scene. Most importantly, human vision can generate the internalized representation of the perceived objects so as to enable other mental activities such as imagination and planning, etc. Therefore, the first consideration in the mind of a human designer should be the answer to the question of what the representation of the concept-physical meanings at the various levels inside machine vision should be.

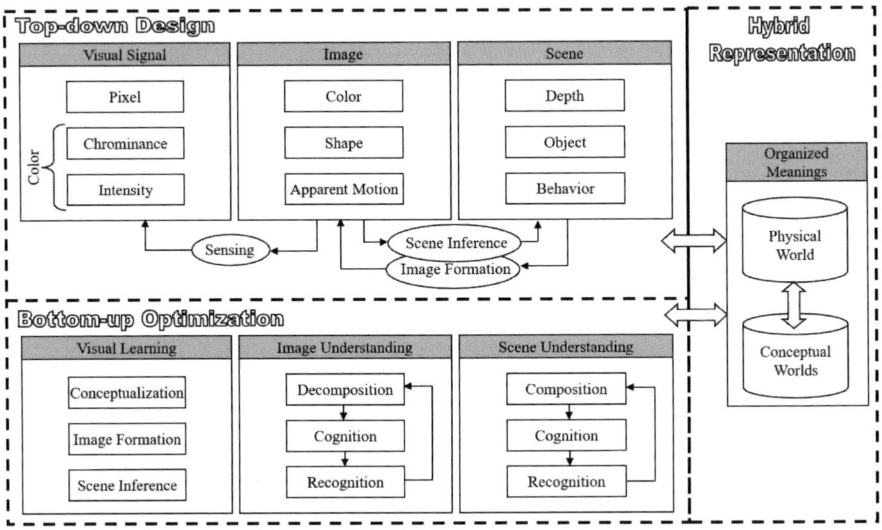

Fig. 6.9 A meaning-centric framework for the design and implementation of cognitive machine vision, in which the basic principle consists of: (a) top-down design in modelling the visual properties and constraints of both an image and a scene, (b) bottom-up optimization in cognising and recognizing the visual properties and constraints in both an image and a scene, and (c) hybrid representation of the visual properties and constraints of both an image and a scene in both the physical world and the conceptual worlds.

6.2.3.3 *Scene-level Meaning*

Refer to Fig. 6.9. A scene usually consists of objects, which act and interact in space and over time. And, geometrically, an object has its own color, shape, pose, identity, category, and behavior.

Hence, if a machine vision has to achieve the mental ability of visual perception, it is necessary to model both the physical meanings and the conceptual meanings of the objects in a scene. And, such modeling should be the result of a top-down design process, which a human designer of machine vision should be responsible for.

6.2.3.4 *Image-level Meaning*

In human vision, lights coming from the objects in a scene are projected onto the retina of an eye. And, such a projection forms an image in the retina of an eye. Similarly, in machine vision, an image will be the projection of lights, which come from the objects in a scene, onto the image plane inside a camera. In general, this process of projection of lights coming from a scene onto the image plane of a camera is called *image formation.*

When an image is formed, the objects in the image become two-dimensional. However, the two-dimensional objects in an image still preserve both the physical meanings and the conceptual meanings in terms of color, shape, pose, identity, category, and apparent motion. As a result, a human designer of machine vision must provide the clear answer to the question of how to model both the physical meanings and the conceptual meanings in an image. And, an important task in machine vision is to infer the meanings of the objects in a scene from the meanings of the objects in an image. Such a process is called *scene inference.*

6.2.3.5 *Signal-level Meaning*

Again, in human vision, the lights focused on the retina of an eye are converted into visual impulses (i.e. electric signals) by the photo-receptors of the eye. And, these visual impulses are then transmitted to the brain through optic nerve.

Similarly, in machine vision, the image sensors placed behind the image plane of a camera will convert the optically-focused lights into electric signals, which are then digitized and transmitted to a processing unit (e.g. a microprocessor, or an array of microprocessors).

And, inside the processing unit of machine vision, an image is represented as a matrix of pixels, each of which stores a color. Interestingly, a color is a cognitive entity because it has not only the physical meanings (e.g. the tri-stimulus values of a color) but also the conceptual meanings (e.g. the conceptual label such as red, green, blue, yellow, pink, etc). Therefore, it is necessary to design an appropriate model to represent both the physical meanings and the conceptual meanings at the level of visual signals.

6.2.3.6 *Bottom-up Optimization*

We know that the ultimate purpose of machine vision is to transform visual signals into the cognitive state of knowing the color, shape, pose, identity, category, and behavior of the objects in a scene. And, such a transformation is obviously a bottom-up process. Most importantly, in order to achieve such a transformation, a human designer of machine vision must answer these crucial questions of:

- What are the innate principles, or algorithms, which enable the bottom-up transformation from visual signal to visual cognition and recognition?
- How to guarantee the success and robustness of the bottom-up transformation from visual signal to visual cognition and recognition?

Clearly, if we have the answer to the first question, the bottom-up transformation from visual signal to visual cognition and recognition should be operated in an optimized manner in order to achieve the success and robustness.

Interestingly enough, human vision achieves such a requirement of robustness by making use of the principle of parallelism and selection. In other words, human vision runs, in parallel, all the possible paths of the bottom-up transformation. And, at a lower level within the chain (or network) of the bottom-up transformation, only the best results are selected, and are allowed to be further transformed at a higher

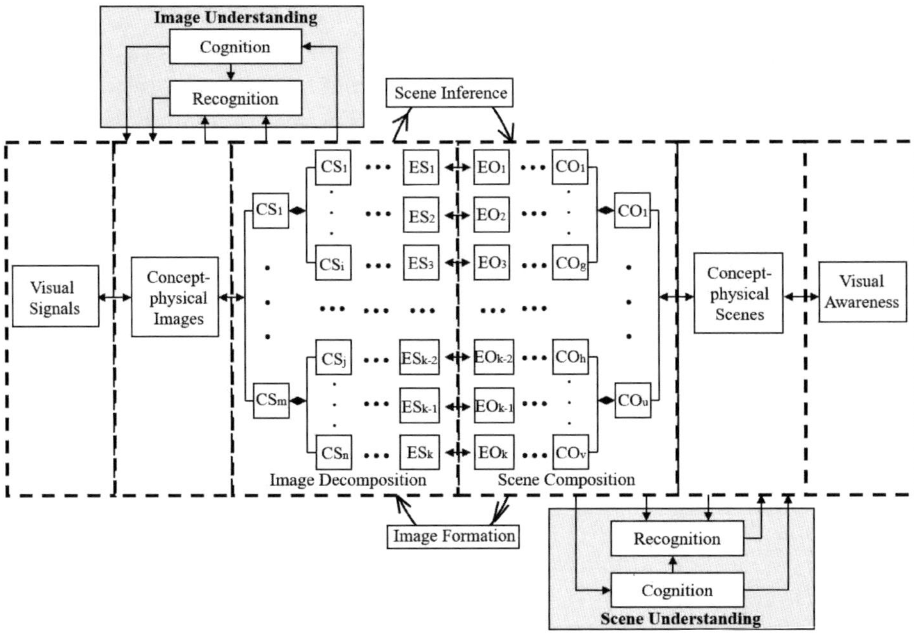

Fig. 6.10 Bottom-up optimization, in which visual signals are transformed into visual awareness. And, such a transformation are built on top of: (a) visual learning, (b) image understanding, and (c) scene understanding. During the process of bottom-up optimization, an image is decomposed into a hierarchical layout of elementary shapes (ES) and composite shapes (CS). And, the corresponding scene is constructed as a result of a hierarchical composition of elementary objects (EO), and composite objects (CO).

level. It makes sense to believe that this property of human vision should inspire human designers to look for appropriate computing architecture of machine vision. However, as this topic is beyond the scope of the book, we will not further elaborate on this issue.

Now, let us look at Fig. 6.10. We can see that the transformation from visual signals to visual awareness will go through two important steps, namely: (a) image decomposition, and (b) scene composition. The former aims at decomposing an image into a hierarchical layout of elementary shapes, and composite shapes. The latter is to construct a scene as a result of a hierarchical composition of elementary objects, and composite objects.

Most importantly, the success of the transformation from visual signals to visual awareness depends on the following three basic building blocks: (a) visual learning, (b) image understanding, and (c) scene understanding.

6.2.3.7 *Visual Learning*

In cognitive machine vision, visual learning plays three major roles, namely: (a) the learning of conceptual meanings (or models), (b) the learning of image formation, and (c) the learning of scene inference.

6.2.3.8 *Conceptualization*

Human vision is the result of natural evolution. And, through the long history of natural evolution, the innate principles and solutions underlying human vision have already been incorporated into the blueprint of human vision. However, it is not conceivable to build the contents of all the possible conceptual meanings about a scene, or an image, into human vision at the design stage. Instead, it is the actual fact that the innate principle of visual learning has already been put in place, which enables human vision to autonomously acquire the possible conceptual meanings of a scene, or an image, through the social interaction.

For instance, the sensation of a specific color has not only the physical meanings (e.g. tristimulus values), but also the conceptual meanings (e.g. a word such as red, green, blue, etc). Interestingly enough, when the color at each pixel of a physical image has been attributed with a conceptual label (e.g. a word, or a symbol), such a physical image is then called as a *concept-physical image*.

As machine vision is the result of design by human designers, it would be tedious if all the possible conceptual meanings about a scene have to be pre-built in advance at the design stage. On the other hand, it would be advantageous to design the innate principle of visual learning, which will enable machine vision to autonomously learn the conceptual meanings of a scene through human-machine interaction.

Also, the conceptualization of meanings should include the aspect of the visual learning of conceptual models. For instance, generative models such as equations of lines, surfaces, and complex shapes could be learnt by machine vision, instead of being pre-built in advance.

6.2.3.9 *Image Formation*

So far, no one exactly knows the innate principles and solutions underlying human vision, which transform the visual signals captured by human eyes into the cognitive state of knowing the color, shape, pose, identity, category, and behavior of the objects in a scene. This is because only the designer of human vision knew about these principles and solutions.

Nevertheless, a person with a normal vision could experience the process, in which various images of a scene, or an object, are perceived by his or her eyes. And, such experiences are certainly useful to help achieve better understanding of an image, or a scene. Here, one important role of visual learning is to acquire the appearance models of the objects in a scene. Obviously, it is not possible to pre-build such appearance models in advance at the design stage.

Similarly, a machine such as a vehicle or robot may not be aware of the engineering principle governing the formation of images inside its vision. But, it may be advantageous to incorporate the principle of visual learning into machine vision, which will enable a machine to autonomously learn various appearance models of the objects in a scene. In other words, it is useful for a machine to learn how the objects in a scene are projected into the image plane of a camera, and how images are formed inside a camera. And, it is our belief that the compositional rules of an image in terms of elementary shapes and composite shapes are first learnt, and then are applied to guide the processes of cognition and recognition in image understanding.

6.2.3.10 *Scene Inference*

It is a common fact that the images captured by human eyes from a scene only give the partial views of the scene. However, human vision is able to create the sense of a relatively-complete view of a scene. For instance, when we see the three legs of a table which has four legs, we do not have the impression that the table's fourth leg is missing. This observation strongly suggests that human vision has a certain level of awareness about an object's composition in a scene. Most importantly, such an awareness of scene composition is not pre-built in advance at the design stage, but is incrementally learnt through autonomous interaction in a scene (e.g. playing with, and around, objects in a scene, etc).

As the knowledge about scene composition is important to scene understanding, a human designer of machine vision must equip a machine with a mental capability, which will enable a machine to incrementally learn the possible compositions of a scene in terms of components, parts and entities. In other words, the purpose of such visual learning is for a machine to gain the experience of undertaking *scene inference*, which inversely maps an *image decomposition* into the corresponding *scene composition*. In our opinion, it is likely that the compositional rules of a scene in terms of components, parts and entities are first learnt, and then are applied to guide the processes of cognition and recognition in scene understanding.

6.2.3.11 *Image Understanding*

Refer to Fig. 6.9. Image understanding depends on a recursive loop, which consists of: (a) image decomposition, (b) cognition of concept-physical meanings, and (c) recognition of concept-physical meanings. And, the convergence of such a recursive loop will result in the final, and stable, understanding of an image, or a photo.

6.2.3.12 *Image Decomposition*

Refer to Fig. 6.10. The input to image understanding is a static image, or a photo. Therefore, the first step toward achieving image understanding is to decompose a static image (or a photo) into a hierarchical layout of elementary shapes and composite shapes. For instance, the proximal pixels having the same physical meanings (e.g. the same physical property) could be grouped together to form an elementary shape.

However, in order to achieve the robustness of grouping pixels into elementary shapes, it is useful to make use of the conceptual meanings of pixels. This requirement implies that the processes of decomposition, cognition and recognition should be operated in a recursive manner. In other words, the (partial) results of cognition and recognition should be feed back to the process of decomposition in order to produce better results of image decomposition.

Subsequently, at a higher level, several elementary shapes, which satisfy a geometrical constraint such as a circle, a rectangle, or a polygon, etc, could be put together to form a composite shape. Here, the selection of a geometrical constraint depends on the intention and result of cognition and recognition. Hence, it is necessary to operate the processes of decomposition, cognition and recognition in a recursive manner.

Finally, at the highest level, several composite shapes, which satisfy a topological constraint such a face, a human body, a car, or a plane, etc, could be put together to form a complex composite shape. As a topological constraint is predicted by the process of recognition, this fact also supports the claim that the processes of decomposition, cognition and recognition must be operated in a recursive manner.

6.2.3.13 *Cognition of Concept-physical Meanings*

Literally, cognition means to know, or be aware of, something which is usually unknown in advance. And, the two important tasks underlying visual cognition are: (a) visual computing, and (b) visual learning. The former aims at computing the physical meanings behind the visual signals. For instance, at each pixel, it is possible to compute the physical properties of its color. And, the latter is indispensable for the learning of the conceptual meanings associated with visual entities in an image. For instance, a shape could be conceptually described as a circle, a triangle, or a rectangle, etc.

Refer to Fig. 6.10. Under the context of image understanding, the process of cognition is to compute the physical meanings of pixels, and shapes. For instance, the cognised physical meanings of pixels will help achieve better results of grouping pixels into elementary shapes. Subsequently, the cognised physical meanings of elementary shapes will be used to further guide the task of arranging elementary shapes into composite shapes. And, this kind of interaction between decomposition and cognition will continue until the final success of understanding an image. Clearly, such an interaction demonstrates the importance of the recursive loop of decomposition, cognition and recognition.

6.2.3.14 *Recognition of Concept-physical Meanings*

Literally, recognition means to categorize and identify something which is partly, or totally, known in advance. For instance, if a machine has learnt the association between the appearance of an apple and the conceptual meanings of the word "apple", this machine should be able to recognize the conceptual meanings of an apple, if an input image contains the appearance of a similar apple. Interestingly, this example shows that the process of recognition depends on the prior knowledge (i.e. the concept-physical meanings) of visual entities, which are known in advance. Now, the question will be: How to gain such a prior knowledge?

For human vision, there is no doubt that the prior knowledge used for recognition is obtained by the process of cognition only. In other words, the physical meanings are computed by human vision, while the conceptual meanings are learnt through the interaction in an environment. This is because human vision is not re-programmable after the birth of a human being. However, a human designer of machine vision has the freedom to program, or re-program, the prior knowledge used for recognition by a machine. Therefore, a machine could obtain the prior knowledge used for recognition in two ways, namely: (a) cognition, and (b) programming. But, in general, it is worth noting that it is not possible for a machine to autonomously achieve recognition, if the machine is not able to undertake cognition.

Refer to Fig. 6.10. Under the context of image understanding, it should be relatively easy to undertake the visual task of recognizing the concept-physical meanings of the visual entities (e.g. pixels, or shapes) in a two-dimensional image space. This is because 2D-to-2D matching behind the process of recognition is a tractable problem.

However, we know that human vision is able to transform the visual signals coming from a two-dimensional image space into the cognitive state of knowing the colors, shape, poses, identities, categories, and behaviors of the objects in a three-dimensional scene space. In other words, the input is a 2D image and the output is a 3D scene. Then, we may raise this question: How to achieve this kind of 2D-to-3D matching behind the process of recognition?

6.2.3.15 *Scene Understanding*

Refer to Fig. 6.9. The success of scene understanding depends on a recursive loop, which consists of: (a) scene composition, (b) object cognition, and (c) object recognition. Also, the convergence of such a recursive loop will result in the final, and stable, understanding of a scene.

6.2.3.16 *Scene Composition*

In the scenario of image understanding, the input includes: (a) a static image (or a photo), and (b) the prior knowledge about the visual entities which have been learnt in advance. However, in the scenario of scene understanding as shown in Fig. 6.10, the input is the live images (or videos) captured by the cameras of a machine, or the eyes of a human being. In addition, the prior knowledge about the familiar visual entities is optional. Therefore, the first step towards scene understanding is to compute the depth of the objects in a scene. In chapter 5, we will study a cognitive approach to the visual perception of depth.

After the visual perception of depth, the result will a three-dimensional matrix of spatial locations. And, the subsequent task is to decompose such a three-dimensional matrix into a hierarchical composition of a scene in terms of elementary objects and composite objects. Clearly, such a decomposition depends on the intention, or the expected outcome, at the level of object cognition, or object recognition. This requirement implies that the processes of scene composition, object cognition and object recognition should be operated within a recursive loop.

6.2.3.17 *Object Cognition*

Refer to Fig. 6.10. Under the context of scene understanding, the input to object cognition could be the partial result of scene composition. And, the purpose of object cognition is two folds. First of all, it is necessary to compute the physical meanings of a new object, which a machine does not know in advance. Second, through the process of object cognition, the conceptual meanings of a new object will be learnt by a machine.

As a live image of a new object only gives a partial view of it, it is necessary to cognise a new object from different views in order to obtain as complete meanings of a new object as possible. However, if the different views of a new object come from different scenes at different time periods, then it is necessary to invoke object recognition in order to fuse the results of cognition. Clearly, this requirement supports the principle of recursively operating the processes of scene composition, object cognition and object recognition within a loop.

6.2.3.18 *Object Recognition*

Under the context of scene understanding, the input to object recognition is the partial result of scene composition. Then, through a process of 3D-to-3D matching, it is possible to use the partial knowledge about an object in a scene to identify the cognised model of the object in the memory of a machine. In return, the recognized model of an object will help to fine-tune the result of scene composition. Therefore, it is natural that the processes of scene composition, object cognition and object recognition should be operated within a recursive loop.

6.2.3.19 *Verification of Concept-physical Meanings*

Within the framework of cognitive vision, we advocate a hybrid representation to organize the concept-physical meanings of a scene perceived by a machine in both the physical world and the conceptual worlds. And, such a hybrid representation should be the result of the top-down design by a human designer of machine vision. Moreover, the content inside such a hybrid representation should be incrementally enriched by the visual experience of a machine. Then, one may raise this relevant question: Will such a hybrid representation of the concept-physical meanings make a machine to actually gain the visual awareness about a scene, or the cognitive state of knowing a scene? In other words, the concern here is how to verify, or attest, that a machine has gained the visual awareness, or the cognitive state of knowing a scene.

In order to answer this question, it is necessary to incorporate the mental processes, such as reasoning, decision-making, and action-taking, into a machine's brain, and to let these mental processes act or interact with the organized meanings. In other words, it is only through the observable dynamism of manipulating the internalized meanings that we will be able to assess the degree of visual awareness gained by a machine.

For instance, certain types of visual awareness will trigger some specific sentiments, or emotions, in human beings. Therefore, it will be interesting to test the degree of a machine's visual awareness by observing the sentiments, or emotions, manifested by the machine. However, this method of verification is strongly linked to humanoid robotics, which is beyond the scope of this book.

On the other hand, it is possible to ask a machine to describe what it sees, in a human language. This is because the conceptual description of a scene is only possible, if a machine has understood the concept-physical meanings of a scene.

6.3 Summary

In this chapter, we have studied the nature of human vision, which is the most important sensory system for knowledge acquisition. Vision enables human beings to discover new knowledge from doing experiments, venturing into exploration, and inventing tools and machines. Vision also enables human beings to quickly transfer knowledge through the process of writing and reading. Inspired by the astonishing power of human vision, machine vision has been under intensive research and development. Hence, we have studied some important aspects that machine vision must possess in order to match the performance of human vision for tasks such as color perception, shape perception, and depth perception, which are the three pillars underlying image understanding and scene understanding. These mental capabilities will enable machines of future to successfully achieve knowledge acquisition from visual signals.

6.4 Exercises

(1) Are today's computers aware of the contents of programs inside them?
(2) Are today's computers aware of the algorithms underlying the programs inside them?

(3) Are today's computers aware of the meanings of the results computed by the programs inside them?
(4) Comment on the question of whether or not the detail of human mind (i.e. the behavior of a human brain) can be discovered (or explained) through the observation and analysis of the signals emitted by the nerve cells inside a human brain.
(5) Comment on the question of whether or not the detail of the operating system inside a modern computer can be discovered (or explained) through the observation and analysis of the signals detected from the electronic devices and chips inside such a computer.
(6) Comment on the question of whether or not the pure computation with visual signals will lead to the cognitive state of knowing the concept-physical meanings of a scene.
(7) Find existing answers in the literature to the question of what the visual awareness is.
(8) Explain the purpose and importance of top-down design in cognitive machine vision.
(9) Explain the purpose and importance of bottom-up optimization in cognitive machine vision.
(10) Explain the purpose and importance of hybrid representation in cognitive machine vision.
(11) Explain the difference between cognition and recognition.
(12) Explain the difference between image understanding and scene understanding.
(13) Explain the difference between visual computing (i.e. computation with visual signals) and visual learning.
(14) Why is human vision composite?
(15) Why is human vision developmental?
(16) Why is human vision cognitive?
(17) What are the meanings of a scene?
(18) How to achieve robustness in visual perception?
(19) What is intelligence?
(20) How to perceive depth?
(21) What is visual attention?
(22) What is visual awareness?
(23) What are the important coordinate systems in human vision?
(24) Assume that you have never learnt the Chinese language. When you see "a new product for free" in Chinese, will you be able to extract the conceptual knowledge of "a new product for free" in English from the visual signals sensed by your eyes?
(25) When you see a region in red, your eyes will output visual signals corresponding to the red color, to your brain. You are asked to comment on the question of

whether or not the lights emitting from the red region also carry the conceptual knowledge of "red" in English, or in Chinese.

(26) Search in the Internet and find the explanation about the anatomy and physiology of human eye. And, explain what is fovea?

(27) The philosophy of Realism states that meanings are independent of the mind. Give your comments on this statement (e.g. is it partially, or totally, true?).

Chapter 7

Knowledge Acquisition from Color

7.1 Introduction

All entities in the universe are the manifestation of energy. Most importantly, the dynamics in the universe is animated by energy transfers from one domain to another domain (e.g. from mechanical energy to electrical energy). Among the various forms of energy transfer, radiation has a particular importance in vision.

In fact, all entities in the physical world emanate in one way or another. The electromagnetic waves emanated from an entity are the movements of the particular energy carriers called *photons*. In physics, we know that the emanation depends on not only the material property, but also the geometrical property, of an entity's surface. Therefore, the emanation makes it possible to sense the appearance of an entity's surface.

It is interesting to take note that an electromagnetic wave is characterized by two physical quantities, namely: (a) the intensity, and (b) the wavelength (or frequency). Within the spectrum of wavelengths between 390 and 780 nanometers, the electromagnetic waves are called the *visible lights*. This is because the wavelengths within this interval can be sensed by human eyes. In addition, it is worth noting that lights at different wavelengths will cause different visual sensations in the consciousness, or psychology, of a human being. And, the visual sensations of this type are commonly known as *colors*.

Under a normal condition of illumination, human eyes will see a colorful world. If a machine such as a vehicle, or a humanoid robot, will co-exist with human beings in a made-for-human environment, it is necessary for it to have a similar perceptual capability in terms of color perception. Hence, the first ability of a machine's visual perception should be the perception of color, which is the first step toward the acquisition of new knowledge from visual signals in the physical world. Here, we do not downplay the importance of other types of signals which could also contribute to the task of knowledge acquisition. However, the focus here is intentionally placed on knowledge acquisition from visual signals.

In general, visual perception involves: (a) detection, (b) modeling, (c) cognition and (d) recognition. Therefore, in this chapter, we will study the biologically-inspired fundamentals which could enable a machine to sense, model, cognise and recognize colors. So, we begin with the study of the basics underlying color imaging in machine vision.

7.2 Transformation From Light to Color Image

In the literature, most textbooks on computer vision, or machine vision, usually devote one whole chapter to discuss about the various aspects of image formation (or camera design). In this book, we adopt a different approach, in which the relevant aspects of image formation (or camera design) are tightly associated with the relevant topics which are to be exposed in different chapters. In fact, this spirit is in line with the philosophy of top-down design in modeling, because a particular aspect of image formation (or camera design) should be dictated by one, or more, particular need in vision.

In machine vision, two important aspects of image formation (or camera design) are: (a) the sensing of color (i.e. to detect colors of the incoming lights), and (b) the sensing of depth (i.e. to compute the positions from which lights emanate toward cameras). In this chapter, we study the first aspect.

7.2.1 *Light Gathering*

Human beings are living in an infinitely large universe. However, our eyes have a fixed small size. Therefore, the first requirement of designing an artificial eye is to gather lights coming from a large scene into an area of a small eye. On the other hand, a scene is normally a three-dimensional space, where the emanations of lights may occur at different locations and in different directions. Hence, the second requirement of designing an artificial eye is to sense the directions, locations, or both of them, with regard to light emanations.

In theory, it would be very difficult to directly sense the locations of light emanations. This is because there are an infinite number of locations in a same direction of an incoming light ray. And, it is impossible to place an infinite number of photo-detectors, or photo-receptors, inside a small eye, each of which is especially sensitive to lights coming from one particular location. Although it is possible to make use of a focusing device to discern a location along a direction of an incoming light ray, it poses a big challenge to incorporate a large number of focusing devices inside a small eye, each of which could take care of the process of focusing in one particular direction of an incoming light ray. By examining a human eye, one could notice that the compromise in formulating the requirement of eye design is to first sense the directions of incoming light rays, and then to derive the locations of emanations indirectly.

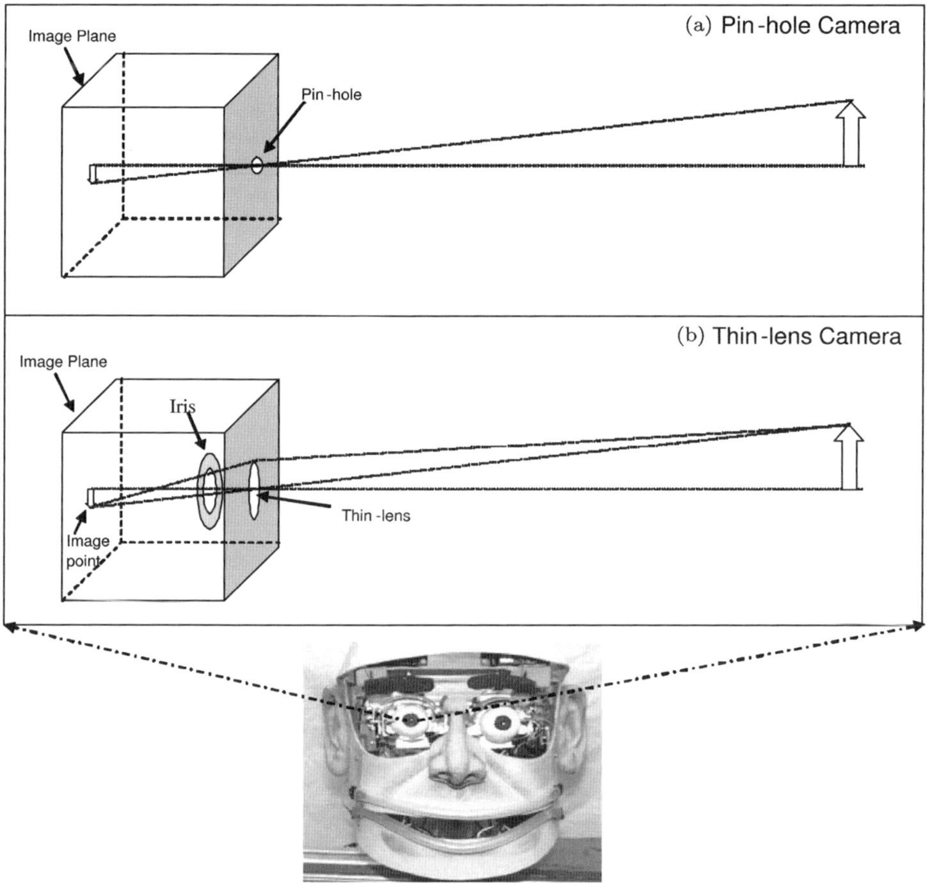

Fig. 7.1 The incoming lights can be gathered by either a pinhole camera in which the incoming lights go through a small hole, or a thin-lens camera in which the thin lens focuses the incoming lights and projects them onto the image plane.

As a matter of fact, the simplest solution to sense the directions of emanations is to design a pin-hole camera as shown in Fig. 7.1(a). In a pin-hole camera, light rays coming from a location in a scene pass through a tiny hole, and form a spot on the image plane. If a photo-detector, or photo-receptor, is placed behind this spot, the intensity and wavelength of an incoming light could be converted into the corresponding electric signals. However, there are two problems with a pin-hole camera, namely:

- Weak response to intensity:
  In order to form shape images, the pin-hole has to be as small as possible. However, smaller the pin-hole, lesser the amount of photons will come inside the

camera. Hence, a pin-hole will cause a poor response to energy emanated from a scene.

- Blurring effect due to refraction:
  In optics, we know that lights will deviate from their straight paths when they pass through a narrow gap (e.g. grid). This phenomenon is called *refraction.* This means that a small pin-hole may cause refraction of incoming lights. And, refraction will deteriorate the sharpness of images.

Interestingly enough, the shortcomings of a pin-hole can be overcome by substituting the pin-hole with a thin-lens, as shown in Fig. 7.1(b). With a thin-lens in place of the pin-hole, the light rays coming from a location in a scene can still pass through the center of the lens. However, a light ray coming from another direction will be re-directed by the lens, which will make it converge to a common point called the *image point.* When an image point is in focus, the thin-lens will make all the incoming lights from a same location (in a scene) converge to this image point. Because of this property, the geometrical model of a thin-lens is equivalent to the geometrical model of a pin-hole. This aspect will be discussed in detail when we study the visual perception of depth (i.e. positions from which light rays emanate towards cameras).

Refer to Fig. 7.1(b). When a thin-lens is put in place for the purpose of substituting the pin-hole, the effective openness of the thin-lens is under the control of a device called *diaphragm*, or *iris*, which is playing the role similar to a human eye's iris. In other words, the iris (or diaphragm) of a camera is to control the amount of lights entering the camera. For instance, when the light intensity is too high, the iris should open less. And, when the light intensity is weak, the iris should open more.

7.2.2 *Color Sensing*

With a thin-lens, a machine's eye is able to capture the incoming lights from a same location in a scene, and focus them onto an image point for the purpose of sensing lights. As a light's wavelength will cause the psychological sensation of color in the mind of human beings, light sensing also refers to *color sensing.* And, color sensing should measure both the intensity and wavelength of a light.

People may believe that it should be an easy task to sense a color, as long as there is a photo-detector, which can measure the wavelengths within a certain range. In practice, it is not so easy because the quantity of intensity is coupled with the quantity of wavelength. In other words, the relationship between the output of a color sensor and a color itself is not monotonic. For instance, let $f(L, I)$ be the output of a photo-detector, or photo-receptor. And, the output is a function of both wavelength L and light-intensity I. With a fixed value of I, the relationship between $f(L, I)$ and L may be monotonic. But, when I varies, the relationship between $f(L, I)$ and L will vary as well.

As a photo-detector, or photo-receptor, will only output one value at a time instance, it is generally impossible to discern the wavelength corresponding to a

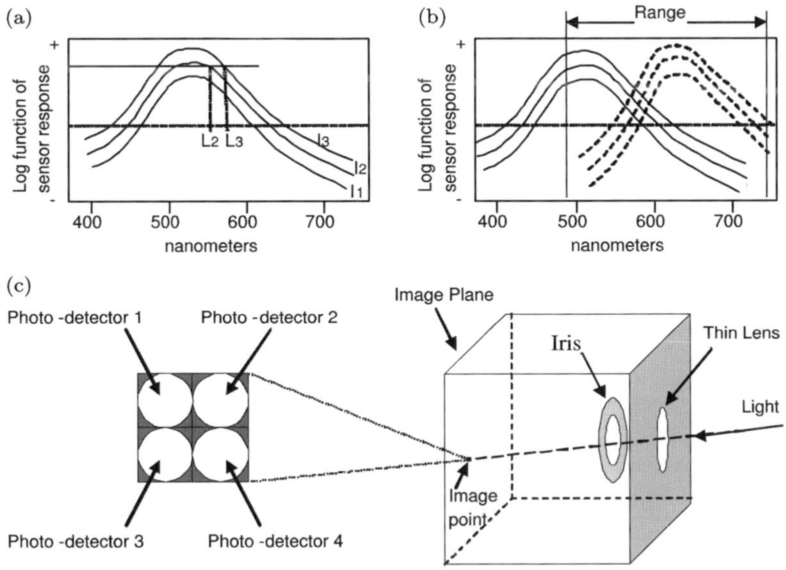

Fig. 7.2 A photo-detector, or photo-receptor, will output a value, which is a function of both light's wavelength and light's intensity. When light's intensity changes, the relationship between the output (e.g. the log function of sensor's response) and wavelength will also change. As shown in (a), it is not possible to discern a wavelength with a single sensor. As shown in (b), in order to discern a wavelength and produce the psychological sensation of color, it is necessary to use, at least, two sensors of different characteristics. In practice, one can place four sensors of different characteristics behind each image point, as shown in (c).

particular sensor-output. Figure 7.2(a) shows some hypothetical responses of a photo-detector, or photo-receptor, as a function of intensity (I) and wavelength (L). Given a specific level of response as indicated by the solid horizontal line, there are at least two possible wavelengths, such as L_2 and L_3, which conform to this level of response. Therefore, in general, one type of photo-detector, or photo-receptor, is not sufficient to discern colors of incoming lights from a particular direction. However, if an environment maintains a stable illumination (i.e. the intensity almost remains constant), it is possible to sense lights of different wavelengths with one type of photo-detectors, or photo-receptors. In biology, there are natural creatures, in which their eyes have only one type of photo-receptors.

If it is necessary to have more than one type of photo-receptors for color sensing, the question will be: Are two photo-detectors, or photo-receptors, sufficient enough so as to discern colors? The answer is affirmative as shown in Fig. 7.2(b). When the characteristics of two photo-receptors are not the same, we have two different output functions, such as $f_1(L, I)$ and $f_2(L, I)$. In other words, two output values from $f_1(L, I)$ and $f_2(L, I)$ permit unique solutions to two unknown L (i.e. wavelength) and I (i.e. light-intensity). Surprisingly enough, some natural creatures (e.g. maritime mammals) materialize this solution for the design of their eyes.

In theory, the solution of using two photo-detectors of different characteristics will only work within their common range of wavelengths as indicated in Fig. 7.2(b), which may not cover the full range of wavelengths of the visible lights. Fortunately, this problem can be easily solved by adding one, or more, photo-detector of different characteristics. And, it seems that a good solution is to have, at each point on the image plane, three photo-detectors of different characteristics in order to sense the colors of visible lights. Interestingly enough, the nature has adopted this solution for the design of human eyes. As a matter of fact, there are three types of cones in a human eye, which are responsible for color sensing. A human eye also has rods, which are responsible for intensity sensing and night vision. In addition, rods have a peak response around the green color. This explains why human eyes are much more responsive to green color than any other colors.

Inspired by the properties of human eyes, a similar solution could be adopted for the design of machines' artificial eyes. As shown in Fig. 7.2(c), a general solution is to place four photo-detectors, or photo-receptors, of different characteristics behind each image point in order to sense the colors of incoming lights which are focused at this image point. In particular, at each image point, there are at least one photo-detector with the peak response to red, another one with the peak response to green, and a third one with the peak response to blue. Not surprisingly, the fourth photo-detector is normally chosen to be sensitive to green, like what a rod of human eye does. The combination of outputs from these three (or four) photo-detectors will allow a machine to re-produce, or perceive, the colors that human beings daily experience. In literature, the values, which respectively measure the responses to red, green and blue, are called *tristimulus values*.

7.2.3 *Color Synthesis*

Color has been extensively studied in the literature. One obvious motivation behind this study is to understand color vision, or color perception. However, another important motivation is to synthesize colors for other purposes such as display (e.g. TV), printing (e.g. color magazine) and painting (e.g. color artworks).

Here, it is important to take note that color perception has nothing to do with color synthesis. For instance, a set of three photosensitive cells of different characteristics is sufficient enough to discern the colors of visible lights. But, it is not necessary for them to be specifically sensitive to the triplet of (red, green, blue), or the triplet of (cyan, yellow, magenta). In theory, the three photosensitive cells behind an image point can be sensitive to any triplet of colors, as long as the combined response-range covers the whole range of wavelengths of the visible lights.

Interestingly enough, the study of color synthesis shows that there are two methodologies for mixing colors in order to create new ones. These two methodologies are:

- Additive Color Mixing Systems:
  For example, the triplet of (red, green, blue) permits the synthesis of any color. As the equal mixture of red, green and blue will produce the white color, this system is called an *additive color mixing system*. And, most of today's computer

screens and TV sets adopt additive color mixing systems to synthesize and display colors.

- Subtractive Color Mixing Systems:
  For instance, the triplet of (cyan, yellow, magenta) also permits the synthesis of any color. But, the equal mixture of cyan, yellow and magenta will produce the black color. This is why it is called a *subtractive color mixing system*. In printing industry, subtractive color mixing systems are widely used for the production of colors on printing materials such as (white) papers.

It is worth noting that under color synthesis, there are extensive studies on various color models, and color image processing techniques. However, in machine vision, what we are interested in are the suitable models for color sensing as well as color perception. For this reason, we will focus our discussion on RGB and L*u*v color models in the next section. Also, it is important to take note that in literature, the values attributed to red, green and blue, which allow the synthesis of any arbitrary color, are called *trichromatic values*.

7.2.4 *Color Image Formation*

In today's color cameras used for machine vision, an image plane is purposely designed by human designers, and is composed of an array of photosensitive cells, which are tightly aligned together to form a tiny rectangular plane. In particular, a photosensitive cell is specifically designed with a photo-detector and a dedicated color filter. Such design, or combination, allows a photosensitive cell to output a signal which is a function of light's intensity and a particularly chosen color such as red, green or blue. As a result, the three, or four, photo-sensitive cells placed behind an image point of a machine's eye are sufficient enough to sense the colors of incoming lights which are focused at this image point.

Mathematically, an image point on an image plane can be treated as an element of a two-dimensional matrix. In machine vision, a two-dimensional matrix representing an image (i.e. picture) is called a *picture map* or *pixmap* in short. And, an element of a pixmap is called a *picture element* or *pixel* in short.

Refer to Fig. 7.3. At each pixel, there are at least three corresponding photosensitive cells which output three values representing the red, green and blue, respectively. Therefore, the spatial dimension of an image is two. But, the photometric dimension of an image is three. This means that an image should be treated as a combination of three pixmaps (R, G, B) corresponding to the red, green and blue, respectively. If we denote (r, c) (i.e. row index r and column index c) the spatial location of a pixel, image I of the size of $n \times m$ (i.e. n rows and m columns) can be described by

$$I = \{R(r,c), G(r,c), B(r,c) \mid r \in [1,n],\ c \in [1,m]\}. \tag{7.1}$$

Alternatively, we can also model a pixel with two sets of coordinates, namely: (a) the image coordinates (i.e. (r, c)), and (b) the color coordinates (e.g. (R, G, B)).

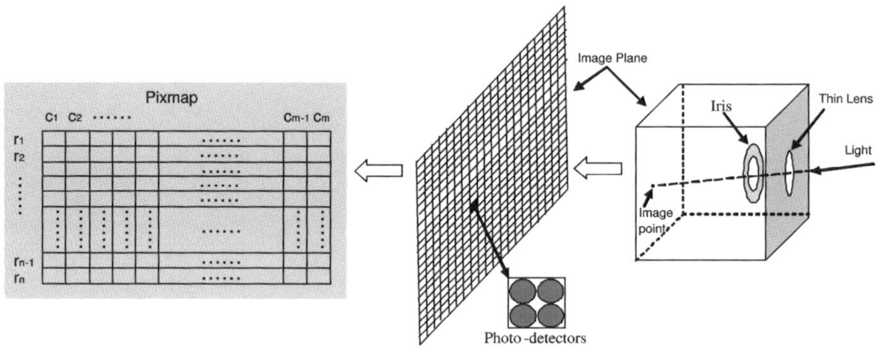

Fig. 7.3 A pixmap is a two-dimensional matrix representing an image, in which each element corresponds to one image point where there are at least three photosensitive cells for the sensing of colors of incoming lights.

Such representation is useful because we can further extend it to include other sets of coordinates such as the coordinates in a three-dimensional scene. Hence, in terms of coordinates, image I can be described by

$$I = \{(r, c, R_{r,c}, G_{r,c}, B_{r,c}) | r \in [1, n],\ c \in [1, m]\}. \tag{7.2}$$

7.3 Representation of Color

In machine vision, lights are first converged through the optical lens and then sensed by the photosensitive cells inside a machine's eyes. At each image point, there are at least three photosensitive cells of different characteristics. Most importantly, at an image point, the outputs from a set of three photosensitive cells of different characteristics permit both the perception and re-production (e.g. display, or printing) of colors.

At sensor level, it is clear that a color is measured by three quantities. But, the perception of a specific color is conceptual at cognitive level. Hence, the following questions arise:

- How to physically represent colors?
- How to represent the conceptual meanings of colors?

In this section, we will study both physical and concept-physical color models, which could be designed in advance for a machine's vision system.

7.3.1 *Color's Properties in Physical World*

As the outputs from three photosensitive cells behind an image point are independent, we can define a three-dimensional space, in which the principal axes serve as the references to these three outputs, respectively. Mathematically, we call such a space a *color space*.

7.3.1.1 *RGB Color Space*

In 1802, Thomas Young has postulated a hypothesis which stated that there are three types of photo-receptors in an human eye and that these three types of photo-receptors have the peak responses around the red, green and blue (or violet), respectively. In 1959, Young's hypothesis has been proven by a team led by the Nobel laureate, George Wald. This discovery shows that at sensor level, human vision adopts a color space represented by three axes corresponding to red, green and blue. This color space is now commonly called the *RGB color space.*

Inspired by the properties of human eyes, the designers of machines' eyes have adopted a similar top-down design approach. In fact, the solution behind the design of machines' eyes purposely chooses three types of color filters for the photosensitive cells, which are specifically transparent to red, green and blue, respectively. In this way, the output from an image point at the image plane inside a machine's eye can be represented by a point in the RGB color space. In literature, the red, green and blue are also called three *primary colors.*

Based on the above-mentioned solution of design, an image point inside a machine's eye will give rise to three values corresponding to three primary colors such as red, green and blue. And, the three outputs are functions of intensities (I) and wavelengths (L) of incoming lights. In the RGB color space, these three outputs are called *color coordinates.* For simplicity of notation, we denote $R(I, L)$ (or R in short) the output from the photosensitive cell for red, $G(I, L)$ (or G in short) the output from the photosensitive cell for green, and $B(I, L)$ (or B in short) the output from the photosensitive cell for blue.

In human vision, the visual signals are digital in nature (i.e. series of electrical pulses) because human brain's nervous systems are apt to process digital signals. Similarly, a machine's brain is a digital system, which can only handle digital signals. Therefore, the analogue color coordinates must be digitized into the corresponding digital formats. In machine vision, a device which converts analogue visual signals into the digital ones is called an *image digitizer.* As a result of image digitization, the RGB color space in consideration should be a discrete color space, as shown in Fig. 7.4.

Example 7.1. In most of today's digital color cameras, a primary color such as red, green or blue is encoded in one byte (i.e. 8 bits). This means that a primary color's value has 256 levels ranging from 0 to 255. Therefore, if a primary color is encoded in one byte, the maximum resolution is 256 at sensor level. For the purpose of processing or perception, we could intentionally reduce the color resolution. For example, let us consider the color plane when the primary color of red is equal to 100 (i.e. $\mathrm{R} = 100$). We reduce its resolution to three levels, namely: 32×32, 16×16 and 8×8. The results are shown in Fig. 7.5. We can see that the reduction of color resolution by a factor of 8 will not cause significant lost of color sensation.

<><><><><>

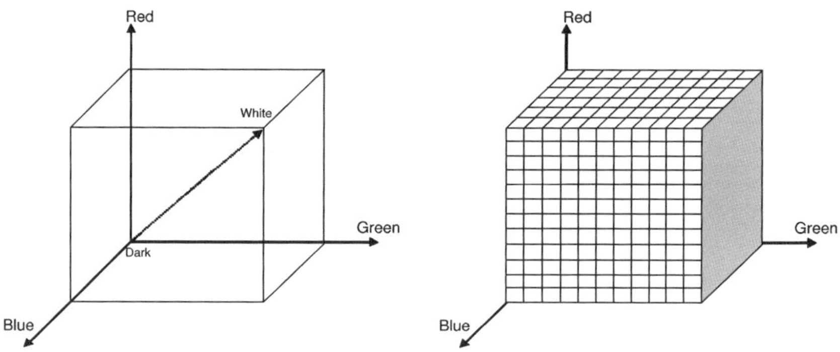

Continuous RGB Color Space
(a)

Discrete RGB Color Space
(b)

Fig. 7.4 At the sensor level of human, or machine, vision, the RGB color space is discrete in nature, in which the primary colors of red, green and blue take discrete values.

Example 7.2. We have an input image, in which each color is represented in a discrete RGB color space. And, each axis in the RGB color space has 256 levels (i.e. each primary color of red, green or blue is encoded with one byte, or 8 bits). Now, we apply a factor of reduction of 8 to reduce each primary color from 256 levels to 32 levels. The result is shown in Fig. 7.6. We can see that the difference between the original and reduced colors is not perceivable.

<><><><><>

7.3.1.2 *L*u*v Color Space*

In physics, the signal which encodes colors should be a one-dimensional signal. This is because the sensations of colors only depend on wavelengths. However, at the sensor level of human (or machine) vision, colors are represented in a three-dimensional RGB color space. Then, this question arises: Could we reduce a three-dimensional color space into a two- or one-dimensional color space?

In theory, the answer should be affirmative. However, in practice, there are many attempts of transforming the RGB color space into another three-dimensional space, in which the intensity (i.e. energy-related quantity) is separated from the chrominance (i.e. wavelength-related quantity).

It is interesting to take note that in the late of nineteenth century, Ewald Hering (1834–1918) has postulated a theory, in which he advocated three opponent processes, namely: red versus green, blue versus yellow, and black versus white. The evidence behind Hering's theory is that in the minds of human beings, the sensations of red and green (or blue and yellow) cannot occur at the same time and at a same spot. In other words, human vision does not create the sensation of reddish-green, or bluish-yellow. Most importantly, the mixture of red and green (or blue and yellow) in a right proportion will completely nullify the color sensation in the minds of human beings. This process is called an *opponent process*.

Fig. 7.5 The color plane when R = 100: (a) the original resolution of 256×256, (b) the resolution of 32 × 32, (c) the resolution of 16 × 16, and (d) the resolution of 8 × 8.

Refer to Fig. 7.7. Hering's theory suggests that at the processing level of human vision, colors are represented in a space which is different from the RGB color space. More specifically, there should exist a three-dimensional color space, in which the three principal axes represent the transitions from red to green, blue to yellow, and black to white, respectively.

Inspired by Hering's theory on human vision, a lot of studies has been done in order to derive such a color space. In 1976, the Commision International d'Eclairage (CIE) has introduced two color spaces named as L*a*b and L*u*v, respectively. The key properties of these two color spaces are as follows:

- Intensity Separation:
  In both color spaces, the intensity of a color is represented by one principal axis, denoted as L. Therefore, the chrominance of a color only depends on two coordinates such as (a, b) or (u, v).

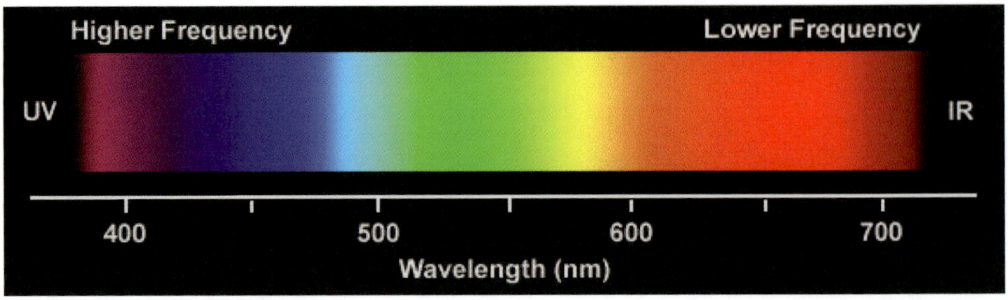

(a) Original Image

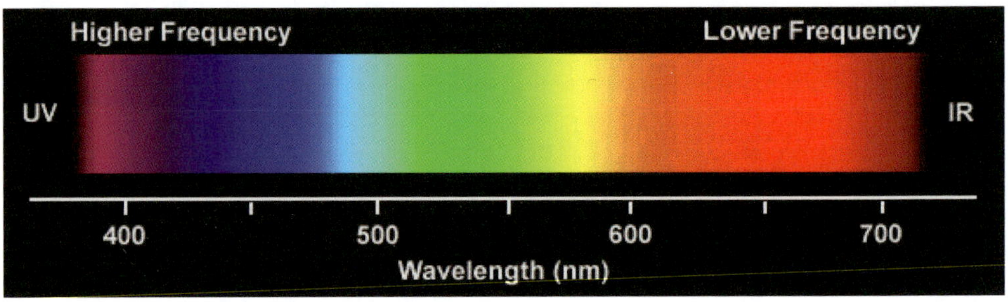

(b) Indexed Image (RGB)

Fig. 7.6 Example of color reduction, in which the original image is shown in (a), and the image of reduced colors is shown in (b).

- Distinct Color Clusterings:
  In the RGB color space, there are no distinct clustering of various colors. However, in the L*a*b, or L*u*v, color space, similar colors are clustered together.
- Perceptual Uniformity:
  Because of the distinct color clustering, equal distances within a color cluster will give rise to the same sensation of color differences in the minds of human beings. A color space having such a property is also called a perceptually-uniform color space.

Another interesting property of the L*a*b and L*u*v color spaces is that both of them are the transformed version of the RGB color space. And, the transformation follows these three general steps:

- Rotation:
  Refer to Fig. 7.4. It is easy to see that the intensity could be represented by an axis along the diagonal direction from black to white. Hence, the rotation of one principal axis in the RGB color space toward such an axis in the diagonal direction will result in the L axis, which represents the intensity of colors.

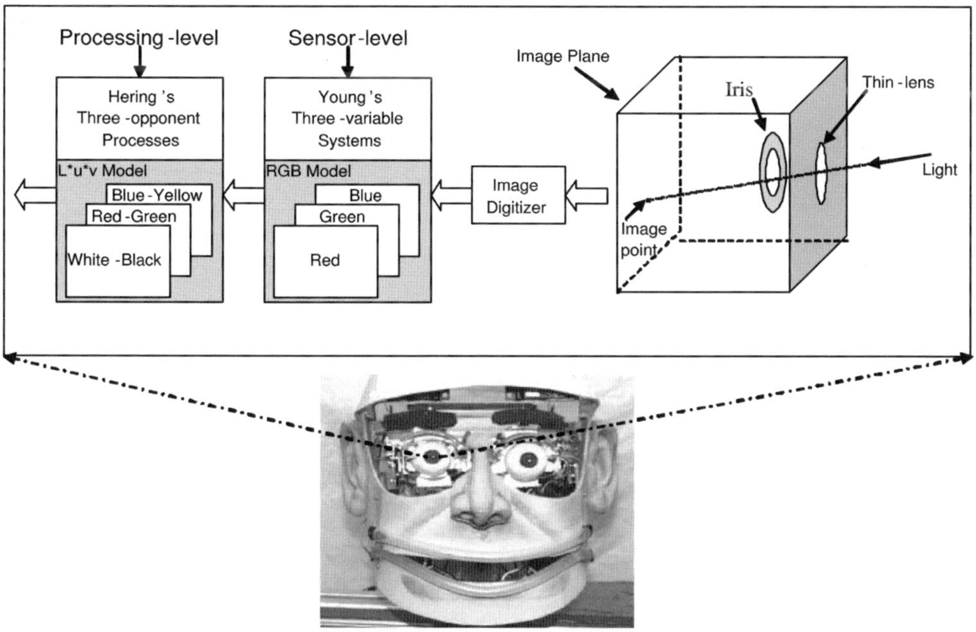

Fig. 7.7 Color representation in the early stage of human vision. At the sensor level, colors are represented in the discrete RGB color space. However, at the processing level, Hering's theory suggests that colors are transformed into a new space, such as the L*u*v color space, where the opponent processes occur.

- Morphing:
  In order to form distinct clustering of colors, it is necessary to undertake non-uniform scaling with regard to the RGB color space. As a result, a cube in the RGB color space will no more be a cube after the transformation into a new color space.
- Selection of Principal Axes:
  Mathematically, the combined effect of rotation and morphing is called the *affine transformation*. After the affine transformation of the RGB color space, a new color space is defined by a set of specifically-selected principal axes. Among these axes, one (i.e. L) represents the intensity of colors.

According to the CIE standard, the L*a*b color space is formed as follows:

- Affine Transformation:
  Let (R, G, B) denote the color coordinates in the RGB color space, and (X, Y, Z) the transformed coordinates in a three-dimensional space. Now, we choose the following affine transformation matrix

$$M_{\text{Lab}} = \begin{pmatrix} 0.607 & 0.174 & 0.200 \\ 0.299 & 0.587 & 0.114 \\ 0.000 & 0.066 & 1.116 \end{pmatrix} \tag{7.3}$$

to transform (R, G, B) into (X, Y, Z) in the following way:

$$\begin{pmatrix} X \\ Y \\ Z \end{pmatrix} = M_{\text{Lab}} \bullet \begin{pmatrix} R \\ G \\ B \end{pmatrix}. \tag{7.4}$$

If (R_w, G_w, B_w) are the coordinates corresponding to the white color, then the transformed coordinates by M_{Lab} will be

$$\begin{pmatrix} X_w \\ Y_w \\ Z_w \end{pmatrix} = M_{\text{Lab}} \bullet \begin{pmatrix} R_w \\ G_w \\ B_w \end{pmatrix}. \tag{7.5}$$

And, the result will be $(250.155, 255.000, 301.410)$ if (R_w, G_w, B_w) equal $(255.0, 255.0, 255.0)$.

- Selection of Principal Axes:
  The affine transformation by M_{Lab} will result in a new three-dimensional space. In this new space, we choose three principal axes as follows:

$$\begin{cases} L = 116 \bullet \left(\dfrac{Y}{Y_w}\right)^{1/3} - 16 \\ a = 500 \bullet \left[\left(\dfrac{X}{X_w}\right)^{1/3} - \left(\dfrac{Y}{Y_w}\right)^{1/3}\right] \\ b = 200 \bullet \left[\left(\dfrac{X}{X_w}\right)^{1/3} - \left(\dfrac{Z}{Z_w}\right)^{1/3}\right] \end{cases} \tag{7.6}$$

  And, the three principal axes L, a and b define the so-called L*a*b color space.

In the RGB color space of machine vision, the color coordinates are bounded by the minimum and maximum values (e.g. between 0 and 255). Hence, the color coordinates in the L*a*b space are also bounded by their respective minimum and maximum values, which are

- $L_{\text{max}} = 116.0$, when (R, G, B) equal $(255, 255, 255)$ (i.e. white)
- $L_{\text{min}} = -16.0$, when (R, G, B) equal $(0, 0, 0)$ (i.e. black)
- $a_{\text{max}} = 96.1$, when (R, G, B) equal $(255, 0, 255)$ (i.e. pink)
- $a_{\text{min}} = -137.7$, when (R, G, B) equal $(0, 255, 0)$ (i.e. green)
- $b_{\text{max}} = 170.4$, when (R, G, B) equal $(255, 0, 0)$ (i.e. red)
- $b_{\text{min}} = -78.5$, when (R, G, B) equal $(0, 0, 255)$ (i.e. blue)

Example 7.3. At the sensor level of a digital color camera, a color is represented by three coordinates in the RGB color space. If the resolution of each primary color is 256 (i.e. one byte per primary color), all the perceivable colors by the digital camera will fall inside a cube of the size of $256 \times 256 \times 256$. When the color cube in the RGB color space is transformed into the L*a*b color space, it will not remain

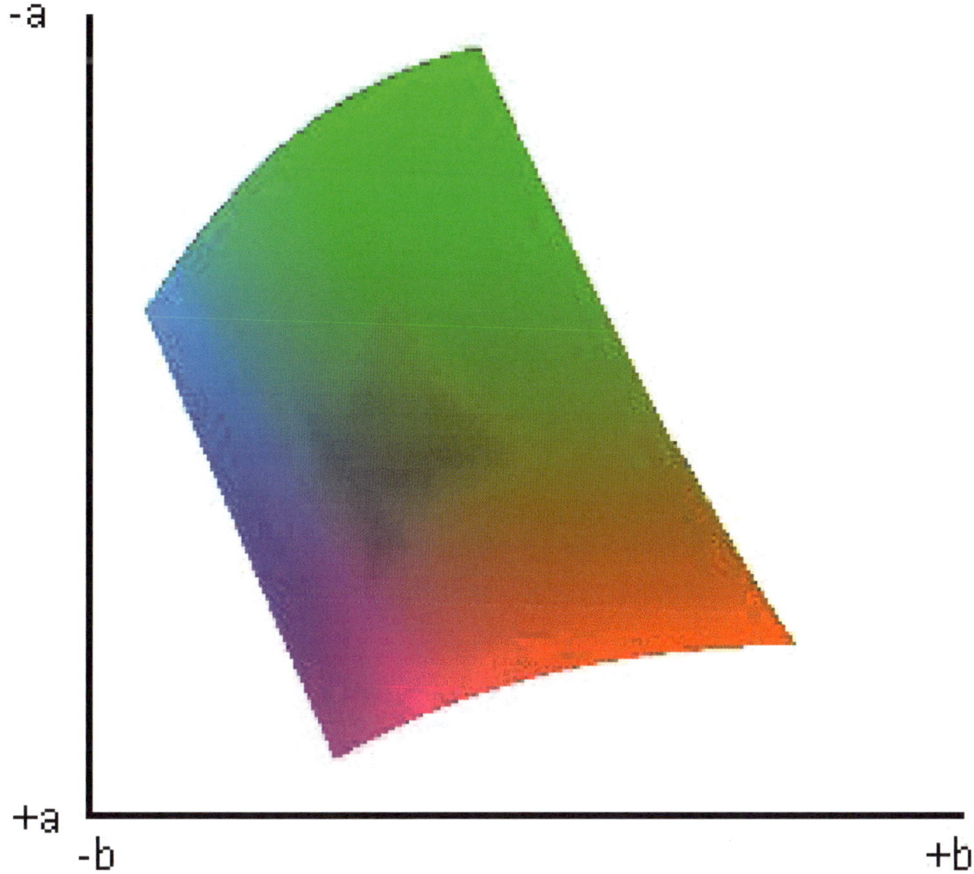

Fig. 7.8 The color plane when L = 75 in the L*a*b color space. We can see that it is not a rectangular shape.

as a cube. In Fig. 7.8, we show the color plane when L is equal to 75. It is easy to see that it is not a rectangular shape. But, we can easily discern the distinct clusterings of red, green, blue and pink in this color plane.

<><><><><>

Similarly, the L*u*v color space can be formed in the following way:

- Affine Transformation:
  We choose the following affine transformation matrix

$$M_{\text{Luv}} = \begin{pmatrix} 0.412 & 0.358 & 0.180 \\ 0.213 & 0.715 & 0.072 \\ 0.019 & 0.119 & 0.950 \end{pmatrix} \tag{7.7}$$

to transform coordinates (R, G, B) into coordinates (X, Y, Z). Thus, we have

$$\begin{pmatrix} X \\ Y \\ Z \end{pmatrix} = M_{\text{Luv}} \bullet \begin{pmatrix} R \\ G \\ B \end{pmatrix}. \tag{7.8}$$

If (R_w, G_w, B_w) are the coordinates corresponding to the white color, the transformed coordinates of the white color will be

$$\begin{pmatrix} X_w \\ Y_w \\ Z_w \end{pmatrix} = M_{\text{Luv}} \bullet \begin{pmatrix} R_w \\ G_w \\ B_w \end{pmatrix}. \tag{7.9}$$

When (R_w, G_w, B_w) equal $(255, 255, 255)$, (X_w, Y_w, Z_w) will be $(242.366, 255.000, 277.632)$.

- Selection of Principal Axes:
  The affine transformation by M_{Luv} will also result in a new three-dimensional space. In this new space, we have certain freedom to choose a set of three principal axes in order to form a new color space. In practice, one selection is as follows:

$$\begin{cases} L = 20 \bullet \sqrt{Y} \\ u = 130 \bullet \sqrt{Y} \bullet \left(\dfrac{4X}{X + 15Y + 3Z} - \dfrac{4X_w}{X_w + 15Y_w + 3Z_w} \right) \\ v = 130 \bullet \sqrt{Y} \bullet \left(\dfrac{9Y}{X + 15Y + 3Z} - \dfrac{9Y_w}{X_w + 15Y_w + 3Z_w} \right) \end{cases} \tag{7.10}$$

  In fact, these three principal axes L, u and v define the so-called *modified* L*u*v color space, which is slightly different from the CIE's standard.

As we mentioned above, the color coordinates in the RGB color space of machine vision are bounded by the minimum and maximum values. Thus, the corresponding coordinates in the L*u*v space are also bounded by their respective minimum and maximum values, which are

- $L_{\text{max}} = 319.4$, when (R, G, B) equal $(255, 255, 255)$ (i.e. white)
- $L_{\text{min}} = 0.0$, when (R, G, B) equal $(0, 0, 0)$ (i.e. black)
- $u_{\text{max}} = 242.1$, when (R, G, B) equal $(255, 0, 0)$ (i.e. red)
- $u_{\text{min}} = -127.9$, when (R, G, B) equal $(0, 255, 0)$ (i.e. green)
- $v_{\text{max}} = 169.1$, when (R, G, B) equal $(255, 255, 0)$ (i.e. yellow)
- $v_{\text{min}} = -179.5$, when $(R, G, B$ equal $(58, 0, 255)$ (i.e. blue)

The above values show that the L, u, and v axes nicely reflect the Hering's theory of color opponent processes. And, it is interesting to know that in this modified L*u*v color space, the ranges of the coordinates are in the same magnitude. We will see the importance of this property when we will discuss the use of multi-scale color indices to represent the physical meanings of colors.

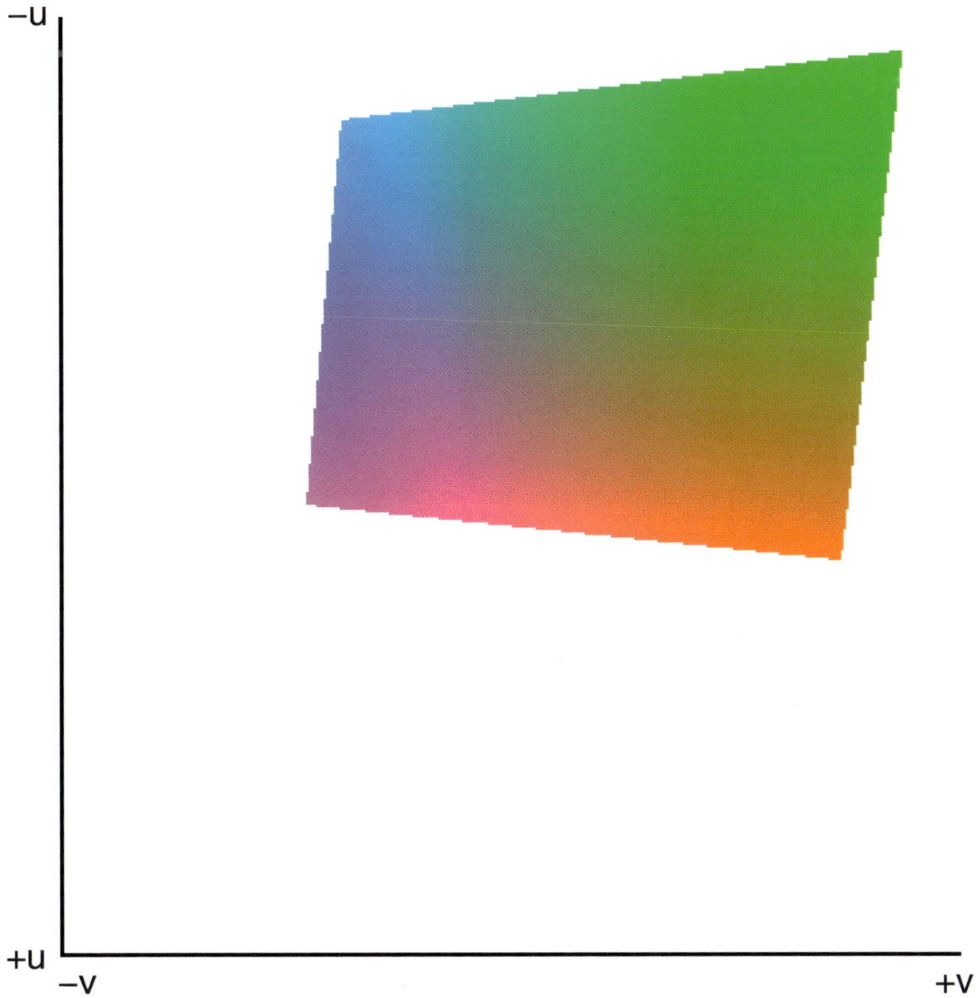

Fig. 7.9 The color plane when L = 240 in the modified L*u*v color space. We can see the distinct clusterings of red, green, blue and pink.

Example 7.4. As matrix M_{Luv} represents an affine transformation, a cube in the RGB color space will not be transformed into a corresponding cube in the L*u*v color space. For instance, in Fig. 7.9, we show the color plane when L is equal to 240. It is clear that the color plane is not exactly rectangular. However, this example shows that colors in the L*u*v color space also form distinct clustering.

<><><><><>

Example 7.5. In the L*u*v color space, we examine the color plane, at which L equals 240. We apply the factor of 8 to reduce the colors. The original colors

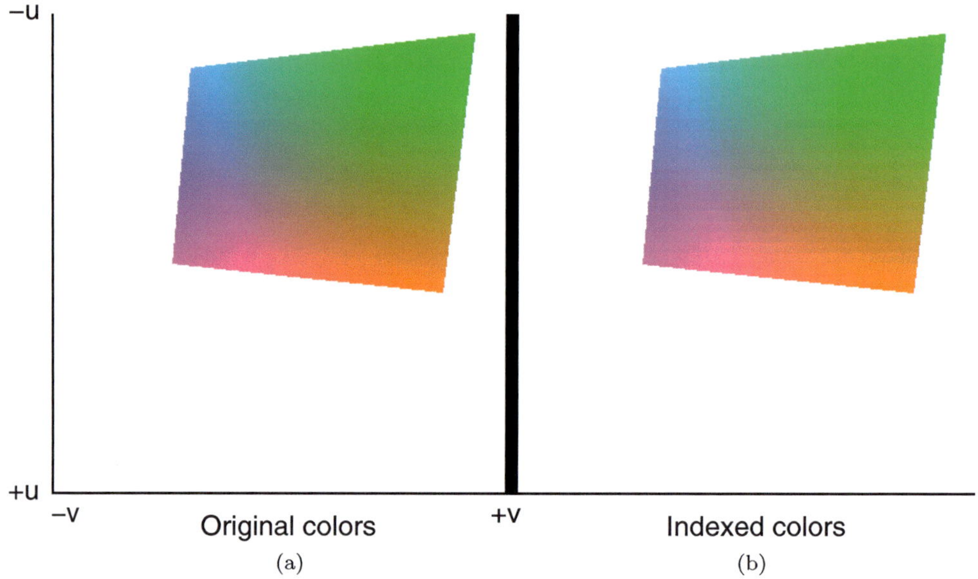

Fig. 7.10 Example of results of indexed colors in the L*u*v space, in which the original colors are shown in (a), and the corresponding reduced colors are shown in (b).

in the plane are shown in Fig. 7.10(a). And, the corresponding reduced colors are shown in Fig. 7.10(b). Again, we can see that the difference between the original and indexed colors is not perceivable.

<><><><><>

7.3.1.3 *Multi-scale Color Indices*

At sensor level, colors can be sensed by the photo-detectors (i.e. image sensors) of different characteristics. And, the outputs from image sensors are the direct measurements of the physical meanings (i.e. properties) of colors. Then, this question arises: What is the exact nature of a color's physical meanings?

We know that a color is represented by a point in a color space. Hence, the coordinates of a point in a color space are the unique representation of a color's identity. In other words, two identical colors will refer to a same point in a color space.

On the other hand, we also know that in the physical world, two different entities may possess a certain degree of similarity in one aspect or other. This implies that similar entities will form a group which is commonly known as a *category*. And, this phenomenon also applies to colors. In other words, similar colors will usually form a category. For example, a cluster of points within a certain volume in a color space will chromatically appear similar to a certain degree of tolerance. In this sense, a cluster can be called a *color category*. Therefore, a color will carry two types of physical meanings, namely: (a) color identity, and (b) color category.

In theory, the outputs of image sensors are the direct measurements of colors' identities in the physical world. Then, the following question arises: How to measure colors' categories in the physical world?

We know that a color's category corresponds to a specific clustering of points within a certain volume of a color space. As a clustering of points may expand differently in a color space, this implies that there are color categories of different scales in a color space. Then, this question will surface out: How to determine, or design, color categories in the physical world? Interestingly enough, the answer lies in the top-down design of multi-scale color indices which will be an unified approach to represent the identities and categories of colors in the physical world.

Multi-scale Color Indices in RGB Color Space

If an eye, or a camera, contains the photo-detectors (i.e. image sensors) which are particularly sensitive to red, green and blue, an arbitrary color will be measured by a set of three photo-detectors and be represented by a point in the RGB color space. As we mentioned before, the RGB color space is a three-dimensional discrete space, in which each coordinate has a specific resolution. For instance, if a primary color is encoded with one byte (i.e. 8 bits), a color coordinate's resolution will be 256 (i.e. 256 levels or values). Therefore, a color's discrete coordinates in the RGB color space can be considered as a representation of the color's identity in the physical world.

If it is easy to measure the identities of colors in the physical world, it is not so obvious to design the categories of colors in the physical world. Here, we present a simple solution to the top-down design of colors' categories in the physical world.

We know that the outputs from image sensors are digitized into discrete values, which vary within a finite range. This means that the colors sensible to an eye, or a camera, are within a fixed volume in a color space. And, this fixed volume is called the *visible volume* in which all colors can be sensed by an eye, or a camera. Conceptually, the visible volume corresponds to the color category at the lowest scale. And, the color category of the lowest scale is also called the category of *all-visible-colors*. In other words, all colors which can be sensed by an eye, or a camera, belong to this category.

Now, let $[R_{\min}, R_{\max}]$ denote the range of the red primary color, $[G_{\min}, G_{\max}]$ the range of the green primary color, and $[B_{\min}, B_{\max}]$ the range of the blue primary color. Then, the category of all-visible-colors in the RGB color space will correspond to the volume defined by

$$\{[R_{\min}, R_{\max}], [G_{\min}, G_{\max}], [B_{\min}, B_{\max}]\}$$

where the suffix "min" indicates the minimum value of a primary color, and the suffix "max" the maximum value of a primary color.

And, it becomes clear that any sub-division of the above volume will give rise to color categories at the higher scales. If we denote s the level of scale (i.e. $s = 1, 2, \ldots$), a simple solution to create a hierarchy of sub-categories of colors is to uniformly divide the visible volume as follows:

$$\left\{ \frac{[R_{\min}, R_{\max}]}{2^s}, \frac{[G_{\min}, G_{\max}]}{2^s}, \frac{[B_{\min}, B_{\max}]}{2^s} \right\}, \quad s = 1, 2, \ldots \tag{7.11}$$

Here, we call this operation the *subdivision of visible volume*.

In Eq. 7.11, the dividing factor of 2 is chosen. In practice, the dividing factor could be other numbers such as 3, 4, etc. We can see that at scale 1, each axis in the RGB color space has two intervals. Therefore, there are $2 \times 2 \times 2$ categories at scale 1. Similarly, at scale 2, each axis in the RGB color space has four intervals (i.e. 2^2). Hence, there are $4 \times 4 \times 4$ categories at scale 2.

If a primary color is encoded with one byte, the categories of colors at scale 8 (i.e. $2^8 = 256$) become the identities of colors at sensor level. As a result, the multi-scale categories unify the representation of the physical meanings of colors. Now, the question will be: What are the multi-scale index coordinates, or indices, of a color within the hierarchy of color categories?

Let $(R_{r,c}, G_{r,c}, G_{r,c})$ denote the sensor outputs at image location (r, c). The multi-scale index coordinates, or indices, could be computed as follows:

$$\begin{cases} I_R(r,c,s) = ceil\left(\dfrac{R_{r,c} - R_{\min}}{R_{\max} - R_{\min} + 1} \bullet 2^s \right) \\ I_G(r,c,s) = ceil\left(\dfrac{G_{r,c} - G_{\min}}{G_{\max} - G_{\min} + 1} \bullet 2^s \right) \\ I_B(r,c,s) = ceil\left(\dfrac{B_{r,c} - B_{\min}}{B_{\max} - B_{\min} + 1} \bullet 2^s \right) \end{cases}, \quad s = 1, 2, 3, \ldots \tag{7.12}$$

where function *ceil*(.) returns the largest integer of its argument, and (I_R, I_G, I_B) are the index coordinates, or indices, at image location (r, c) and at scale s.

Example 7.6. We have an input image as shown in Fig. 7.11(a). In the image, we randomly select two image points, which are marked by '+'. In the RGB color space, the multi-scale color indices at these two selected points are shown in Fig. 7.11(b).

<><><><><>

Refer to Eq. 7.11. The sub-division of the maximum volume is uniform. This will cause the problem of boundary of division. In other words, two nearby colors could be arbitrarily separated by a boundary of division. If this happens, these two colors will belong to two different categories while they should belong to a same category. For the example shown in Fig. 7.11, the colors at the two selected image points belong to the same category up to scale 3 only.

Image of Original Colors (RGB)

(a)

Scale1: (1,1,2)	Scale1: (1,1,2)
Scale2: (2,2,3)	Scale2: (2,2,3)
Scale3: (3,4,6)	Scale3: (3,4,6)
Scale4: (5,7,11)	Scale4: (5,8,11)
Scale5: (9,14,22)	Scale5: (9,15,22)
Scale6: (18,28,44)	Scale6: (18,29,44)
Scale7: (36,56,88)	Scale7: (35,57,88)
Scale8: (71,112,176)	Scale8: (69,113,176)

Multi–Scale Index Coordinates (RGB Space)

(b)

Fig. 7.11 Example of multi-scale color indices in the RGB color space. The original image is shown in (a). And, the selected image points are marked with "+". The multi-scale color indices at the selected image points are shown in (b).

With the structure of hierarchical categories, the problem of boundary of division can be easily solved. The solution is to use the notion of *augmented category*, in lieu of *nominal category*. More precisely, we call the categories obtained by sub-division the *nominal categories*. And, a nominal category together with its immediate neighbors at a same scale is called an *augmented category*. Refer to Fig. 7.11 again. Now, the colors at the two selected image points belong to the same augmented category up to scale 7. We will come back to this issue when we will study the topic of color recognition at the later part of this chapter.

*Multi-scale Color Indices in L*u*v Color Space*

According to Hering's theory of color opponent processes (i.e. red versus green, blue versus yellow, and white versus black), a color's intensity is separated from its chrominance at the processing level in human vision. For machine vision, we could do the same by transforming the color coordinates in the RGB color space into the corresponding coordinates in the L*u*v color space. This process is called *intensity separation*. And, Eqs. 7.7 to 7.10 fully describe the process of intensity separation in the L*u*v color space.

As we mentioned before, the conversion from the RGB color space to the L*u*v color space follows an affine transformation. This means that the finite visible volume in the RGB color space will be transformed into the corresponding visible volume in the L*u*v color space, which is finite as well. As the visible volume in the L*u*v color space represents the color category at the lowest scale, the sub-division of this visible volume will generate a hierarchy of color categories at the higher scales.

For the sake of simplicity, let $[L_{\min}, L_{\max}]$ denote the range of intensity L, $[u_{\min}, u_{\max}]$ the range of coordinate u, and $[v_{\min}, v_{\max}]$ the range of coordinate v, where the suffix "min" indicates the minimum value of a coordinate, and the suffix "max" the maximum value of a coordinate. Without loss of efficiency, we could use the volume defined by

$$\{[L_{\min}, L_{\max}],\ [u_{\min}, u_{\max}],\ [v_{\min}, v_{\max}]\}$$

to approximately represent the color category of all-visible-colors in the L*u*v color space. In this way, the sub-division of this visible volume will give rise to color categories at the higher scales in the L*u*v color space.

Similarly, if we denote s the level of scale (i.e. $s = 1, 2, ...$), a simple solution to create a hierarchy of sub-categories of colors is to uniformly divide the visible volume as follows:

$$\left\{\frac{[L_{\min}, L_{\max}]}{2^s}, \frac{[u_{\min}, u_{\max}]}{2^s}, \frac{[v_{\min}, v_{\max}]}{2^s}\right\}, \quad s = 1, 2, \ldots \tag{7.13}$$

Now, let $(L_{r,c}, u_{r,c}, v_{r,c})$ denote the sensor level's color coordinates at image location (r, c), in the L*u*v color space. Then, in the L*u*v color space, the multi-scale index coordinates, or indices, could be computed as follows:

$$\begin{cases} I_L(r,c,s) = ceil\left(\dfrac{L_{r,c} - L_{\min}}{L_{\max} - L_{\min} + 1} \bullet 2^s\right) \\ I_u(r,c,s) = ceil\left(\dfrac{u_{r,c} - u_{\min}}{u_{\max} - u_{\min} + 1} \bullet 2^s\right) \\ I_v(r,c,s) = ceil\left(\dfrac{v_{r,c} - v_{\min}}{v_{\max} - v_{\min} + 1} \bullet 2^s\right) \end{cases}, \quad s = 1, 2, 3, \ldots \tag{7.14}$$

where function *ceil*(.) returns the largest integer of its argument, and (I_L, I_u, I_v) are the index coordinates, or indices, at image location (r, c) and at scale s.

Example 7.7. We have an input image as shown in Fig. 7.12(a). The original colors are in the RGB color space. We first transform the color coordinates from the RGB color space into the L*u*v color space. Then, we randomly select two image points, which are marked by '+'. In the L*u*v color space, the multi-scale color indices at these two selected points are shown in Fig. 7.12(b). It is interesting to observe that these two colors belong to a same nominal category up to scale 3. However, they belong to a same augmented category up to scale 4.

<><><><><>

7.3.2 *Color's Properties in Conceptual Worlds*

In a color space, a point represents a color perceived at any location in the image plane of an eye, or a camera. In other words, a point in a color space is the physical representation, or a psychological sensation, of a specific color in the minds of human beings. However, in the consciousness of human beings, the psychological

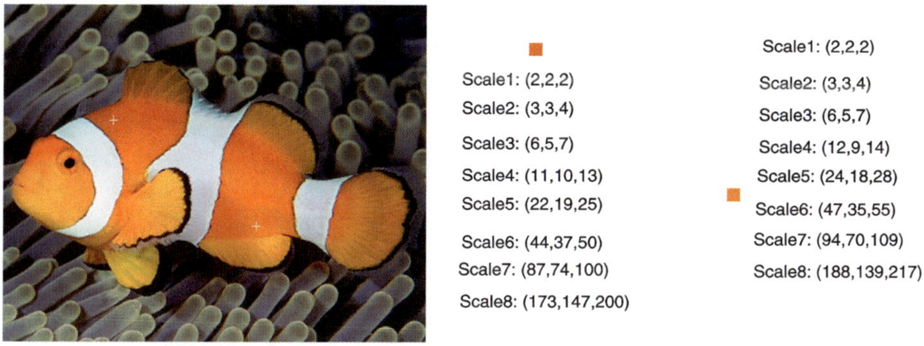

Image of Original Colors (RGB) — (a)

Multi–Scale Index Coordinates (Luv Space) — (b)

Fig. 7.12 Example of multi-scale color indices in the L*u*v color space. The original image is shown in (a). And, the selected image points are marked with "+". The multi-scale color indices at the selected image points are shown in (b).

sensation of a specific color (e.g. white, red, green, or blue, etc) can be communicated out conceptually (i.e. the use of human languages), instead of physically. This observation implies that a color space must have extra dimensions in order to account for conceptual meanings.

7.3.2.1 *Multi-scale Color Labels*

In the RGB color space, the physical meanings of a color at image location (r, c) could be represented by multi-scale color indices $\{I_R(r, c, s), I_G(r, c, s), I_B(r, c, s)\}$ where s indicates the scale. Similarly, in the L*u*v color space, a color's physical meanings at image location (r, c) could be represented by multi-scale color indices $\{I_L(r, c, s), I_u(r, c, s), I_v(r, c, s)\}$. When scale s takes the largest permissible value $s_{\max}$ (e.g. 8), index coordinates $\{I_R(r, c, s_{\max}), I_G(r, c, s_{\max}), I_B(r, c, s_{\max})\}$, or $\{I_L(r, c, s_{\max}), I_u(r, c, s_{\max}), I_v(r, c, s_{\max})\}$, represent a color's identity. Otherwise, they represent a color's nominal category at scale s. Therefore, it is clear that the multi-scale color indices unify the representation of the physical meanings of colors in terms of identity and category in the physical world.

Besides the physical meanings, we know that colors do carry conceptual meanings which are represented by words (or labels) in a human language. According to Berlin and Kay (1969), colors are conceptually partitioned into eleven (11) basic categories in the context of English language. Among these eleven basic categories, three are achromatic and eight are chromatic. For instance, the three achromatic categories are black, white and gray. And, the eight chromatic categories are red, yellow, green, blue, purple, orange, pink and brown. Other color categories could be formed by a combination of these basic categories. For instance, psychologically, one could observe yellowish-red, greenish-yellow, and bluish-green, etc. However, if one considers another human language, the conceptual meanings of colors will be represented by different labels or words.

Now, we examine the discrete RGB color space. It is interesting to note that at scale 1, there are eight (i.e. $2 \times 2 \times 2$) categories. And, each of these color categories has a corresponding name in English (i.e. black, white, red, green, blue, yellow, pink, and cyan). At a higher scale, it is also possible to associate a label (or word) to a color category. Then, within a hierarchy of multi-scale color categories, this interesting question arises: Could the conceptual meanings of a color be computed from the visual signals directly?

Obviously, the answer is not affirmative because words (or labels) representing color identities (or categories) in a particular human language can only be learnt through human-machine (or machine-machine) interaction. This means that visual learning is indispensable in machine vision. And, visual learning is a bottom-up process, which depends on the types of human-machine interface, such as conversational dialogue, or gesture, etc. We will discuss the detail of visual learning of a color's conceptual meanings in the later part of this chapter.

7.3.3 *Data Structure of Organized Meanings*

Human vision is cognitive in nature. Similarly, machine vision should also be cognitive. This implies that machine vision must possess an organized memory, in which the meanings in both the physical world and the conceptual worlds are stored. In practice, the concept-physical meanings of colors can be implemented in the form of organized data structure as shown in Fig. 7.13(a).

For instance, in a generic data structure of colors, these properties and constraints must be included:

- Color coordinates measured at sensor level.
- Multi-scale color indices (i.e. the physical meanings).
- Multi-scale color labels (or words) (i.e. the conceptual meanings in one, or more, human language).

With an organized data structure of colors in place, the next question will be: What are the innate mental processes which enable a machine to autonomously understand the colors perceived at the image planes?

As shown in Fig. 7.13(b), the general mental processes should include: (a) intensity separation, (b) subdivision of visible volume, and (c) visual learning. However, in the following sections, we will study specific mental processes under the topics of color cognition and recognition, which constitute the core of color perception, or understanding.

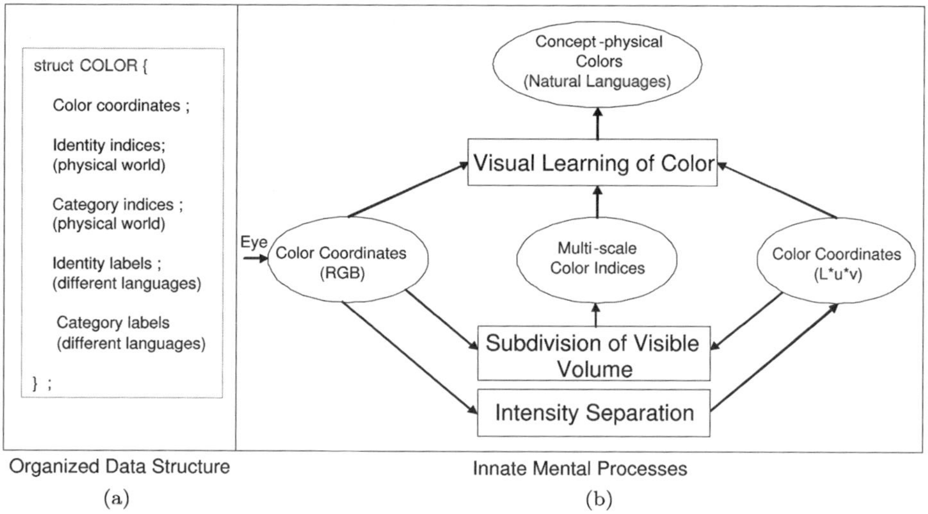

Fig. 7.13 A concept-physical model of colors, in which color representation in the L*u*v space is obtained by the process of intensity separation. And, the multi-scale color indices are the result of subdivision of visible volume in a color space. Finally, the process of visual learning will add another dimension to the colors. This new dimension includes conceptual labels in terms of color identities and categories in a human language.

7.4 Cognition of Color

Literally, color cognition means to know the meanings of colors, which are not familiar (or known) to a machine beforehand. Refer to Fig. 7.14 and Fig. 7.13. As the concept-physical model of colors can be designed in advance, color cognition should embrace both top-down design in modeling and bottom-up optimization in experiencing. And, the ultimate purpose of color cognition is to determine the identities and categories of sensed colors in both the physical world and the conceptual worlds (i.e. texts in the form of human languages).

As meanings exist in both the physical world and the conceptual worlds, color cognition will imply two mental tasks such as: (a) visual computation of a color's physical meanings (i.e. in the physical world), and (b) visual learning of a color's conceptual meanings (i.e. in the conceptual worlds).

7.4.1 *Visual Computing*

Colors can be sensed by a machine's eyes. At sensor level, colors are represented in the RGB color space. Subsequently, the designer of machine vision could put in place a transformation which transforms the RGB color space into another color space such as the L*u*v space. Therefore, at processing level, it is possible to implement innate solutions which compute the identities and categories of sensed colors in the physical world.

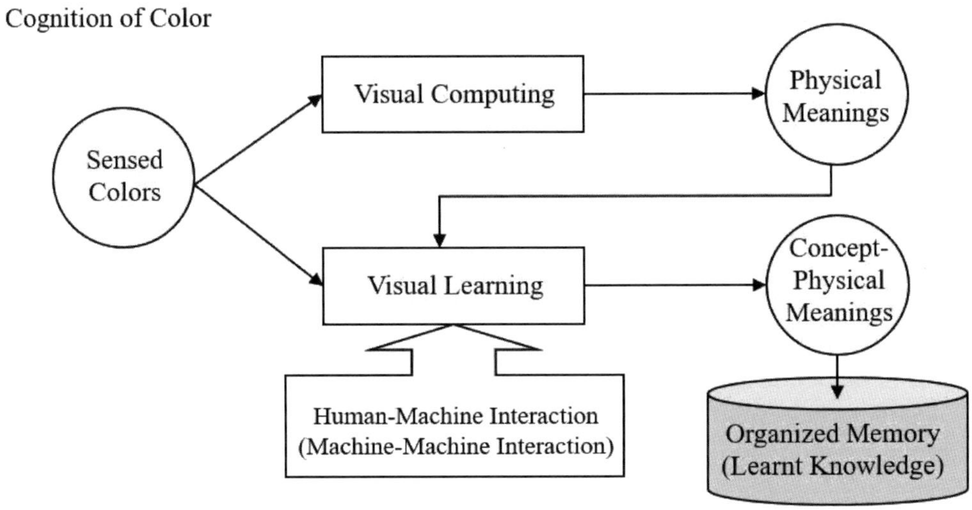

Fig. 7.14 Color cognition, in which two mental tasks are: (a) visual computing of a color's physical meanings and (b) visual learning of a color's conceptual meanings. And, the outcome of color cognition will be stored in an organized memory.

7.4.1.1 *Color Identity*

When a color is sensed by a machine's eyes, its identity indices can be directly measured. For example, let $(R_{r,c}, G_{r,c}, B_{r,c})$ be the coordinates of a color at pixel location (r, c) (i.e. r^{th} row and c^{th} column). If $s_{\max}$ denotes the maximum permissible scale, then the color's identity indices in the RGB color space will be

$$\{I_R(r, c, s_{\max}), I_G(r, c, s_{\max}), I_B(r, c, s_{\max})\}. \tag{7.15}$$

For instance, if each primary color is encoded with one byte, then we have

$$\begin{cases} s_{\max} = 8 \\ R_{\min} = G_{\min} = B_{\min} = 0 \\ R_{\max} = G_{\max} = B_{\max} = 255. \end{cases} \tag{7.16}$$

By applying Eq. 7.12, we obtain

$$\begin{cases} I_R(r, c, s_{\max}) = R_{r,c} \\ I_G(r, c, s_{\max}) = G_{r,c} \\ I_B(r, c, s_{\max}) = B_{r,c}. \end{cases} \tag{7.17}$$

Example 7.8. We have an input image as shown in Fig. 7.15(a). In the image, we randomly select some image points, which are marked by '+'. Then, we compute the identity indices of colors at the selected image points. And, the result is shown in Fig. 7.15(b) together with the selected colors which are depicted in the small rectangular areas.

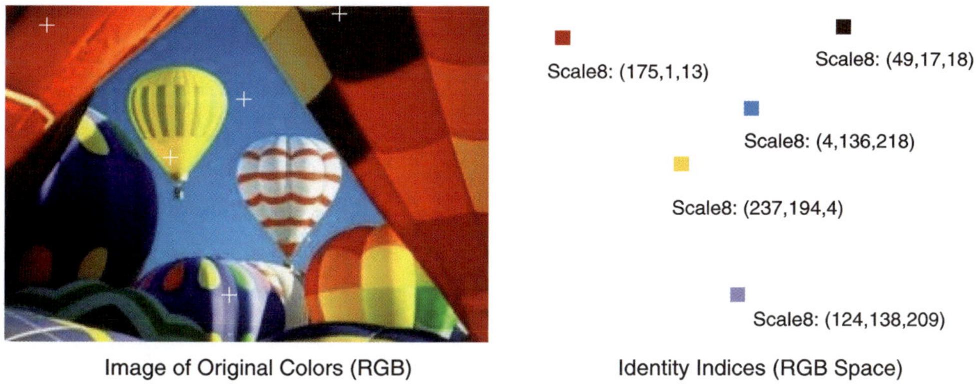

(a) (b)

Fig. 7.15 Example of cognising colors in the RGB color space. The original image is shown in (a). And, the selected image points are marked with '+'. The computed identity indices of colors at the selected image points are shown in (b).

<><><><><>

In the L*u*v space, the chrominance of a color is separated from its intensity. And, the chrominance only depends on two coordinates (i.e. u and v). Now, let $(L_{r,c}, u_{r,c}, v_{r,c})$ be the coordinates of a color at pixel location (r, c) (i.e. r^{th} row and c^{th} column). Similarly, if $s_{\max}$ denotes the maximum permissible scale, then the color's identity indices in the L*u*v color space will be

$$\{I_L(r, c, s_{\max}),\ I_u(r, c, s_{\max}),\ I_v(r, c, s_{\max})\}. \tag{7.18}$$

For instance, if we choose $s_{\max}$ to be 8, then we have

$$\begin{cases} I_L(r, c, 8) = ceil\left(\dfrac{L_{r,c} - L_{\min}}{L_{\max} - L_{\min} + 1} \bullet 256\right) \\ I_u(r, c, 8) = ceil\left(\dfrac{u_{r,c} - u_{\min}}{u_{\max} - u_{\min} + 1} \bullet 256\right) \\ I_v(r, c, 8) = ceil\left(\dfrac{v_{r,c} - v_{\min}}{v_{\max} - v_{\min} + 1} \bullet 256\right). \end{cases} \tag{7.19}$$

Example 7.9. Figure 7.16(a) shows an image, in which a pixel's color is represented in the RGB color space. We first transform the input image into the L*u*v color space. In this way, a pixel's intensity is separated from its chrominance. Then, we randomly select some image points, which are marked by '+'. Then, we compute the identity indices of colors at the selected image points. The result is shown in Fig. 7.16(b) together with the sensed colors which are depicted in the small rectangular areas.

<><><><><>

Image of Original Colors (RGB) Identity Indices (Luv Space)

(a) (b)

Fig. 7.16 Example of cognising colors in the L*u*v color space. The original image is shown in (a). And, the selected image points are marked with '+'. The computed identity indices of colors at the selected image points are shown in (b).

7.4.1.2 *Color Category*

Refer to the discrete RGB color space illustrated in Fig. 7.4. In theory, any clustering of interconnected cells within a reasonable volume could represent a color category in the physical world. In other words, the category indices of a color are not fixed, but dynamically changing.

However, in practice, one could adopt the solution of a hierarchy of multi-scale categories to design the nominal categories of colors in a color space. For instance, Eq. 7.12 and Eq. 7.19 offer a simple solution of sub-dividing the visible volume in order to generate a hierarchy of sub-categories in the RGB, and L*u*v, color spaces, respectively.

Example 7.10. We have a color image as input, as shown in Fig. 7.17(8). At each pixel location, the primary colors are known. For every row r and column c in the input image, we apply Eq. 7.12 to compute multi-scale color indices $(I_R(r,c,s), I_G(r,c,s), I_B(r,c,s))$ corresponding to color coordinates $(R_{r,c}, G_{r,c}, B_{r,c})$ at pixel location (r,c). Then, for every scale s, we apply

$$\begin{cases} R^s_{r,c} = \dfrac{R_{\max} - R_{\min} + 1}{2^s} \bullet I_R(r,c,s) \\ G^s_{r,c} = \dfrac{G_{\max} - G_{\min} + 1}{2^s} \bullet I_G(r,c,s) \\ B^s_{r,c} = \dfrac{B_{\max} - B_{\min} + 1}{2^s} \bullet I_B(r,c,s) \end{cases} \tag{7.20}$$

to compute coordinates $(R^s_{r,c}, G^s_{r,c}, B^s_{r,c})$ which represents the color of the category indexed by $(I_R(r,c,s), I_G(r,c,s), I_B(r,c,s))$ at pixel location (r,c). As a result, at each scale, we have an image of categories' colors. These images are shown in Fig. 7.17. At scale 8, the image of categories' colors is the same as the original image, or the image of identities' colors.

<><><><><>

(1) RGB Image at Color Scale 1

(2) RGB Image at Color Scale 2

(3) RGB Image at Color Scale 3

(4) RGB Image at Color Scale 4

(5) RGB Image at Color Scale 5

(6) RGB Image at Color Scale 6

(7) RGB Image at Color Scale 7

(8) RGB Image at Color Scale 8

Fig. 7.17 Images of colors at the multi-scale categories in the RGB color space. At scale s, the number of colors is $2^s \times 2^s \times 2^s$. And, the image at scale 8 is the same as the original image.

7.4.2 *Visual Learning*

Human vision is developmental in nature, because of the autonomous learning involved. In a similar manner, machine vision should also possess the ability of autonomous learning. This is especially true, because the association of conceptual labels (or words) to sensed colors can only be achieved by a visual learning process, but not pure computation with visual signals. Most importantly, this learning process in machine vision depends on human-machine interaction. Visual learning could also be done through machine-machine interaction if a knowledgeable machine could teach a novice machine in future.

However, from an engineering point of view, one may ask this question: What is the innate principle underlying the association of conceptual labels (or words) to sensed colors? This is the long-lasting problem of symbol grounding in artificial intelligence.

In order to devise a definite answer to this question, one can consider at least two possible approaches, namely: (a) the use of symbol grounding network, and (b) the use of fuzzy logic.

7.4.2.1 *Use of Symbol Grounding Network*

The first approach in visual learning is the so-called *associative memory* or *symbol grounding network*. When we adopt the approach of symbol grounding network to undertake visual learning, the physical meanings of colors are the input, and the conceptual labels (or words) are the output. In this way, the whole purpose of visual learning is to transform the input to the output through the nodes in the middle layer(s) of a symbol grounding network.

In the physical world, the physical meanings of colors are represented by a hierarchy of color indices. In this hierarchy, it may be sufficient enough to do the visual learning of conceptual labels at a certain scale. After that, the conceptual labels could be inherited by the categories at other scales. Then, this question arises: What should be the scale at which it is appropriate to start the visual learning of conceptual labels?

As the answer depends on the nature of a human language, it is difficult to provide a definite solution. Empirically, one can choose scale 2 or 3. This is because the number of color-related words in a human language is in the order of $2^2 \times 2^2 \times 2^2$, or $2^3 \times 2^3 \times 2^3$.

Alternatively, one can choose to start the visual learning at the highest scale. That is to say that the conceptual labels will be associated to perceived colors at sensor level. Interestingly enough, visual learning at the highest scale simultaneously solves the problem of associating conceptual labels to color categories at other scales. For instance, if two colors at sensor level have been assigned conceptual label i, but are similar at scale k (i.e. they belong to a color category at scale k), then it is equivalent to say that conceptual label i is assigned to a color category at scale k, which these two colors belong to.

Now, let us consider the case of visual learning of conceptual labels at the highest scale. And, for the sake of illustration, we choose the L*u*v color space and denote (L, u, v) the coordinates of an input color. In this case, coordinates L, u, and v are the input to a symbol grounding network. If we work with the RGB color space, the input will be R, G, and B. Also, for simplicity of illustration, we discuss the innate principle of associating one conceptual label to a list of colors given as a training set. This process of visual learning is equally applicable to other conceptual labels of colors.

Refer to Fig. 7.18. A symbol grounding network normally consists of three, or more, layers. The first layer is the input layer. The input to this layer is the vector of coordinates of a color sensed by a machine. In a color space, a color is actually a vector. Hence, a color vector is also a color's feature vector.

Through human-machine (or machine-machine) interaction in real-time, a trainer (e.g. a human being or an expert machine) will teach a machine what is an input color's conceptual label. And, a new conceptual label to be learnt by the machine will be stored inside a new node in the middle layer. Now, the question is: What is the innate principle to transform the input to the output through one, or more, node in the middle layer(s)?

With the use of symbol grounding functions, the innate principle underlying the visual learning of a color's conceptual label(s) by a symbol grounding network is very simple. For example, during the process of visual learning, a node in the middle layer will memorize all the input colors' feature vectors, which are to be associated to a same conceptual label of color. From the memorized list of colors' feature vectors, a node in the middle layer will determine its symbol grounding function as follows:

$$p(c_x, c_i) = \frac{1}{\sigma_i \sqrt{2\pi}} e^{\frac{-1}{2\sigma_i^2} \|c_x - c_i\|^2} \tag{7.21}$$

where

- c_x is an input color vector to node i in the middle layer,
- c_i is the mean vector of node i. The mean vector is the averaged sum of all the color vectors learnt by node i. Actually, c_i is the reference color vector of node i.
- $p()$ is the symbol grounding function which calculates the possibility for c_x to share the same conceptual label of reference color c_i.
- σ_i is the standard deviation of the Euclidean distances from reference color c_i to all the learnt color vectors stored inside node i.

The output layer of a symbol grounding network consists of a single node, which is to do the selection of a conceptual label of color as output. The selection is based on a max operator, which selects node i in the middle layer if node i's symbol grounding function outputs the maximum value of possibility among the values of possibility from all the other nodes in the middle layer. If node i has been selected, its conceptual label w_i will be the output from the output layer.

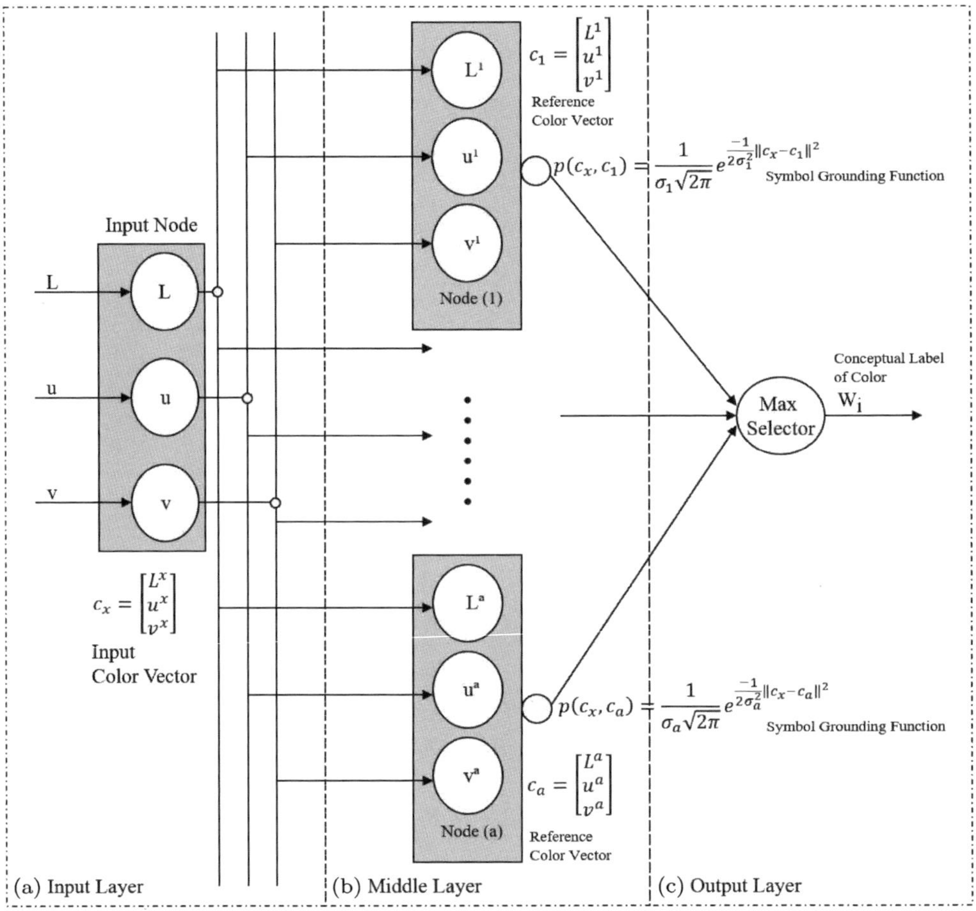

Fig. 7.18 Visual learning with symbol grounding network. In a symbol grounding network, the coordinates of a sensed color are sent to the input node. And, a conceptual label will be given by trainer. If the input of a color is accepted as a valid candidate to be associated to a new label, a new node will be created in the middle layer(s) which stores this new label and also the coordinates of the input color. In this way, each node in the middle layer will store a list of learnt colors which are associated to a same conceptual label. From the list of learnt colors stored inside each node, a symbol grounding function will be determined and it will be used for color recognition.

7.4.2.2 *Use of Fuzzy Logic*

Symbol grounding network is simple and efficient for the learning of associative relationships between values in the physical world and words in a conceptual world. Another way to handle such symbol grounding problem in artificial intelligence is the use of fuzzy logic.

Fuzzy logic was originally developed as an effective tool to cope with the vagueness in linguistics, or soft computing (i.e. computation with words). For example, fuzzy logic could handle the phenomenon of elasticity (or vagueness) of conceptual

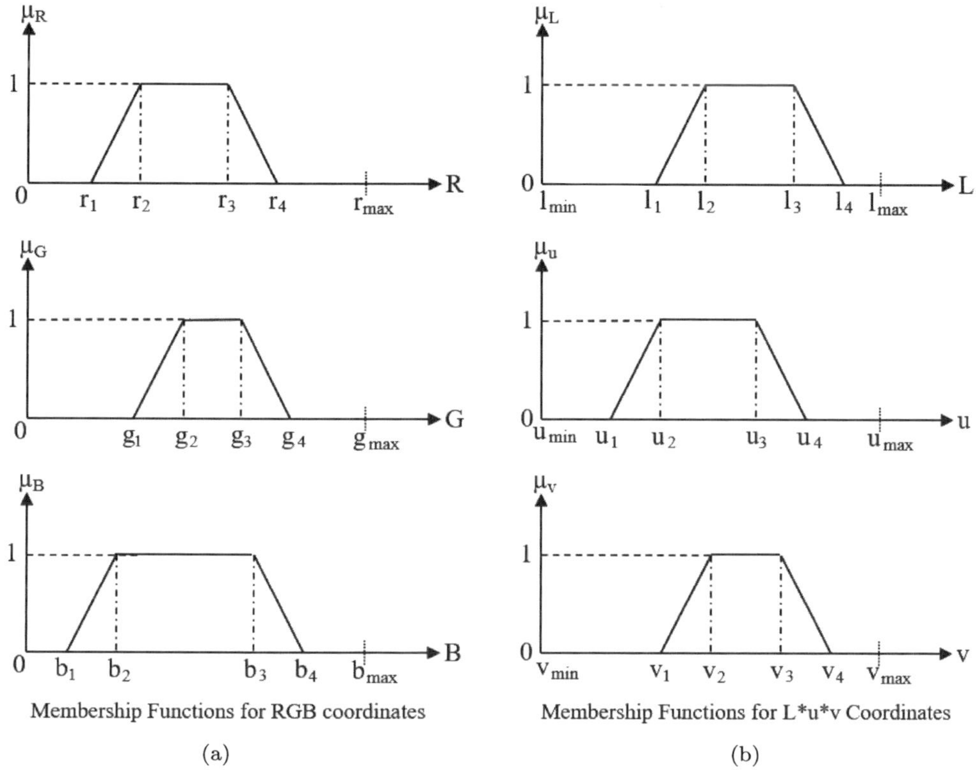

Fig. 7.19 Visual learning with fuzzy logic approach, in which each conceptual label will have three membership functions. And, each membership function will have four parameters such as (l_1, l_2, l_3, l_4) in μ_L.

labels (or words) such as dark-blue, sky-blue, or reddish blue, etc. The elasticity (or vagueness) of words (or conceptual symbols) is usually described by the technical term *fuzziness*. And, the logic of fuzziness, or fuzzy logic in short, actually aims at transforming crisp values in the physical world into fuzzy labels in a conceptual world.

Interestingly enough, fuzzy sets could embrace all the possible conceptual labels (or words) which describe all the color categories in a human language. For example, for each conceptual label (or word) in a fuzzy set, three membership functions could be designed or learnt. These membership functions correspond to three color coordinates (L, u, v) in the L*u*v space (or (R, G, B) in the RGB space), respectively.

Without loss of generality, we consider color coordinates (L, u, v) in the L*u*v space. Also, for simplicity of illustration, a conceptual label can interchangeably refer to a color's identity, or category, in a human language.

Refer to Fig. 7.19. A conceptual label will have three membership functions (μ_L, μ_u, μ_v), each of which is modeled by four parameters, such as (l_1, l_2, l_3, l_4) for membership function μ_L. Hence, a conceptual label will have twelve (12) parameters, which must satisfy the following necessary condition:

$$\begin{cases} l_1 < l_2 < l_3 < l_4 \\ u_1 < u_2 < u_3 < u_4 \\ v_1 < v_2 < v_3 < v_4. \end{cases} \tag{7.22}$$

If we examine a membership function in Fig. 7.19, we can find these three ranges:

- Crisp Range:
  In this range, a membership function outputs the value of 1. It means that a sensed color belongs to this conceptual label. For instance, the crisp range of membership function μ_u is between u_2 and u_3.
- Non-Response Range:
  In this range, it is certain that a sensed color does not belong to this conceptual label. For instance, the non-response range of membership function μ_u includes intervals $(0, u_1)$ and $(u_4, u_{\max})$.
- Fuzzy Range:
  In this range, we cannot totally exclude a sensed color from this conceptual label. For instance, the fuzzy range of membership function μ_L includes intervals (l_1, l_2) and (l_3, l_4).

During the process of human-machine (or machine-machine) interaction in real-time, the purpose of visual learning is to dynamically update the twelve parameters in Eq. 7.22. In practice, the innate algorithm underlying the visual learning of a conceptual label by fuzzy logic could consist of three orderly steps as follows:

- Update of Crisp Range:
  If a trainer (e.g. a human being or an expert machine) tells a machine that sensed color (L, u, v) belongs to a given conceptual label, the visual learning process will update intervals (l_2, l_3), (u_2, u_3) and (v_2, v_3), accordingly. For instance, if $L < l_2$, we let $l_2 = L$. Otherwise, if $L > l_3$, we let $l_3 = L$. If this is the first instance of visual learning, we let $l_2 = l_3 = L$, $u_2 = u_3 = u$ and $v_2 = v_3 = v$.
- Update of Non-Response Range:
  On the other hand, if a trainer (e.g. a human being or an expert machine) tells a machine that sensed color (L, u, v) does not belong to a given conceptual label, the visual learning process will update intervals $(0, l_1)$, $(l_4, l_{\max})$, $(0, u_1)$, $(u_4, u_{\max})$, $(0, v_1)$, and $(v_4, v_{\max})$, accordingly. For instance, if $L < l_2$ and $L > l_1$, we let $l_1 = L$. Otherwise, if $L < l_4$ and $L > l_3$, we let $l_4 = L$.
- Verification of Necessary Condition:
  After the update of crisp range or non-response range, it is important to check whether the necessary condition in Eq. 7.22 still holds. If not, we must re-adjust l_1, l_4, u_1, u_4, v_1, or v_4, accordingly. For instance, if $l_1 > l_2$, we let $l_1 = l_2 - \epsilon$ where ϵ is a system parameter of machine vision. And, if $l_3 > l_4$, we let $l_4 = l_3 + \epsilon$.

7.4.2.3 *Use of Human-Machine Interface*

Besides innate solutions underlying visual learning, it is also necessary to develop an appropriate human-machine (or machine-machine) interface, which could facilitate the autonomous interaction between human beings and machines (or between two machines). In general, a human-machine interface may include the exchange of both audio and visual signals. In particular, a machine should also include a haptic interface in order to support human-machine interaction. As a result, a machine should be designed in a way to be apt to undertake visual learning.

7.5 Recognition of Color

The mental process of color cognition enables a machine to incrementally build up its organized memory, which stores the internalized version of the concept-physical meanings of colors. And, the amount of contents in the organized memory will grow as a function of human-machine (or machine-machine) interaction in real-time. In this process of color cognition, it is clear that there is an interplay between the principles of top-down design and bottom-up optimization. Most importantly, the success of color cognition will enable another important mental process called *color recognition.*

Refer to Fig. 7.20. Color recognition literally means to match a sensed color against the learnt colors in order to know the identity and/or category of the sensed color. In theory, the matching process in color recognition must be based on a criterion of similarity, which judges the similitude of two colors in terms of identity, category, or both. In general, the result of color recognition will be the concept-physical meanings of sensed colors. However, in machine vision, one particular application of color recognition is to classify the pixels of an image into image components, each of which contains colors that are similar in terms of identity or

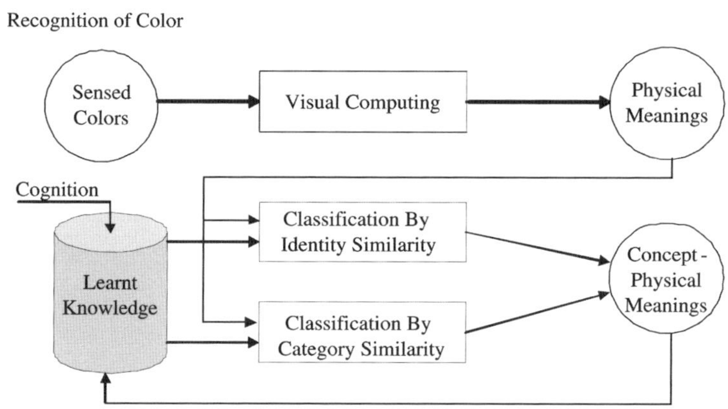

Fig. 7.20 Color recognition, in which the criterion of decision-making can be based on the similarity on identity, category, or both.

category. This approach consists of a common solution to the general problem of *image segmentation* in computer vision, or machine vision. And, the ultimate purpose of image segmentation is to obtain the composition of an image in terms of image components, which is a prerequisite to the general task of *image understanding*. As image segmentation has a paramount importance in vision, a specific task of color recognition is to classify the pixels of an image into image components.

7.5.1 *Use of Symbol Grounding Network*

If a machine has a learnt memory as a result of color cognition, the symbol grounding network, such as the one shown in Fig. 7.21, can be used to first recognize the colors of the pixels in an input image. Subsequently, the pixels, of which the colors are similar in terms of color identity or category, can be grouped together to form an image component.

Refer to Fig. 7.18. Under the context of color recognition, coordinates (L, u, v) of a color are given as input, and the response (i.e. value of possibility) from each node in the middle layer will be calculated according to Eq. 7.21. Color label W_i will be the output if the value of possibility from node i is the maximum one.

Refer to Fig. 7.21. Assume that the symbol grounding network has N nodes in the middle layer. And, each node in the middle layer is capable of recognizing one learnt conceptual label, which represents a color's identity, or category, in a human

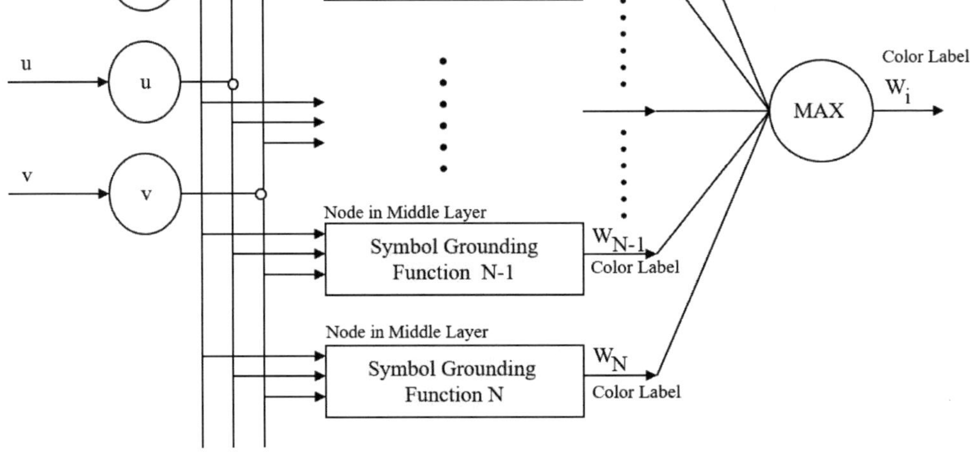

Fig. 7.21 Color recognition by symbol grounding network, in which a sensed color is the input. Color label W_i will be the output if the value of possibility from node i is the maximum one.

language. When a machine undertakes the task of recognizing the colors of pixels in an input image, it will first scan the input image row by row and column by column. At a pixel's location (r, c), color coordinates (L, u, v) are taken as input, which will trigger the responses from these N nodes in the middle layer. Among these responses, the maximum one will indicate the recognized conceptual label. For instance, if label i is the recognized one, then

$$W_{\max} = W_i = \max\{W_k,\ k = 1, 2, \ldots, N\}. \tag{7.23}$$

When all the pixels in an input image have undergone the process of color recognition, these pixels can then be grouped into image components, each of which has a same conceptual label of color. In other words, all the pixels having a same conceptual label of color will be considered to belong to a same image component. If needed, an image component could be further split into sub-components.

7.5.2 *Use of Fuzzy Logic*

After visual learning, the fuzzy sets of conceptual labels of color could also be used for the purpose of doing color recognition. For instance, under the context of color recognition, it is necessary to know what is the degree of certitude for a sensed color to be associated to a learnt conceptual label. In fact, if we consider conceptual label i, the degree of certitude of assigning sensed color (L, u, v) to conceptual label i can be measured by the membership function μ_i, which depends on μ_L, μ_u, and μ_v in the following way:

$$\mu_i(L, u, v) = \mu_L(L) \bullet \mu_u(u) \bullet \mu_v(v). \tag{7.24}$$

Refer to Fig. 7.22. We consider that a machine has a learnt memory as a result of color cognition. Also, in this organized memory, all the conceptual labels of learnt colors form a fuzzy set. If the membership functions of each element in the fuzzy set are known, then this fuzzy set can be used to recognize the colors of pixels in an input image.

For instance, we assume that a machine has learnt M conceptual labels. These labels normally represent color identities (or categories) in a human language. Given an input image, a machine will scan it row by row and column by column. At a pixel's location (r, c), color coordinates (L, u, v) in the L*u*v color space are taken as input to all the membership functions in the fuzzy set of conceptual labels. In response to input (L, u, v), the membership functions in the fuzzy set will produce the following outputs

$$\{\mu_k(L, u, v),\ k = 1, 2, \ldots, M\}. \tag{7.25}$$

Among these outputs, the maximum one

$$\mu_{\max} = \max\{\mu_k(L, u, v),\ k = 1, 2, \ldots, M\} \tag{7.26}$$

indicates the conceptual label which represents the identity (or category) of recognized color at pixel location (r, c).

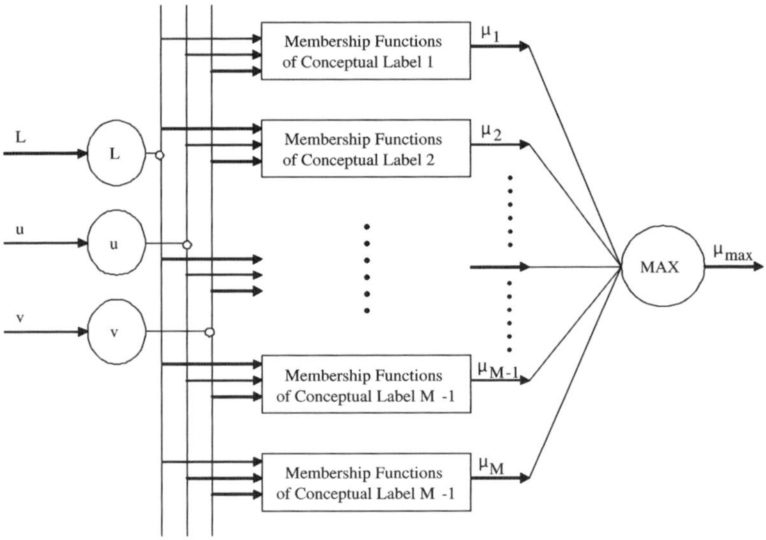

Fig. 7.22 Color recognition by fuzzy logic, in which a sensed color will trigger the responses from the membership functions of learnt conceptual labels in the organized memory. And, maximum output μ_i will indicate the recognized identity (or category) of the sensed color in the conceptual world.

When all the pixels of an input image have undergone the process of color recognition, pixels having similar color identity (or category) could then be grouped together to form image-components.

7.5.3 *Use of Multi-scale Color Indices*

The first objective of doing color recognition is to determine the color identities in an input image. This task is commonly known as *color identification.* And, the second objective of doing color recognition is to determine the color categories in an input image. This task is known as *color categorization.* Therefore, the general process underlying color recognition will include

- Color identification, and
- Color categorization.

However, if a hierarchy of multi-scale color indices in the physical world has been designed in advance, an approach based on the use of multi-scale color indices could unify color identification and color categorization into a single task. This multi-scale approach is explained as follows:

Augmented Multi-scale Color Indices

Without loss of generality, we consider the hierarchy of multi-scale color indices in the RGB color space. Assume that we take a seed color at pixel location (r_0, c_0)

of an input image. Then, this pixel's color coordinates at sensor level will be $(R_{r_0,c_0}, G_{r_0,c_0}, B_{r_0,c_0})$. By applying Eq. 7.12, the multi-scale color indices at pixel location (r_0, c_0) will be

$$\{I_R(r_0, c_0, s), I_G(r_0, c_0, s), I_B(r_0, c_0, s)\}\,, \quad s = 1, 2, \ldots, 8. \tag{7.27}$$

At scale s, color indices $\{I_R(r_0, c_0, s), I_G(r_0, c_0, s), I_B(r_0, c_0, s)\}$ are called *nominal color indices*, which indicate a color category at this scale.

By definition, nominal color indices and its immediate neighbors are called *augmented color indices*. For instance, at pixel location (r_0, c_0), augmented multi-scale color indices will be

$$\{I_R(r_0, c_0, s) \pm 1, I_G(r_0, c_0, s) \pm 1, I_B(r_0, c_0, s) \pm 1\}\,, \quad s = 1, 2, ..., 8. \tag{7.28}$$

Multi-scale Similarity of Color Categories

Now, assume that we examine the color at another pixel location (r, c). And, we would like to know the color similarity between pixel locations (r_0, c_0) and (r, c). In traditional approaches, the similarity test is based on a similarity (or distance) function and a predefined threshold. If the response of the similarity function with regard to two quantities in comparison is within the threshold, it is said that these two quantities are similar. In practice, it is very difficult to predefine a threshold for similarity test. This is why threshold-based algorithms are not robust in real situations.

Interestingly enough, with the use of multi-scale color indices, we could do away with the use of threshold. In fact, the color at pixel location (r, c) is declared to be similar to the color at pixel location (r_0, c_0) at scale s, if and only if the nominal color indices at pixel location (r, c)

$$\{I_R(r, c, s), I_G(r, c, s), I_B(r, c, s)\}$$

are within the augmented color indices at pixel location (r_0, c_0)

$$\{I_R(r_0, c_0, s) \pm 1, I_G(r_0, c_0, s) \pm 1, I_B(r_0, c_0, s) \pm 1\}$$

at scale s.

Pixel Grouping by Similarity of Color Categories

Finally, if an image is given as input, its pixels can be grouped together at different scales of similarity of color categories. For instance, one can randomly chooses some pixel locations, and uses the colors at these locations as seed colors. Then, the input image is scanned row by row and column by column. At each scanned location, the color is compared against the seed colors. In this way, pixels of similar colors at a scale will be grouped together to form image-components at different scales.

Example 7.11. Figure 7.23(1) shows a color image. We select a pixel which is marked by "+". And, the color at this pixel is chosen to be a seed color. Then, we scan the input image row by row and column by column. At each scanned pixel, we test the color similarity at all the scales. If the color at a scanned pixel is similar to the seed color at a scale, the pixel is retained as part of the image-component at that scale. At the end of color recognition, the image-components at different scales are shown in Fig. 7.23(1)–(8) (NOTE: At scale 1, all colors are similar).

<><><><><>

Example 7.12. Figure 7.24(1) shows another color image. Similarly, we select a pixel which is marked by "+". And, the color at this pixel is chosen to be a seed color. Then, we scan the input image row by row and column by column. At each scanned pixel, we test the color similarity at all the scales. If the color at a scanned pixel is similar to the seed color at a scale, the pixel is retained as part of the image-component at that scale. At the end of color recognition, the image-components at different scales are shown in Fig. 7.24(1)–(8).

<><><><><>

7.6 Summary

We are living in a colorful world. The psychological sensation of colors is undoubtedly one of the most fascinating properties of human vision. In this chapter, we have first studied the photometric aspect of image formation in machine vision. We have learnt that human vision is designed in a specific manner by the nature, in order to perceive colors manifested in the physical world. In particular, we have learnt that there are three types of photo-receptors, or photo-detectors, which are responsible for color sensing in human vision (NOTE: The photo-receptors of the fourth type in human eyes are responsible for monochrome and night vision).

Inspired by human vision, it becomes obvious to us that machine vision could also be designed in a specific way in order to sense colors. For instance, we have learnt that there are indeed three types of photo-detectors behind an image point inside a machine's eyes (or cameras). Because of this specific design, the output from an image point at the sensor level consists of three values, which are the photo-detectors' responses to the three primary colors such as red, green and blue.

Subsequently, we have studied the concept-physical model of colors, which could also be designed by a top-down process in modeling the properties and constraints of colors in both the physical world and the conceptual worlds. For instance, we have learnt that the color at an image point at the sensor level is represented by a point in the RGB color space. This confirms Thomas Young's theory of three-variable systems of color. In addition, we have learnt that at the processing level, the intensity and chrominance of a color are separated. And, the result of intensity

(1) Similar Colors at Color Scale 1

(2) Similar Colors at Color Scale 2

(3) Similar Colors at Color Scale 3

(4) Similar Colors at Color Scale 4

(5) Similar Colors at Color Scale 5

(6) Similar Colors at Color Scale 6

(7) Similar Colors at Color Scale 7

(8) Similar Colors at Color Scale 8

Fig. 7.23 An example of classifying pixels of an input image based on multi-scale similarity of color categories. The input image is shown in (1), on which a selected pixel is marked by "+" where a seed color is taken. The clustered pixels of similar color categories at different scales are shown in (2)–(8).

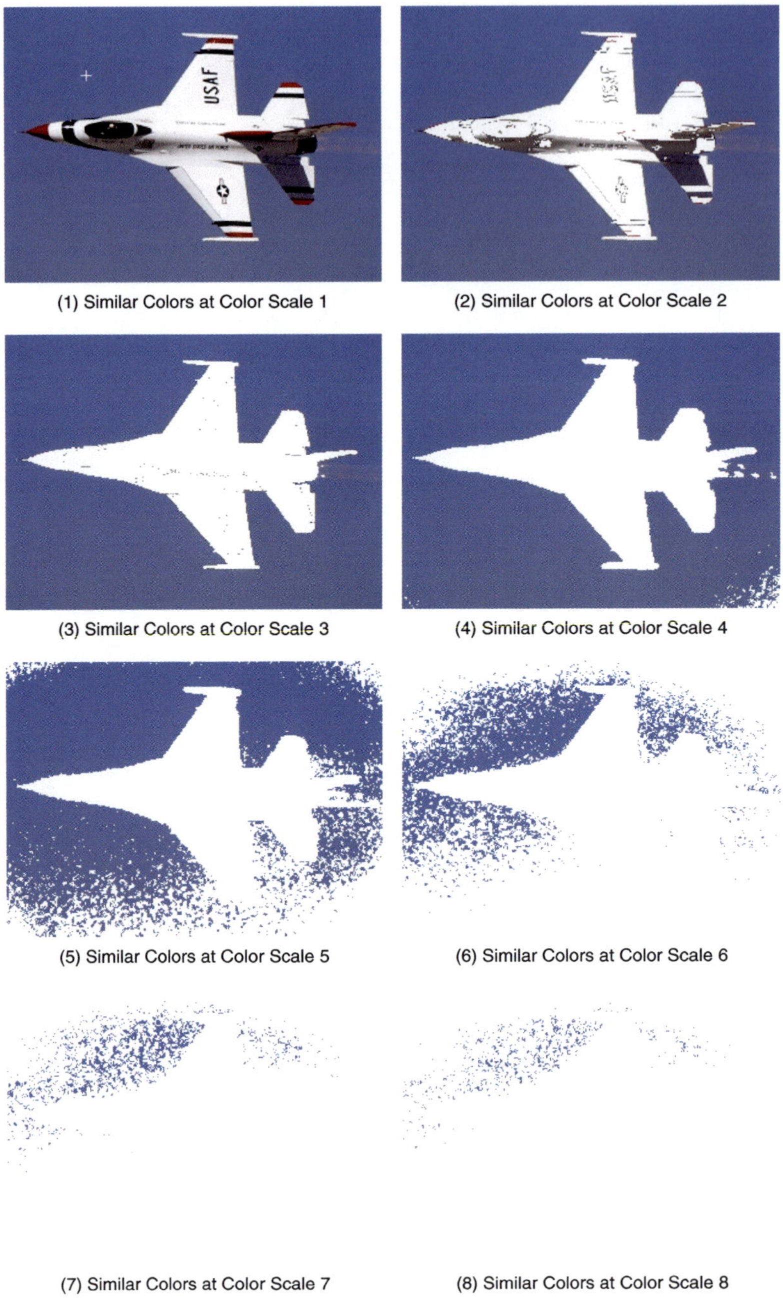

(1) Similar Colors at Color Scale 1
(2) Similar Colors at Color Scale 2
(3) Similar Colors at Color Scale 3
(4) Similar Colors at Color Scale 4
(5) Similar Colors at Color Scale 5
(6) Similar Colors at Color Scale 6
(7) Similar Colors at Color Scale 7
(8) Similar Colors at Color Scale 8

Fig. 7.24 Another example of classifying pixels of an input image based on multi-scale similarity of color categories. The input image is shown in (1), on which a selected pixel is marked by "+" where a seed color is taken. The clustered pixels of similar color-categories at different scales are shown in (2)–(8).

separation could be achieved by transforming the RGB color space into the L*u*v color space, in which the Hering's theory of color opponent processes is verified.

Besides the physical models of colors, we understand as well that the sensations of colors in the mind of human beings could communicate out by words in a human language. This observation has motivated us to study the representation of colors in terms of words or labels in a human language. Most importantly, we are convinced that the conceptual meanings of colors cannot be computed from visual signals, but be learnt through human-machine, or machine-machine, interaction in real-time. This belief has further motivated us to study the aspect of visual learning, which is an important topic in cognitive machine vision.

After the top-down design in modeling the properties and constraints of colors in both the physical and conceptual worlds, we moved on to the study of color cognition, which is a prerequisite to the mental task of color recognition. We know that the two fundamental issues underlying color cognition are: (a) visual computing, and (b) visual learning. The former aims at computing the multi-scale color indices, which represent the physical meanings of colors in the physical world. And, the latter is to learn the conceptual labels (or words), which represents the identities and categories of colors in a conceptual world. Because of top-down design, visual computing of a color's physical meanings consists of simple computation (i.e. multi-scale color indices). However, visual learning of conceptual labels (or words) must invoke a bottom-up optimization process. In particular, we have learnt two innate principles of visual learning, namely: (a) symbol grounding network approach and (b) fuzzy logic approach.

Finally, we understand that the outcome of color cognition is an organized memory, which stores an internalized version of the meanings of colors in both the physical world and the conceptual worlds. The learnt knowledge stored in the organized memory will enable another important mental task, which is known as color recognition. Under the topic of color recognition, we have specifically learnt three basic approaches to the general problem of how to classify the pixels of an input image into image components, each of which contains the pixels having similar colors in terms of color identities or categories.

7.7 Exercises

(1) What is a visible light?
(2) Explain why there are four types of photo-detectors in human eye.
(3) What are the important properties of thin-lens?
(4) What is the purpose of putting a diaphragm in a camera?
(5) What is a pixel?
(6) What are the possible coordinates, which could be associated to a pixel?
(7) Explain why the mixture of two or three primary colors will give rise to the sensation of one color only in the mind of human beings.

(8) Given the color coordinates in the L*a*b color space, derive the equations which permit the computation of the corresponding color coordinates in the RGB color space.
(9) Given the color coordinates in the L*u*v color space, derive the equations which permit the computation of the corresponding color coordinates in the RGB color space.
(10) Give the English names of the color categories at scale 1 in the RGB color space.
(11) List all the possible words in English, which describe colors.
(12) Is it possible to compute the conceptual meanings of colors from visual signals?
(13) Explain the difference between color cognition and color recognition.
(14) Project 1: Implement a program which do the transformation between the RGB and L*a*b color spaces.
(15) Project 2: Implement a program which do the transformation between the RGB and L*u*v color spaces.
(16) Project 3: Implement a program which converts colors, in the RGB color space, into a hierarchy of multi-scale color indices.
(17) Project 4: Implement a program which converts colors, in the L*u*v color space, into a hierarchy of multi-scale color indices.
(18) Project 5: Implement a visual learning program based on neural network.
(19) Project 6: Implement a visual learning program based on fuzzy logic.
(20) Project 7: Implement a human-machine interface which allows a human master to teach a machine (or computer) the conceptual meanings (i.e. identity, or category, labels) of selected colors on an input image.
(21) Project 8: Implement a program which classifies the pixels of an input image into image-components based on color similarity in the RGB, or L*u*v, color space.

Chapter 8

Knowledge Acquisition from Shape

8.1 Introduction

By design at sensor level, human eyes can sense incoming lights. Also, by design at processing level, human vision is able to cognise and recognize colors. The immediate outcome of color perception is that the pixels of an image in human vision have not only color identities, but also color categories. Most importantly, the clustering of proximal pixels of similar color identities, or color categories, create the physical sensation of what is called *shape* in vision. Therefore, shape perception is another pillar underlying both image understanding and scene understanding, which are the two typical scenarios of acquiring knowledge from visual signals.

Interestingly enough, shape is also a concept-physical entity. In the physical world, a shape may correspond to the profile, outline, or boundary, of a physical object. In a conceptual world, most shapes can find their unique designations in a human language (e.g. triangle, square, rectangle, circle, cube, sphere, etc). Hence, it is important for a machine to possess the mental ability to perceive shapes from images. In this chapter, we will study the fundamentals of visual perception of shape under the context of cognitive machine vision (i.e. design-and-optimization paradigm) for knowledge acquisition. In general, we will look into these important issues such as

- Shape Detection
- Shape Representation
- Shape Cognition
- Shape Recognition

8.2 Transformation from Image to Shape

In a machine vision system, the incoming lights to a machine's eye will be converted into numerical values (i.e. digital signals) which are arranged in the form of a two-dimensional matrix. The elements of such a matrix are called *pixels*. At each pixel location, there will be three numeric values which are the measurements of three primary colors such as red, green and blue.

As an image in the eyes of a machine is a two-dimensional matrix of pixels, this implies that an image at the early stage of vision is totally decomposed into the smallest elements called *pixels*. Therefore, it is clear that the first step toward image (or scene) understanding is to compose image components from pixels.

For historical reason, the composition of image components from pixels is commonly called *image segmentation*, or *detection*, in the literature of vision. Strictly speaking, an image at sensor level has already been highly segmented, or fragmented, into pixels. What is necessary in the subsequent step is the composition of image components from pixels.

Within the paradigm of cognitive vision, the ultimate purpose of vision is to achieve the cognitive state of knowing color, shape, pose, and behaviors of objects in a scene. Hence, it goes without saying that visual perception of shape naturally follows after visual perception of color, as shown in Fig. 8.1. And, the very first step in visual perception of shape is to detect the presence of shapes in an image.

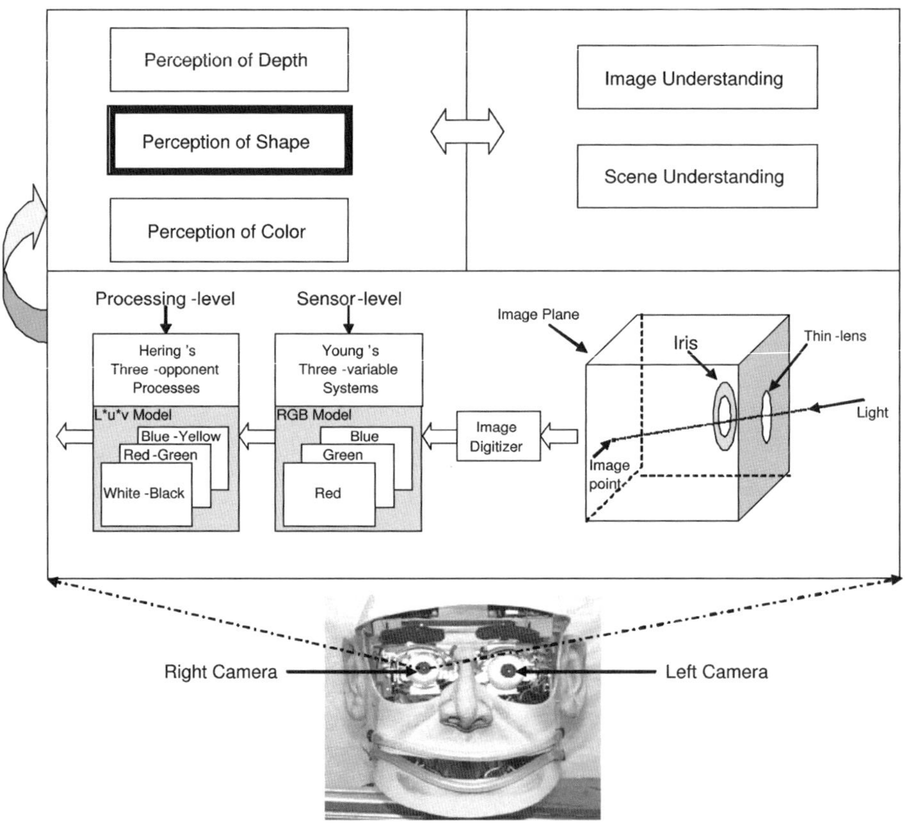

Fig. 8.1 Visual perception of shape under the cognitive vision paradigm, which emphasizes the importance of achieving the cognitive state of knowing color, shape, pose, and behavior of objects in a scene through the mental processes of cognition and recognition.

In this section, we focus our discussion on shape detection. In line with the spirit of design and optimization paradigm, an innate solution to shape detection should be a bottom-up optimization process, which must be operated in conformity to the pre-specified (i.e. pre-designed) definition of shape. Therefore, we present a general definition of shape before studying the basic techniques of shape detection.

8.2.1 *Definition of Shape*

As we mentioned above, the direct cause behind the formation of shape in an image space is the clustering of proximal pixels, which have similar identities, or categories. So far, we have studied the identities and categories of pixels in terms of colors. Therefore, the focus of our discussion in this section is how to group together pixels having similar color identities, or categories. For instance, in Fig. 8.2, we show a zoomed view of a portion of an image. We can see the enlarged pixels. And, the clusterings of pixels having similar colors will give rise to the physical appearance (or impression) of shapes.

After the visual perception of color, the pixels of similar colors in an image will be identified, or categorized, as belonging to (elementary) shapes. Moreover, the further clustering of proximal (elementary) shapes may create more complex

Fig. 8.2 A zoomed view of a portion of an image, in which the enlarged pixels become visible, and the clusterings of pixels having similar colors form shapes.

shapes (e.g. multi-colored shapes). Therefore, without loss of generality, we can postulate a general definition of shape, as follows:

Definition 8.1. A shape is the appearance of a clustering of proximal pixels having the similar identity, or category.

In general, a shape may appear in either a 2D image space, or a 3D scene space. However, a shape from a scene is the reality of its corresponding shape in an image. Conversely, a shape in an image is the manifestation of a real shape from a scene. As we have not yet learnt the topic of visual perception of depth (i.e. positions from which lights emanate towards cameras), it will be premature to discuss about perception of 3D shapes. Therefore, the study in this chapter will mainly focus on the visual perception of shape in a 2D image space.

8.2.2 *Use of Color Histogram*

According to the definition of shape, the objective of shape detection will be to cluster proximal pixels having similar identities, or categories. So far, the available information at a pixel is color. Then, the question becomes how to group proximal pixels together, which are similar in terms of color identities, or color categories.

For instance, when a machine's eye captures an image of a colorful scene, the image plane will contain shapes of specific colors. Naturally, the first question is to know what these specific colors will be. In addition, the next question is to know which pixels belong to a specific color. Obviously, the answer to the second question will lead to the detection of shapes, each of which has a specific color. Also, the answer to the second question depends on the answer to the first question.

In the literature, a common technique to analyze the frequencies of the occurrences of specific colors in an image is the so-called *color histogram*. In terms of algorithm, a color histogram is a matrix, or list, of counters, each of which is assigned to a color indexed by color indices in the RGB, or L*u*v, color space. Without loss of generality, we consider the multi-scale color indices in the RGB color space.

When we have an color image as input, we first compute the multi-scale color indices in the RGB, or L*u*v, color space. Then, we scan the color image row by row and column by column. At pixel location (r, c) (i.e. row r and column c), we will know its multi-scale color indices denoted by

$$\{I_R(r, c, s), I_G(r, c, s), I_B(r, c, s)\}$$

in which s indicates the level of scale. Subsequently, the counter $H[.][.][.]$ associated to the color indexed by this set of color indices is incremented as follows:

$$H[i_R][i_G][i_B] = H[i_R][i_G][i_B] + 1 \tag{8.1}$$

where

$$\begin{cases} i_R = I_R(r, c, s) \\ i_G = I_G(r, c, s) \\ i_B = I_B(r, c, s). \end{cases} \tag{8.2}$$

In other words, Eq. 8.1 counts the number of pixels having color indices $\{I_R(r, c, s), I_G(r, c, s), I_B(r, c, s)\}$ at scale s.

After the full scan of all the pixels in an input image, the frequencies of the occurrences of specific colors are all stored in the three-dimensional matrix of counters H, which is called *color histogram.*

Example 8.1. In order to examine the usefulness of color histogram, we choose a simple example. As shown in Fig. 8.3, a capital letter A in red is placed in a background in gray color.

Under the ideal condition in which there is no noise, this image should have only two specific colors, namely: red and gray. However, due to noises, this image has more than two colors. In order to verify this claim, we first compute the multi-scale color indices in the RGB color space. Then, we compute the color histogram at scale 5. By default, there are $2^5 \times 2^5 \times 2^5$ counters at scale 5. And, we choose to display those counters, of which the values are highest in descending order. The result is shown in Fig. 8.4. We can see that the values in the first two counters are much larger than those in the other counters.

Now, the local maxima in color histogram could be examined. Usually, the values of local maxima in color histogram are called *peaks.* As a color histogram in the RGB color space is a three-dimensional matrix, the neighborhood of a counter will consist of "$3 \times 3 \times 3 - 1$" immediate neighbors as shown in Fig. 8.5. And, the neighborhood is centered at the counter indexed by (i_R, i_G, i_B).

Within the $3 \times 3 \times 3$ neighborhood, if the value of a counter is larger than those of its immediate neighbors, this counter is called a *peak.* For this example, the highest peaks in descending order is shown in Fig. 8.6.

We can see that there are seven peaks at scale 5. Each of these peaks indicates a specific color present in the input image. And, the color indices of these peaks are

$$\begin{array}{lll} I_{\text{peak1}} = (20, 20, 20) & I_{\text{peak2}} = (23, 15, 15) & I_{\text{peak3}} = (24, 13, 13) \\ I_{\text{peak4}} = (26, 10, 10) & I_{\text{peak5}} = (28, 8, 8) & I_{\text{peak6}} = (29, 5, 5) \\ I_{\text{peak7}} = (32, 1, 1). & & \end{array} \tag{8.3}$$

Now, we examine what happens if we reduce the level of scale from 5 to 3. At scale 3, by default, there will be $2^3 \times 2^3 \times 2^3$ counters in the color histogram. However, for this example, only seven of these counters have been incremented. The values of these counters are shown in Fig. 8.7.

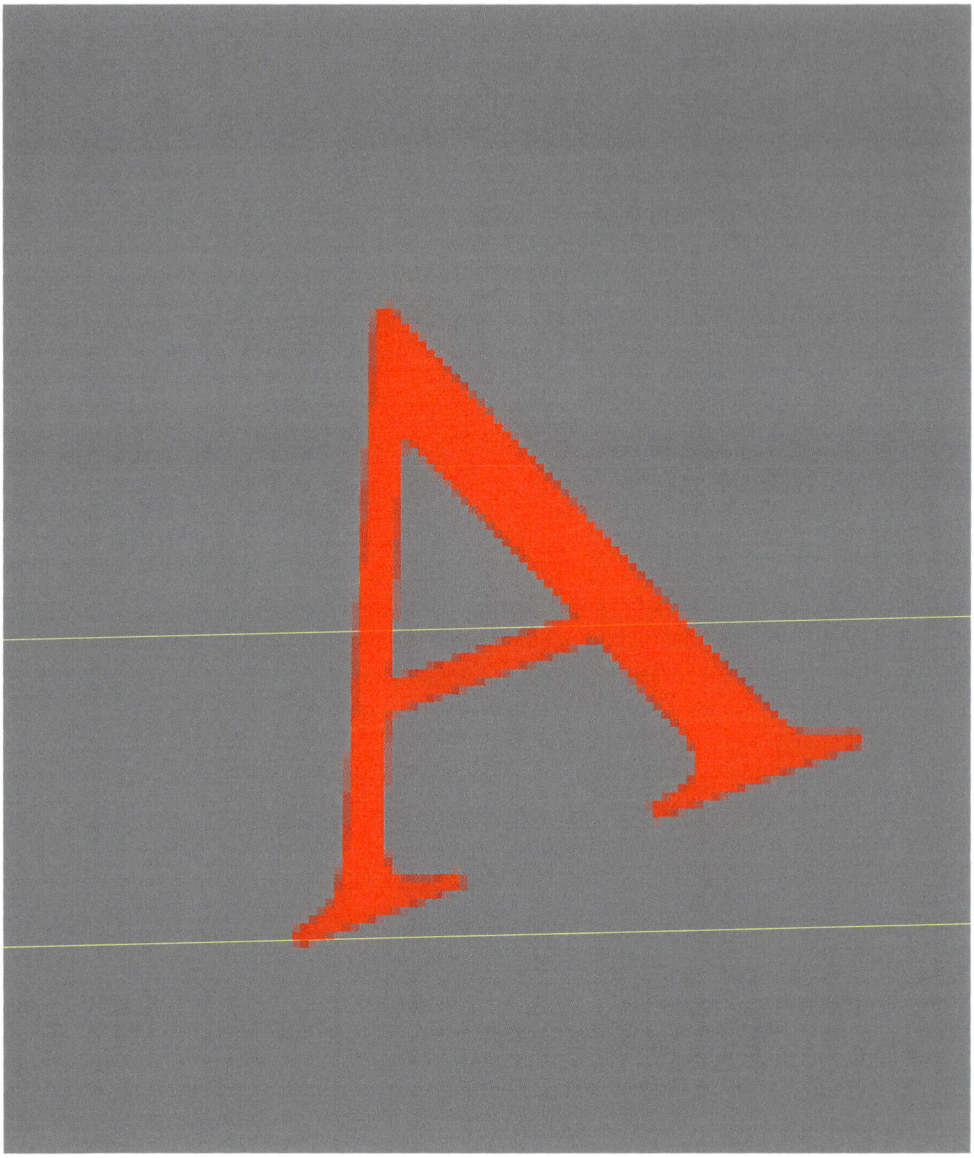

Fig. 8.3 A color image of letter "A" in red which is placed in a background in gray.

Subsequently, we determine the peaks at scale 3. The result is shown in Fig. 8.8. And, the color indices of these two peaks are (5, 5, 5) and (8, 1, 1), respectively. This example shows that a color histogram at an appropriate scale is useful to help discern the specific colors of shapes in an image.

<><><><><>

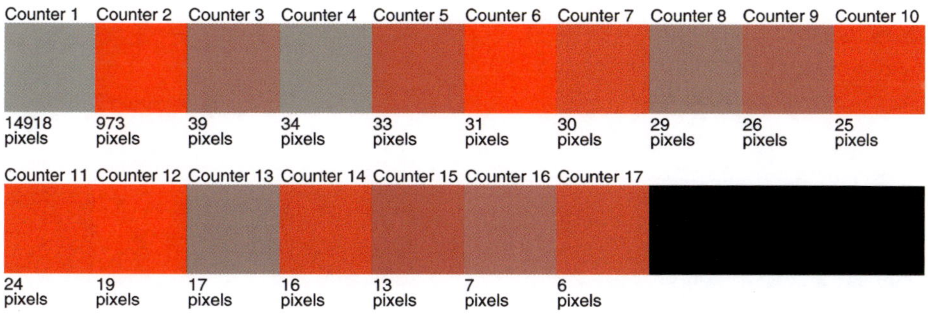

Fig. 8.4 Spectrum of histogram-counters at scale 5. For this example, there are seventeen colors at scale 5. These colors are shown in small squares. Under each colored square, the number indicates the pixel count.

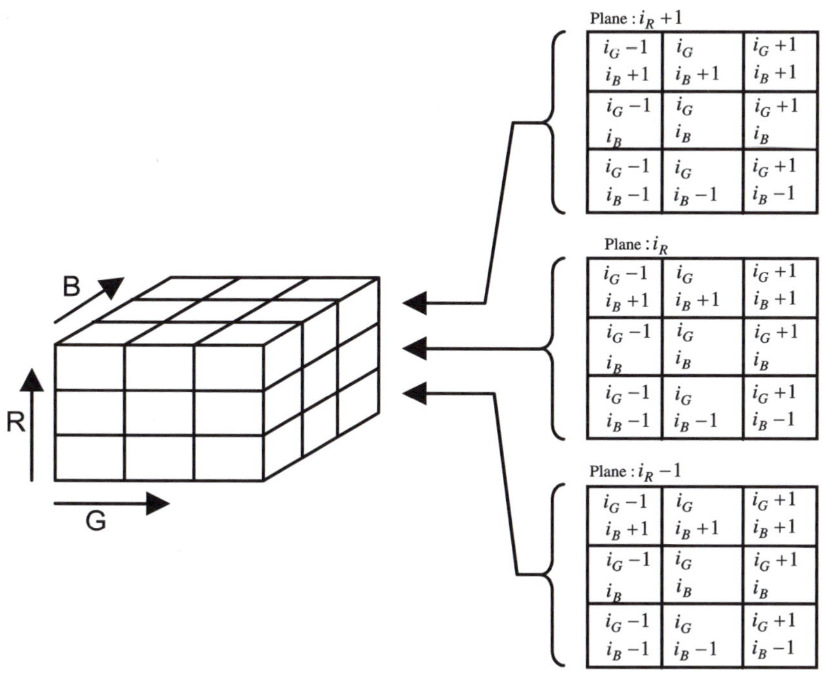

Fig. 8.5 The $3 \times 3 \times 3$ neighborhood centered at location (i_R, i_G, i_B) in a color histogram in the RGB color space at a certain scale.

8.2.3 *Use of Color Region*

The previous example shows that it is possible to simply apply color histogram to first identify specific colors in an input image, and then to group proximal pixels into clusters of these specific colors. The result of the second operation will be the so-called *color regions*.

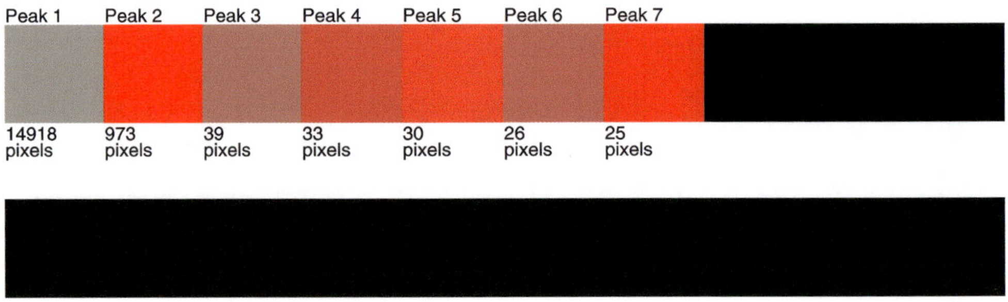

Fig. 8.6 Spectrum of histogram-peaks at scale 5. There are seven peaks at scale 5.

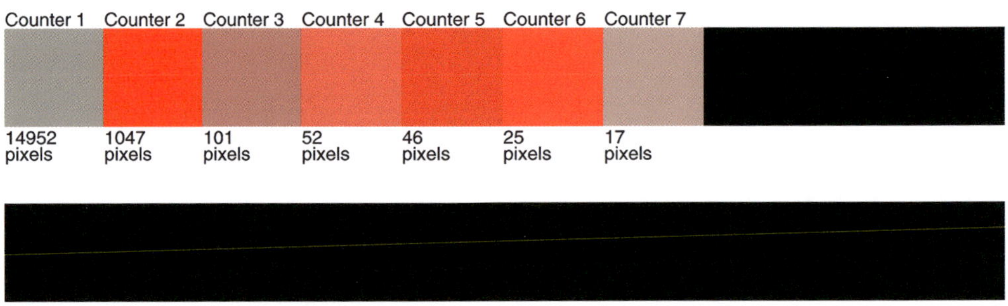

Fig. 8.7 Spectrum of histogram-counters at scale 3. The numbers beneath the colored squares indicate the pixel counts of these counters.

Fig. 8.8 Spectrum of histogram-peaks at scale 3. The numbers beneath the colored squares indicate the pixel counts of these peaks.

Now, the question will be: What is the innate solution to the problem of merging proximal pixels having color identities, or categories, similar to pre-specified colors?

In the literature, there are many studies on solutions and algorithms which aim at partitioning an image into homogeneous regions. These solutions and algorithms are all based on computations of digital signals (e.g. intensities or colors). Because of noises present in digital signals, they suffer the problem of robustness and computational complexity.

Under the context of cognitive machine vision, our concern is on the robust and innate principle, or solution, which enables a machine to partition an image into homogeneous regions, each of which corresponds to a shape.

We have learnt that the color at a pixel can be represented by a set of multi-scale color indices in the physical world. At the highest scale, the color indices represent the color's identity, while the color indices at other scales represent the hierarchical categories of the color. With such a top-down design of multi-scale representation of color indices, it is easy to devise a simple solution to the problem of clustering proximal pixels having similar color category at a chosen scale.

For instance, a practical algorithm will be, as follows:

- Step 1: Determine seed colors of possible shapes.
- Step 2: For each seed color, recursively merge its proximal pixels having the color category, at a pre-specified scale, similar to the one of the seed color. So far, we have learnt that two colors will be said to be similar at a certain scale, if and only if the difference in their color indices at this scale is not greater than one.
- Step 3: Repeat the algorithm until all pixels have been grouped into clusters.

The simplicity of such a solution is expected to make a machine easily adopt it as an innate solution.

In the above algorithm, Step 1 is easy because one can make use of color histogram. With a highly parallel processing architecture, Step 2 is easy as well. This is because the decision of merging any pair of neighboring pixels could be concurrently done by an array of parallel processing units. However, with a computer which operates in a sequential manner, an effective technique of implementation is the so-called *dynamic programming.*

For instance, we assume that we have a seed color at pixel location (r, c) (i.e. row r and column c). We know that in a machine vision system, an image is a two-dimensional matrix. This means that a pixel at (r, c) will have four immediate neighbors at

$$\{(r-1, c), (r+1, c), (r, c-1), (r, c+1)\}$$

as shown in Fig. 8.9a. Alternatively, if one consider the neighborhood of eight immediate neighbors, their locations will be

$$\{(r-1,c-1),(r-1,c),(r-1,c+1),(r,c-1),(r,c+1),(r+1,c-1),(r+1,c),(r+1,c+1)\}$$

as shown in Fig. 8.9b. By dynamic programming, we can develop a function called **clustering-pixel(r, c, s)**, which marks the pixel at (r, c) as part of a new cluster, and recursively merges its immediate neighbors having the similar color category at scale s.

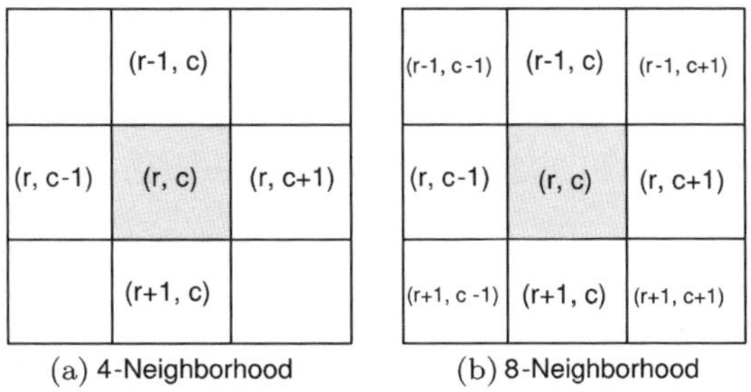

Fig. 8.9 Two common types of neighborhood at pixel location (r, c): (a) 4-neighborhood and (b) 8-neighborhood.

In a pseudo-form of program, function **clustering-pixel(r, c, s)** will look like

```
clustering-pixel(r, c, s)
{
    Mark the pixel at (r, c) as part of the new cluster;

    Get the four immediate neighbors of (r,c):
    (r1, c1), (r2, c2), (r3, c3), (r4, c4);

    %%
    If the color at pixel location (r1, c1) has a similar color
    category (at scale s) to the pre-assigned seed color,
    call function clustering-pixel(r1, c1, s) ;

    %%
    If the color at pixel location (r2, c2) has a similar color
    category (at scale s) to the pre-assigned seed color,
    call function clustering-pixel(r2, c2, s) ;

    %%
    If the color at pixel location (r3, c3) has a similar color
    category (at scale s) to the pre-assigned seed color,
    call function clustering-pixel(r3, c3, s) ;

    %%
    If the color at pixel location (r4, c4) has a similar color
    category (at scale s) to the pre-assigned seed color,
    call function clustering-pixel(r4, c4, s) ;

    return ;
}
```

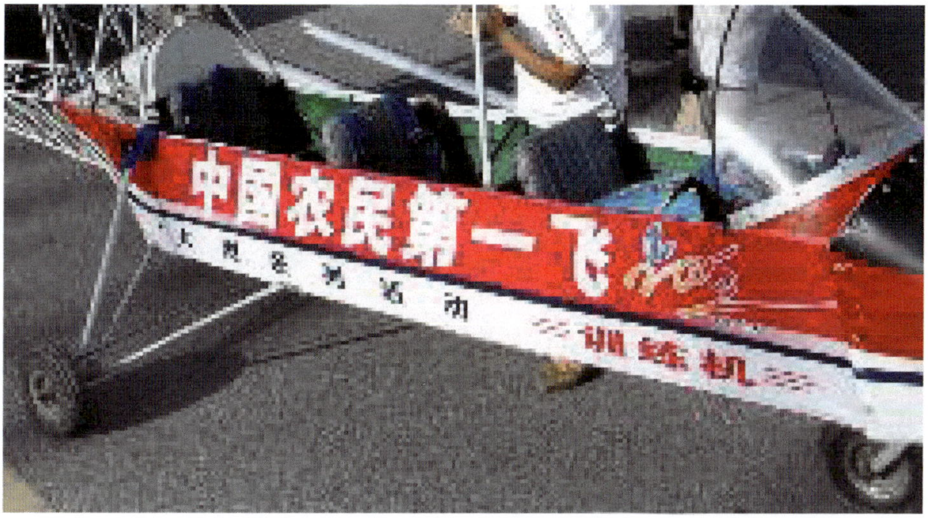

Fig. 8.10 A real image which shows the main body of a small aircraft for training purpose.

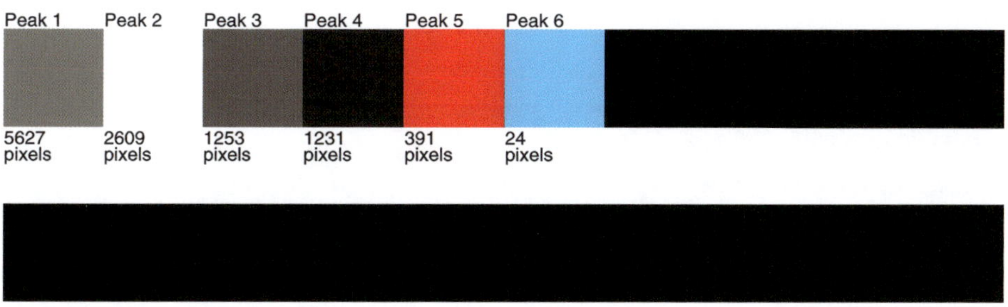

Fig. 8.11 Spectrum of histogram-peaks at scale 3. The numbers beneath the colored squared indicate the pixel counts of these peaks.

Example 8.2. In order to demonstrate the simplicity and robustness of dynamic programming technique with multi-scale color indices, we present here an experiment with a real image of relatively low resolution. This image is shown in Fig. 8.10.

We first compute the multi-scale color indices. Then, we compute the color histogram at scale 3. In the color histogram, we search for the peaks, of which the pixel counts are local maxima. The result of peaks in the color histogram at scale 3 is depicted in Fig. 8.11. There are six peaks in total. We can see that the two dominant peaks are in gray and white, respectively.

Now, we search for shapes in white. For illustration purpose, we manually pick up seven pixels, in which the seed color of white is taken. At each selected pixel location, function **clustering-pixel(r, c, s)** is invoked. And, scale s is set to be 3. The result of seven clusters, or color regions, is shown in Fig. 8.12.

<><><><><>

(a) RGB Image (b) Clusters

Fig. 8.12 Result of color regions. The original RGB color image is shown in (a). The "+" signs indicate the selected pixels where the seed color of white is taken. The clustered pixels around the selected pixels are shown in (b).

Careful readers will notice that seed color in function **clustering-pixel(r, c, s)** remains unchanged during the execution. This property is desirable for detection of color regions, which correspond to relatively flat surfaces such as planar shapes, or characters. However, it is not a good idea to keep a seed color unchanged if a region corresponds to an undulated surface such as the surface of a sphere, or egg.

Interestingly enough, a simple and effective strategy to adaptively change a seed color during the process of clustering proximal pixels is to take the color of a newly merged pixel as a new seed color, and then to continue the process. With this strategy in place, the modified version of function **clustering-pixel(r, c, s)** will simply look like

```
clustering-pixel-modified(r, c, s) {
    Mark the pixel at (r, c) as part of the new cluster;

    Get the color at pixel location (r, c) as the new seed color.

    Get the four immediate neighbors of (r,c):
    (r1, c1), (r2, c2), (r3, c3), (r4, c4);

    %%
    If the color at pixel location (r1, c1) has a similar color
    category (at scale s) to the color at (r, c), call function
    clustering-pixel-modified(r1, c1, s) ;
    %%
    If the color at pixel location (r2, c2) has a similar color
    category (at scale s) to the color at (r, c), call function
    clustering-pixel-modified(r2, c2, s) ;

    %%
```

```
    If the color at pixel location (r3, c3) has a similar color
    category (at scale s) to the color at (r, c), call function
    clustering-pixel-modified(r3, c3, s) ;

    %%
    If the color at pixel location (r4, c4) has a similar color
    category (at scale s) to the color at (r, c), call function
    clustering-pixel-modified(r4, c4, s) ;

    return ;
}
```

(NOTE: One could also put an additional control in place in order to keep the variation of color indices of possible seed colors within a pre-specified range during the execution).

Example 8.3. In order to demonstrate the effectiveness of the modified version of function **clustering-pixel(r, c, s)**, we choose an image which contains many colored eggs, as shown in Fig. 8.13. We can see that the resolution of this real image is relatively low.

Fig. 8.13 A real image which shows many colored eggs in a basket.

(a) RGB Image (b) Clusters

Fig. 8.14 Result of color regions. The original RGB color image is shown in (a). The "+" signs indicate the selected pixels where the initial seed colors are taken. The clustered pixels around the selected pixels are shown in (b).

As usual, we first compute the multi-scale color indices. In order to save space, we skip the result of color histogram. And, for illustration purpose, we manually pick up six pixels, in which the initial seed colors are taken. Then, at each selected pixel location, function **clustering-pixel-modified(r, c, s)** is invoked. And, scale s is set to be 5. The result of six clusters, or color regions, is shown in Fig. 8.14.

<><><><><>

8.2.4 *Use of Color Edge*

The clustering of pixels having similar color identities, or categories, is a direct approach to the problem of shape detection. This is because shapes will directly appear from the outcome of pixel clustering. And, this direct approach makes use of similarities of pixels in terms of their colors.

Alternatively, a shape usually corresponds to a sub-image, which has a finite dimension in an image plane. This means that a shape is normally delimited by its boundary, or edge. Therefore, one may think of the possibility of first detecting the pixels which delimit shapes in an image plane, and then linking these pixels in order to form closed boundaries of shapes. In the literature, the operation of detecting pixels which separate regions in an image is called *edge detection*. And, the outcome of edge detection will provide useful information for further detection of shapes.

In the literature, there have been extensive research efforts on edge detection in the past fifty years. And, most of these efforts have been devoted to the detection of intensity edge. In other words, most efforts are focused on the computation of intensity discontinuity in a gray-scale (or monochrome) image, where edges appear. For instance, the simplest detector of intensity-edge is the so-called *Sobel edge detector*, while a most complex and sophisticated detector of intensity-edge is the *Canny edge detector*. Although some researchers have investigated the issue of color edge detection, their proposed approaches are not much different from those methods being used for intensity-edge detection. Because of this situation, there is a belief that color-edge detection does not bring out much advantage compared to intensity-edge detection.

From our study on the modeling of color, we know that a color normally consists of two parts, namely: (a) chrominance, and (b) luminance (i.e. intensity). In the RGB color space, when the three primary colors, such as red, green and blue, are equal, a color image will become a gray-scale image (i.e. monochrome image). Clearly, a gray-scale image is a special case of color image.

On the other hand, with color image as input, the definition of color edge is much less fuzzier (i.e. vague) than the definition of intensity edge. For instance, a color edge could be defined as follows:

Definition 8.2. A pixel is a color edge if and only if: (a) at least one of its eight immediate neighbors has a different color than itself, and (b) at least one of its eight immediate neighbors has a similar color to itself.

In the above definition, the first condition simply means that a color edge must be at the boundary of a color region, or shape. And, the second condition is to avoid considering a single isolated pixel as a color edge.

However, if we have to define intensity edge, the simplest way is to state it as follows:

Definition 8.3. The pixels of an image where discontinuities in intensity occur are called *edges*.

In order to determine discontinuities in intensity, one must first have a continuous function which describes the surface profile of intensity across an image. Then, it will only become possible to compute the first, or second, order derivatives of this continuous function, which are indispensable for the determination of discontinuities in intensity. Clearly, it is not a simple matter to accurately compute the discontinuities in intensity in a gray-scale image, and then to make decision on where these discontinuities exactly occur in the image plane.

Under the context of cognitive machine vision, we advocate the consideration of color edge for the unique reason that the solution to color-edge detection will be very simple. Because of such simplicity, we could easily make the solution innate to a machine.

For instance, if we have a color image as input, we first scan the image row by row and column by column. At pixel location (r, c) (i.e. row r and column c), we get its eight immediate neighbors. Then, we test the number of its immediate neighbors, of which the colors are similar to the pixel's color in terms of identity, or category, at a certain scale. If this number is within the interval $[1, 7]$, the pixel at

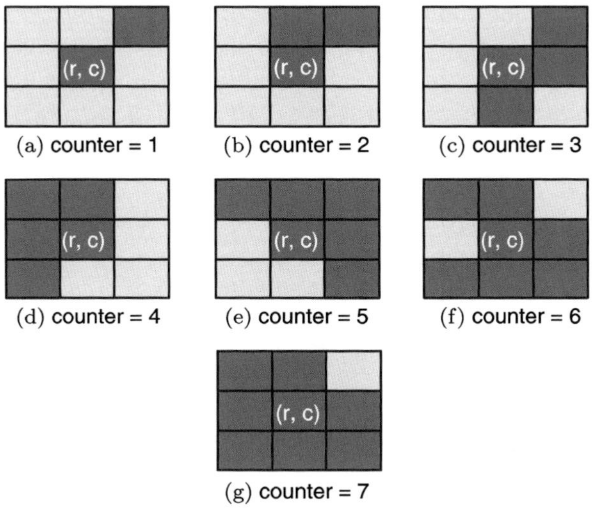

Fig. 8.15 Seven scenarios in which a color edge is detected at pixel location (r, c): (a) one neighbor of similar color, (b) two neighbors of similar color, (c) three neighbors of similar color, (d) four neighbors of similar color, (e) five neighbors of similar color, (f) six neighbors of similar color.

(r, c) is declared to belong to a color edge. In order to demonstrate the simplicity of color-edge detection with multi-scale color indices, Fig. 8.15 illustrates these seven cases, in which the counter will be equal to 1, 2, 3, 4, 5, 6, or 7, respectively.

Interestingly enough, the above operation could be concurrently carried out at all pixels of an input image. In other words, this solution to color-edge detection could be effectively implemented in a highly parallel architecture with a large amount of computing units.

However, if we use a today's computer with a single unit of microprocessor inside, it is also easy to implement the above solution by using a sequential function. For example, we can name this function as **detect-color-edge(image, s)**, where argument "image" is a pointer to an input color image, and argument "s" indicates the level of scale at which the color similarity is being tested. In a pseudo-form of program, function **detect-color-edge(image, s)** will look like

```
detect-color-edge(image, s)
{
    Scan the input image row by row and column by column;

    At currently scanned pixel location (r, c), retrieve its
    color indices at scale s;

    At currently scanned pixel location (r, c), get its eight
    immediate neighbors and their color indices at scale s;
```

(a) RGB Image (b) Color Edges

Fig. 8.16 An example of color-edge detection. The original RGB color image is shown in (a). And, the detected pixels which belong to color edges are shown in (b).

```
    Prepare a counter, and set its initial count to zero;

    For each immediate neighbor, if the neighbor's color is
    similar to the pixel's color at scale s, increment the
    counter by one;

    If the value of the counter is within the interval [1, 7],
    set the pixel at (r, c) as belonging to a color edge;

    Continue the scanning;

    return ;
}
```

Example 8.4. Figure 8.16(a) shows a real color image, which contains many curved edges of various contrasts. We first compute the multi-scale color indices. Then, we invoke function **detect-color-edge(image, s)**, where argument "s" is set to be 3. This means that the test of color similarity is done at scale 3. The output is shown in Fig. 8.16(b). Here, the colors of pixels belonging to color edges are taken from the original image. We can see that two adjacent regions (or shapes) are separated by two edges, which belong to these two regions respectively.

<><><><><>

In order to see the performance of color-edge detection in comparison to the well-known Canny algorithm for intensity-edge detection, we present one more example of result below.

(a) RGB Image

(b) Color Edges

Fig. 8.17 Another example of color-edge detection. The original RGB color image is shown in (a). And, the detected pixels which belong to color edges are shown in (b).

Example 8.5. Figure 8.17(a) shows a real color image, in which a basket contains many colored eggs. As usual, we first compute the multi-scale color indices. Then, we invoke function **detect-color-edge(image, s)**, where argument "s" is set to be 3. The output is shown in Fig. 8.17(b). We can see that the pixels in a textured area (e.g. the sub-image of the basket) appear as edges.

For the purpose of comparison, we use the standard version of Canny edge detector, which is provided in the MATLAB software, to compute the intensity edges from the same image. As the Canny edge detector takes a gray-scale image as input, we first convert the original color image into the corresponding gray-scale image. Figure 8.18(a) shows the gray-scale image. The output from the Canny edge detector is shown in Fig. 8.18(b).

By comparison, we can say that the color edge detector discussed here performs equally well. However, the Canny edge detector is computationally much more complex than our color edge detector.

<><><><><>

8.2.5 *Use of Color Texture*

In the previous sections, the techniques for detecting shapes through color regions, or edges, assume that there are uniform distributions of similar colors within distinct areas in an image plane. However, in real life, the appearance of a shape may not look uniform in color distribution. For instance, in Example 8.5, careful readers will notice that the sub-image of the basket is not uniform in color, but textured. Therefore, another avenue to detect shapes is to identify textured areas in an image.

In the literature, there have been extensive studies on detection and analysis of texture in an image. But, it is interesting to take note that texture is not a cognitive

Fig. 8.18 Result of intensity-edge detection by Canny edge detector. The gray-scale image is shown in (a), which is the input to the Canny edge detector. The output of intensity edges is shown in (b).

entity in an image. For instance, human vision never assigns conceptual meanings to textures. Hence, it is not necessary to dedicate a whole chapter to study texture under the context of cognitive machine vision.

Although texture detection and analysis have been broadly investigated in the past decades, it is still true to say that there is no definite consensus on the definition of texture. This is because the physical appearance of colored, or monochrome, textures in an image can be due to very different causes. For instance, the uniform distribution of small, or large, image-components (e.g. image-components like "+", "o", "L", "T", "Y", or "//") will give rise to the appearance of textures in an image. However, a certain kind of random distributions of small, or large, image-components (e.g. leafs of a tree, or grasses) may also give rise to the appearance of textures in an image.

For instance, some researchers treat images as two-dimensional signals. Then, they define texture as a two-dimensional signal which exhibit specific high spatial-frequencies, and specific orientational characteristics. Based on this definition, techniques for texture detection are mainly based on the use of pass-band filters such as Gabor filters, or wavelet transforms.

In addition, some researchers define texture as a region composed of closely interwoven image-components, which are also called *texture elements*. Based on this definition, some techniques aim at analyzing the distribution and relation of colors, or gray-scale levels, in order to discern the presence of colored, or monochrome, textures in an image. Also, there are some other techniques, which focus on the analysis of texture elements and their placements (or arrangements) in order to identify textured regions in an image.

As the physical appearance of colored, or monochrome, textures in an image depends on the distribution of certain types of image-components (or texture elements), we face two difficulties in defining what color textures in an image should be.

The first difficulty is that a distribution of image-components depends on the spatial scale. For example, at a much larger scale, an image-component in a texture may appear as a distinct shape. Therefore, the relativity in spatial scale makes the definition of texture difficult. The second difficulty in defining texture is that a textured region in an image may contain one type, or multiple types, of image-components. Therefore, it is not easy to stipulate a crisp definition of texture.

Inspired by the appearance of color edges and their distribution in an image, an interesting way of defining color texture is as follows:

Definition 8.4. A specific distribution of color edges within an area in an image is called a *color texture*, or *textured color-region*.

In the above definition, we suggest to consider color edges as the indicator of texture elements. In this way, the detection of above-defined textures can cope with texture elements, which may be of various types and in different sizes. This is because a texture element will have its own edges at a certain scale of color indices. However, there is a fuzzy term in this definition, which is the word *specific*. One may ask: What is the specificity of a distribution of color edges in an image?

Here, we are interested in the spatial distribution of color edges in an image. Hence, we must consider these two important parameters, namely: (a) the periodicity, and (b) the repetitiveness. For instance, the texture elements in an area of an image could be either periodically-, or sporadically-arranged. So, it is necessary to specify such a characteristic of distribution. If we denote t_p the periodicity of a texture, the inverse of t_p actually indicates its spatial-frequency.

On the other hand, a spatial distribution is also characterized by the extent of repeated arrangement of texture elements in some directions. This means that the repetitiveness of texture elements in some directions is another important parameter. For instance, we could choose to consider the horizontal and vertical directions in an image plane (NOTE: It is trivial to add other directions). In either of these two directions, periodicity t_p indicates an interval, in which there is at least one texture element. In practice, the repetitiveness of texture elements in the horizontal, or vertical, direction could be measured by the number of repeats of interval t_p. And, in each of the repeats, there is at least one texture element inside.

Now, we could stipulate a more precise definition of color texture, as follows:

Definition 8.5. A *color texture*, or *textured color-region*, is an area in an image, in which there is a spatial distribution of color edges. And, this spatial distribution is governed by two specific parameters, namely: a) the periodicity, and b) the repetitiveness.

According to the above definition, the detection of color texture will consist of identifying those color edges, which belong to the above-defined textures. Here, we propose a simple algorithm. Hopefully, the simplicity of this algorithm could make it innate to machine vision. The outline of this algorithm is as follows:

An image of color edges is given as the input, which is also called a *color edgemap* in the literature. Then, the algorithm scans the edgemap row by row and column by column. When it reaches a color edge, the algorithm looks at its four directions, namely: (a) the east, (b) the west, (c) the north, and (d) the south. In each of these four directions, the algorithm examines t_n intervals of t_p, which are consecutively repeated. For instance, if the pixel location of the color edge is (r, c) (i.e. at row r and column c), t_n intervals of t_p in the east, west, north and south directions will be

$$east = \{(r, c - k \times t_p), (r, c - (k-1) \times t_p - 1)\}, \quad k = 1, 2, \ldots, t_n, \qquad (8.4)$$

$$west = \{(r, c + (k-1) \times t_p + 1), (r, c + k \times t_p)\}, \quad k = 1, 2, \ldots, t_n, \qquad (8.5)$$

$$north = \{(r - k \times t_p, c), \ (r - (k-1) \times t_p - 1, c)\}, \quad k = 1, 2, \ldots, t_n, \qquad (8.6)$$

$$south = \{(r + (k-1) \times t_p + 1, c), (r + k \times t_p, c)\}, \quad k = 1, 2, \ldots, t_n, \qquad (8.7)$$

respectively. If there is at least another color edge in any one of the intervals in both the east and west directions, the currently-scanned color edge is declared to be a horizontally-textured edge. On the other hand, if there is at least another color edge in any one of the intervals in both the north and south directions, the currently-scanned color edge is declared to be a vertically-textured edge. If a color edge under the scan is a horizontally- and vertically-textured edge, it is treated as part of a texture.

In this algorithm, the status on whether a color edge belongs to a texture does not affect the status of other color edges. This property means that this simple algorithm could be effectively implemented in a highly parallel architecture for data processing. However, with a mono-processor computer, this algorithm could also be implemented as a sequential function.

For instance, let **detect-color-texture(edge, tp, tn)** denote such a function, in which argument "edge" is a pointer to an edgemap, "tp" the input parameter defining the maximum periodicity of all the admissible textures, and "tn" the minimum number of repeats of interval in the east, west, north and south directions from a color edge under the scan. In a pseudo-form of program, function **detect-color-texture(edge, tp, tn)** will look like

```
detect-color-texture(edge, tp, tn)
{
    Scan the input edgemap row by row and column by column;

    When it reaches a color pixel, retrieve its location (r, c);

    %%
    In both the east and west directions, check whether
    there is at least one other color edge in any one of
    the intervals. If yes, the color edge under the scan
    is said to be a horizontally-textured edge;
```

```
    %%
    In both the north and south directions, check whether
    there is at least one other color edge in any one of
    the intervals. If yes, the color edge under the scan
    is said to be a vertically-textured edge;

    %%
    If the color edge under the scan is a horizontally-
    and vertically-textured edge, it is said to be part of
    a texture;

    Continue the scanning;

    return ;
}
```

Example 8.6. In Example 8.5, we have noticed that the pixels in the textured area (e.g. the sub-image of the basket) appear as edges. We use the edgemap in Example 8.5 as input, and invoke function **detect-color-texture(edge, tp, tn)**, in which we set parameters (t_p, t_n) to be $(3, 4)$ (i.e. the maximum periodicity t_p of admissible textures is 3, and the minimum number of repeats of interval t_n from a color edge under the scan is 4). The result from function **detect-color-texture(edge, tp, tn)** is shown in Fig. 8.19(b).

<><><><><>

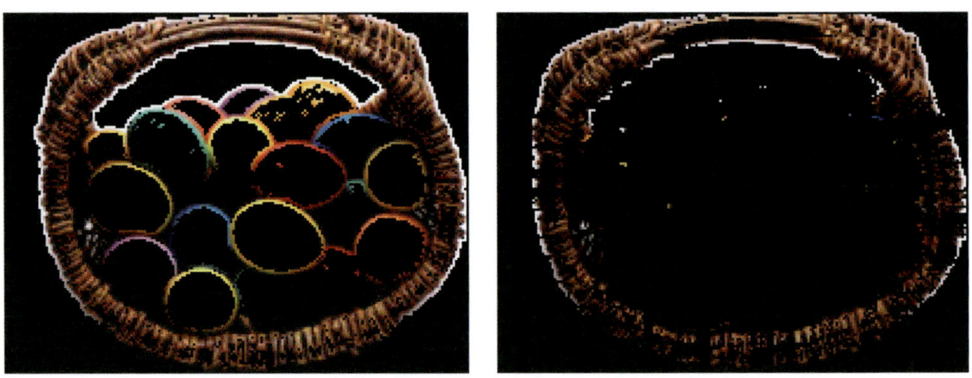

(a) Color Edges (b) Color Textures

Fig. 8.19 An example of color-texture detection. The input of edgemap is shown in (a). And, the detected edges which belong to color textures are shown in (b).

As we have mentioned earlier, textures depend on the spatial scale in an image plane. In other words, a texture which appears at a larger scale may disappear at a smaller scale. In order to appreciate this phenomenon, we present one more example below:

Example 8.7. Figure 8.20(a) shows a color image, in which there are several eggs in white, but one egg in multiple colors. Before color-texture detection, we must obtain color edges as input. For this purpose, the multi-scale color indices are first computed at every pixel in the image. Then, at scale 3, color edges are determined by the algorithm studied in the previous section. The result of color edges is shown in Fig. 8.20(b). We can see that almost all pixels in the multi-colored egg appear as color edges.

For the sake of comparison, we have also applied the Canny edge detector from MATLAB, to detect intensity edges. The result of intensity edges detected by Canny edge detector is shown in Fig. 8.21(b).

Now, we use the result in Fig. 8.20(b) as input, and invoke function **detect-color-texture(edge, tp, tn)**, in which we set parameters (t_p, t_n) to be $(60, 2)$. The result of color edges which are identified to belong to color textures is shown in Fig. 8.22(b). We can see that the whole image is identified to appear as a textured area.

In order to see the effect of scaling down parameter t_p (i.e. maximum periodicity of admissible textures), we reset parameters (t_p, t_n) to be $(3, 4)$. And, the result of color edges which are identified to belong to color textures is shown in Fig. 8.23(b). Now, only the sub-image of the multi-colored egg appears as a textured area. Although there are some isolated edges in Fig. 8.23(b), these noisy edges could be easily removed by a post-processing (e.g. the dimension of a valid texture must be greater than $t_p \times t_n$).

<><><><><>

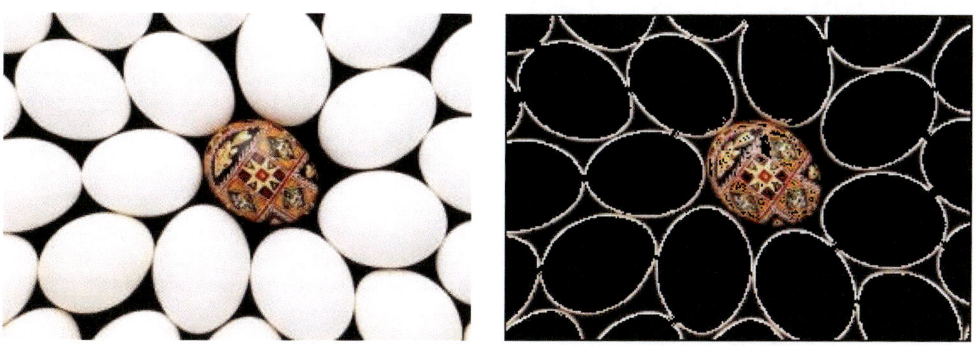

(a) RGB Image (b) Color Edges

Fig. 8.20 Input to color-texture detection: (a) a color image which contains many eggs in white, but one in multiple colors, and (b) color edges as a result of color edge detection which makes use of multi-scale color indices.

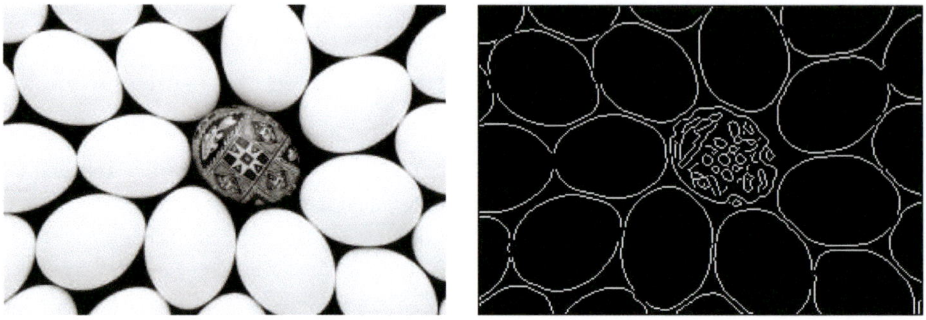

(a) Gray–scale Image (b) Edges (Canny Edge Detector)

Fig. 8.21 Result of intensity edges by Canny edge detector: (a) gray-scale image as input, and (b) intensity edges as output.

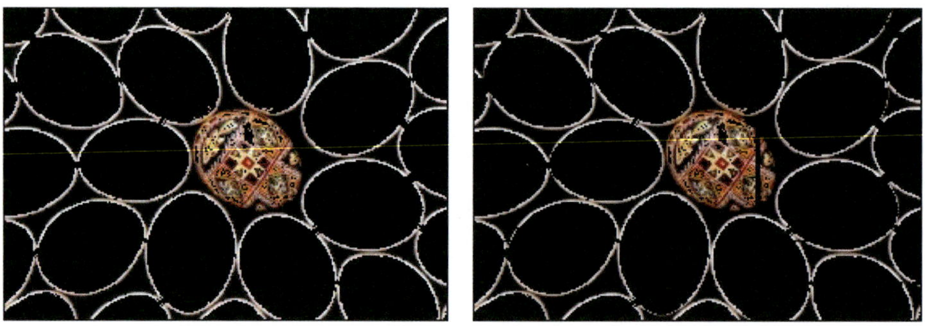

(a) Color Edges (b) Color Textures

Fig. 8.22 Result of color-texture detection when (t_p, t_n) are set to be $(60, 2)$: (a) color edges as input, and (b) color edges which are identified to belong to color textures.

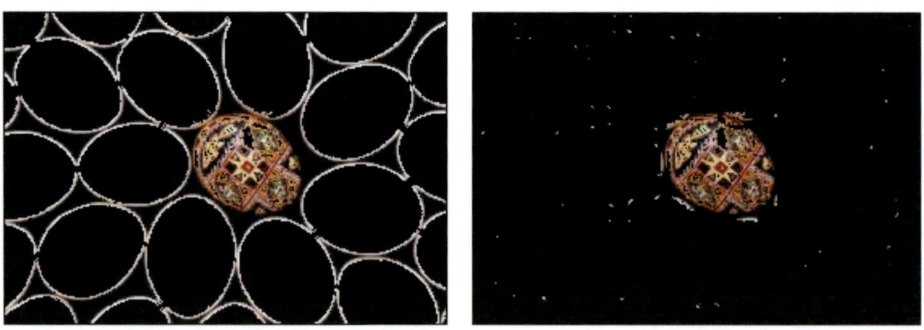

(a) Color Edges (b) Color Textures

Fig. 8.23 Result of color-texture detection when (t_p, t_n) are set to be $(3, 4)$: (a) color edges as input, and (b) color edges which are identified to belong to color textures.

8.3 Representation of Shape

We know that a shape is the appearance of a clustering of proximal pixels having similar identities, or categories. And, the detection of shapes in an image could make use of visual information related to color, edge, or texture. Also, we know that like a color, shape is another type of cognitive entity in an image. This is because a shape usually has the meanings in both the physical and conceptual worlds (e.g. a rectangle, a square, and a triangle, etc).

As a cognitive entity in an image, a shape could be cognised in the first instance, and then be recognized subsequently, by an autonomous machine. Hence, a relevant question will be: What should be the concept-physical model of shape? In this section, we first discuss about the physical properties of shape, and then present a concept-physical representation of shape.

8.3.1 *Properties of Shape in Physical World*

The spatial clusterings of proximal pixels in an image plane will exhibit three types of physical properties, namely: (a) layout (or appearance), (b) pose, and (c) dimension.

8.3.1.1 *Layout or Appearance of Shape*

It is easy to understand that the pixels in a cluster can be configured (i.e. arranged) in very different manners. And, each of such spatial configurations (or arrangements) will give rise to the sensation of a specific appearance, or layout. For instance, Fig. 8.24 shows four different configurations of pixels, each of which appears as a distinct shape in an image.

As the appearance of a shape solely depends on the layout of the shape's pixels in an image, it is important to design innate representations which could accommodate all kinds of shapes. We will come back to this issue shortly.

8.3.1.2 *Scale of Shape*

In geometry, a spatial layout, or appearance, may occur at different scales. For instance, Fig. 8.25 shows the appearance, or spatial layout, of letter "A" at five different scales. We can see that the sensation of shape at these five scales is the same.

When scale changes, the size of a shape normally changes as well. For instance, when a shape is scaled up, its size will normally increase. But, we cannot say that two shapes are at the same scale, if their sizes are equal to one another. For instance, Fig. 8.26 shows two shapes, which are composed of the same segments (e_1, e_2, e_3, e_4). Although their sizes are the same, they are at two different scales.

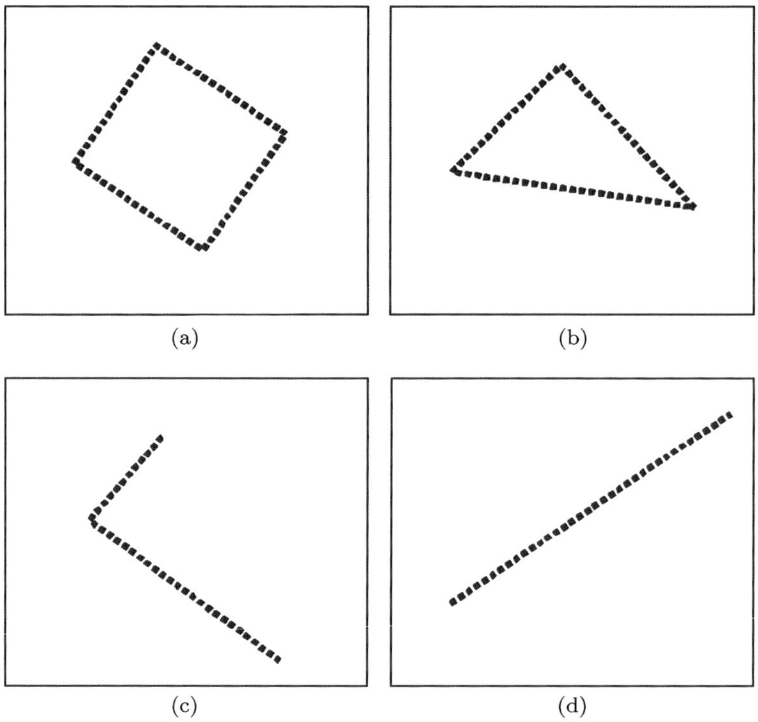

Fig. 8.24 Illustration of pixel configurations, which give rise to sensations of different layouts, or shapes: (a) a rectangular shape, (b) a triangular shape, (c) a L-shape, and (d) a line segment.

Fig. 8.25 Illustration of the appearance of letter "A" at five different scales. At these scales, the sensation of shape remains unchanged.

8.3.1.3 *Pose of Shape*

Once an image is taken (i.e. captured), the instance of a layout, or shape, in it can be treated as a rigid body. In robotics, we know that any rigid body in a space must have its own local frame (i.e. a coordinate system), which nicely represents

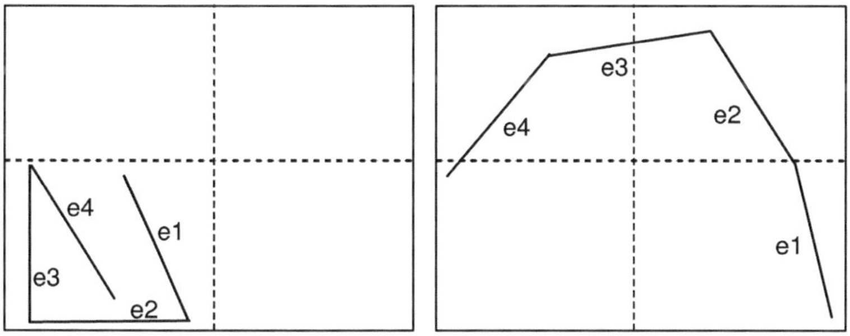

Fig. 8.26 Illustration of the difference between scale and size. The two shapes, one in (a) and the other in (b), are of the same size, because they are composed of the same four segments (e_1, e_2, e_3, e_4). But, they are not at the same scale.

the rigid body's position and orientation. For instance, the origin of a local frame marks the position of a rigid body, while the principal axes of a local frame indicate the orientation of the rigid body with respect to a common reference frame.

In geometry, a rigid body's position and orientation are called a *pose*. As the instance of a shape can be treated as a rigid body in an image, it is natural to adopt the same idea of associating a local frame to a shape. For instance, Fig. 8.27(a) shows a shape with a local frame (i.e. coordinate system).

In computer vision, there are many discussions on what should be translation-, and/or rotation-invariant features. Interestingly enough, with the concept of local frame, all features of a shape computed within its own local frame are absolutely invariant. For instance, we show four different poses of a same shape in Fig. 8.27. Clearly, any feature determined within the local frame for one pose will remain the same for another pose. This is simply because the geometry of a shape with respect to its own local frame remains unchanged.

With the presence of the local frame attached to a shape, it is also easy to obtain scale-invariant features. For instance, one simple solution is to scale a shape up, or down, to a nominal size, before extracting features from the shape.

8.3.2 *Representation of Shape's Physical Properties*

A shape in an image is a manifestation of a real shape from a scene. Therefore, a shape will have its own meanings in terms of properties and constraints in the physical world. Then, the following question arises: What should be the representation of a shape's physical properties?

Under the topic of pattern, or object, recognition in computer vision, there have been extensive investigations on representations of shapes, patterns, or objects in an image. For instance, various statistical, structural, geometrical, or photometric models have been proposed in the literature. However, there is no single, and

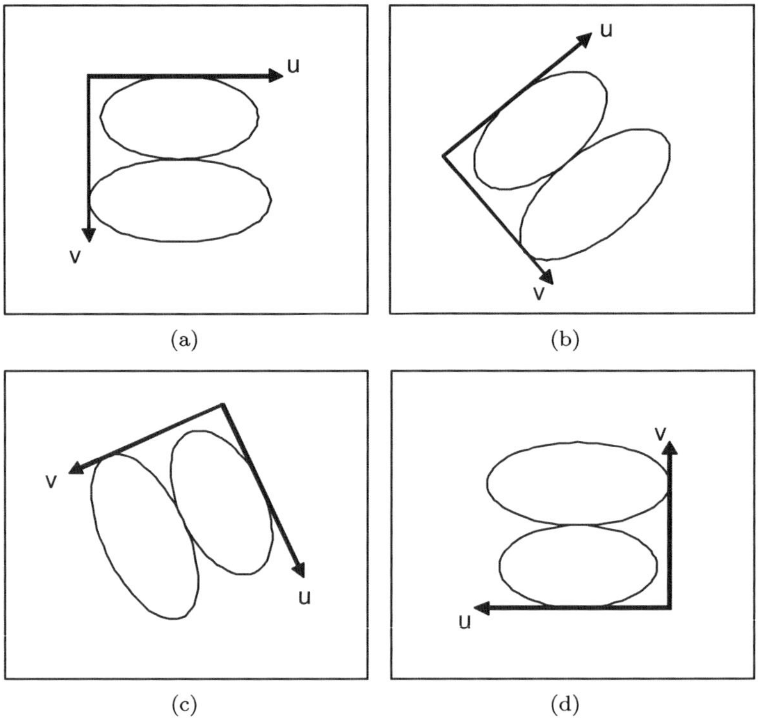

Fig. 8.27 Illustration of various poses of a same shape. When a local frame is associated to a shape, all features computed within the local frame are absolutely invariant.

definite, solution which stands out to satisfactorily answer the question of how to represent shapes, patterns, or objects in an image.

In this book, we generalize the notion of shape, and use it to broadly refer to what are usually called *pattern* and *object* in an image (or scene). In particular, we advocate that the representation of a shape's physical properties should be hybrid, and take into account the following important aspects of a shape:

- Appearance
- Geometry
- Structural Composition

8.3.2.1 *Appearance Model of Shape*

A shape corresponds to a cluster of proximal pixels in an image. Hence, the simplest, and most direct, way of representing a shape is to use a two-dimensional matrix, in which each element could store a binary (i.e. 0 or 1), intensity, or trichromatic, value. Since such a matrix could be displayed as a sub-image, this representation is called an *appearance model*, or *template*, in the literature.

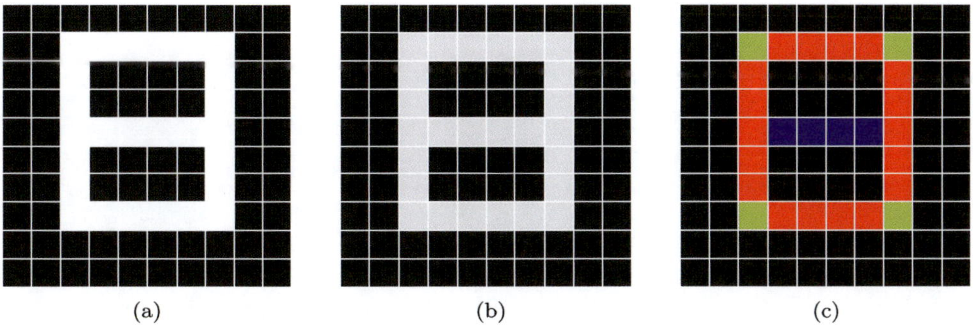

Fig. 8.28 A shape in an image can be represented by a two-dimensional matrix, such as: (a) a matrix of binary values, (b) a matrix of intensity values, or (c) a matrix of trichromatic values.

For instance, Fig. 8.28 shows that a shape of character "8" could be represented by one of these three matrices: (a) a matrix of binary values (i.e. 0 or 1), (b) a matrix of intensity values, or (c) a matrix of trichromatic values.

It is interesting to take note that the appearance model for a shape in an image could be easily extended to the case of a shape in a scene. In fact, the simplest way of representing a shape in a scene is to use a three-dimensional matrix of binary, intensity, or trichromatic, values.

On the other hand, a good representation must be unique and consistent. Then, one may raise the following question: Could the appearance model of a shape be unique and consistent?

In theory, the appearance model of a shape in an image will be unique, if and only if:

- it is translation-invariant,
- it is rotation-invariant,
- it is scale-invariant, and
- the colors of the shape remain unchanged.

As shown in Fig. 8.29, with respect to an image's coordinate system, a same shape may appear very differently. However, if we assign a local coordinate system to a shape, a shape's appearance will be translation- and rotation-invariant with respect to this coordinate system. If we further scale a shape to a nominal size in its local coordinate system, its appearance will also be scale-invariant. In the later part of this chapter, we will study in detail the way of cognising and recognizing appearance models of shapes, which are translation-, rotation-, and scale-invariant.

It is interesting to take note that the appearance model of a shape could be simplified into a representation by its boundary. For instance, the appearance model of a circle could take all the pixels enclosed by the circle. Alternatively, it could simply take the pixels which belong to the boundary of the circle. Another example is the body image of a standing person, which could be simplified into a representation by the body's silhouette.

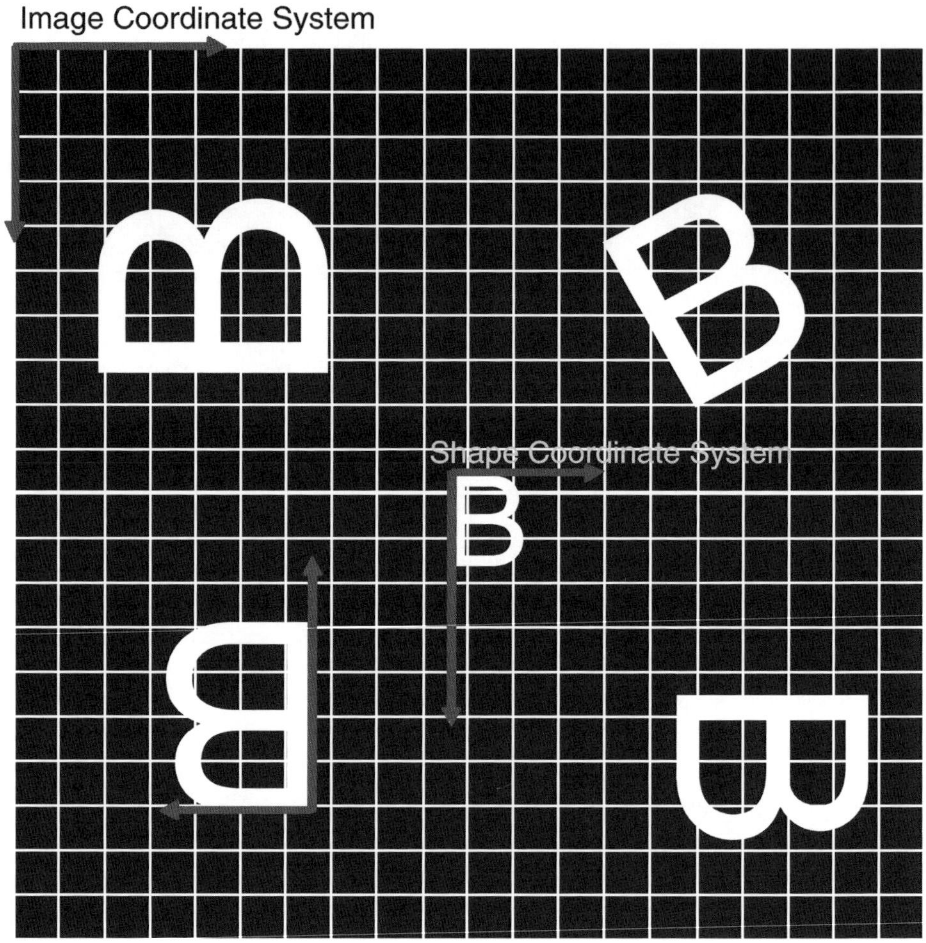

Fig. 8.29 In general, the appearance model of a shape is not unique with respect to the image coordinate system. However, it could become unique with respect to its own local coordinate system, which is also called shape coordinate system.

Another interesting property about appearance model is that we can state a sufficient condition for two shapes to be the same (or similar). For instance, the following sufficient condition is valid:

Sufficient Condition: If the appearance models of two shapes are perfectly superimposed, then these two shapes are the same.

8.3.2.2 *Feature Model or Vector of Shape*

A shape is a cognitive entity in an image. This is because a shape not only has its own identity, but also categories, which it belongs to. For instance, a specific rectangular shape has its own unique appearance (i.e. identity), and also belongs

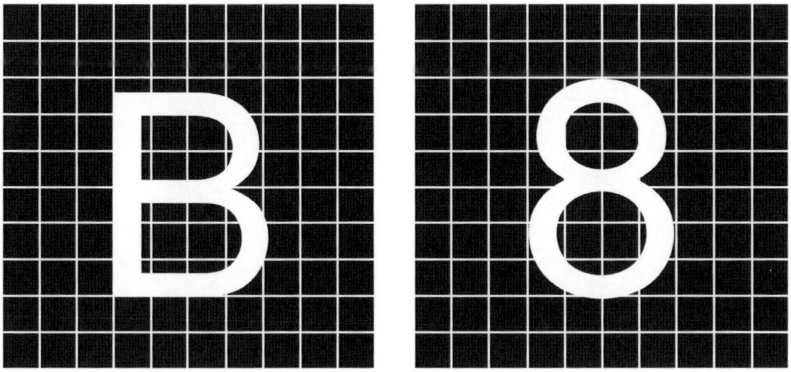

(a) (b)

Fig. 8.30 Example of two similar shapes represented by their appearance models, in which it is not straightforward to meaningfully tell their similarity.

to the category of rectangles, or polygons. As shapes could be categorized, this implies that similar shapes will form categories. In other words, we can talk about the similarity, or resemblance, between two shapes.

When two shapes are represented by their appearance models, it is not straightforward to meaningfully tell their similarity. For instance, Fig. 8.30 shows two shapes which are represented by their appearance models. It is clear that based on the judgment of human vision (i.e. our vision), these two shapes are similar to a certain extent. However, on the basis of appearance model, it is not so straightforward to tell how these two shapes are similar. For example, one may attempt to consider the degree of overlap between them. But, the degree of overlap between the surfaces of two similar shapes does not tell anything meaningful about their difference, or similarity, in geometry, or appearance. For instance, the appear model of letter "O" may not differ too much from the appear model of letter "C". Also, the appear model of letter "Q" is quite close to the appear model of letter "O".

Appearance model is the simplest way of representing a shape in an image. But, it is not quantitative enough for shape categorization. Therefore, we must consider a quantitative model, which metrically characterizes the geometrical properties of a shape. In the literature, a representation which consists of geometrical properties of a shape is called *feature model or vector.* And, the common geometrical properties of a shape, or pattern, include:

Area

We can consider that a pixel of a shape occupies an area of one unit in an image plane. In this way, the sum of areas of a shape's pixels indicates the area occupied by the shape itself.

Perimeter

A shape usually has an outline, or boundary. The length of a shape's boundary is called the *perimeter.*

Width, or Horizontal Range

A pixel of a shape has two coordinates: one along the horizontal axis and the other along the vertical axis. Hence, in terms of coordinates, a shape can be treated as a set of inter-connected coordinates. Then, along the horizontal axis, the difference between the greatest and smallest coordinates is called the *horizontal range*, or *width*, of a shape.

Height, or Vertical Range

Similarly, along the vertical axis, the difference between the greatest and smallest coordinates is called the *vertical range*, or *height*, of a shape.

Enclosing Rectangle

Along the horizontal axis, we draw two vertical lines: one at the smallest coordinate, and the other at the greatest coordinate, of a shape's pixels. Similarly, along the vertical axis, we draw two horizontal lines: one at the smallest coordinate, and the other at the greatest coordinate, of a shape's pixels. It is easy to verify that there four lines will form a rectangle. And, this rectangle is called the *enclosing rectangle* of a shape.

Minimum Enclosing Rectangle

A shape may be placed at different orientations in an image plane. At every possible orientation, a shape has an enclosing rectangle. Among these enclosing rectangles, the one having the smallest area is called the *minimum enclosing rectangle.* It is interesting to take note that the minimum enclosing rectangle may indicate a nominal pose of a shape, at which one can assign a local coordinate system to the shape.

Example 8.8. We have a triangle in an arbitrary pose as shown in Fig. 8.31(a). We first determine the minimum enclosing rectangle, and align the shape so that the minimum enclosing rectangle is parallel to the image. Then, we extract the sub-image enclosed by the minimum enclosing rectangle. This sub-image is shown in Fig. 8.31(c). As the shape may be in any arbitrary pose in the image plane, there are three other possible results as follows:

- the sub-image after 90 degrees of counterclockwise rotation,
- the sub-image after 180 degrees of counterclockwise rotation,
- the sub-image after 270 degrees of counterclockwise rotation.

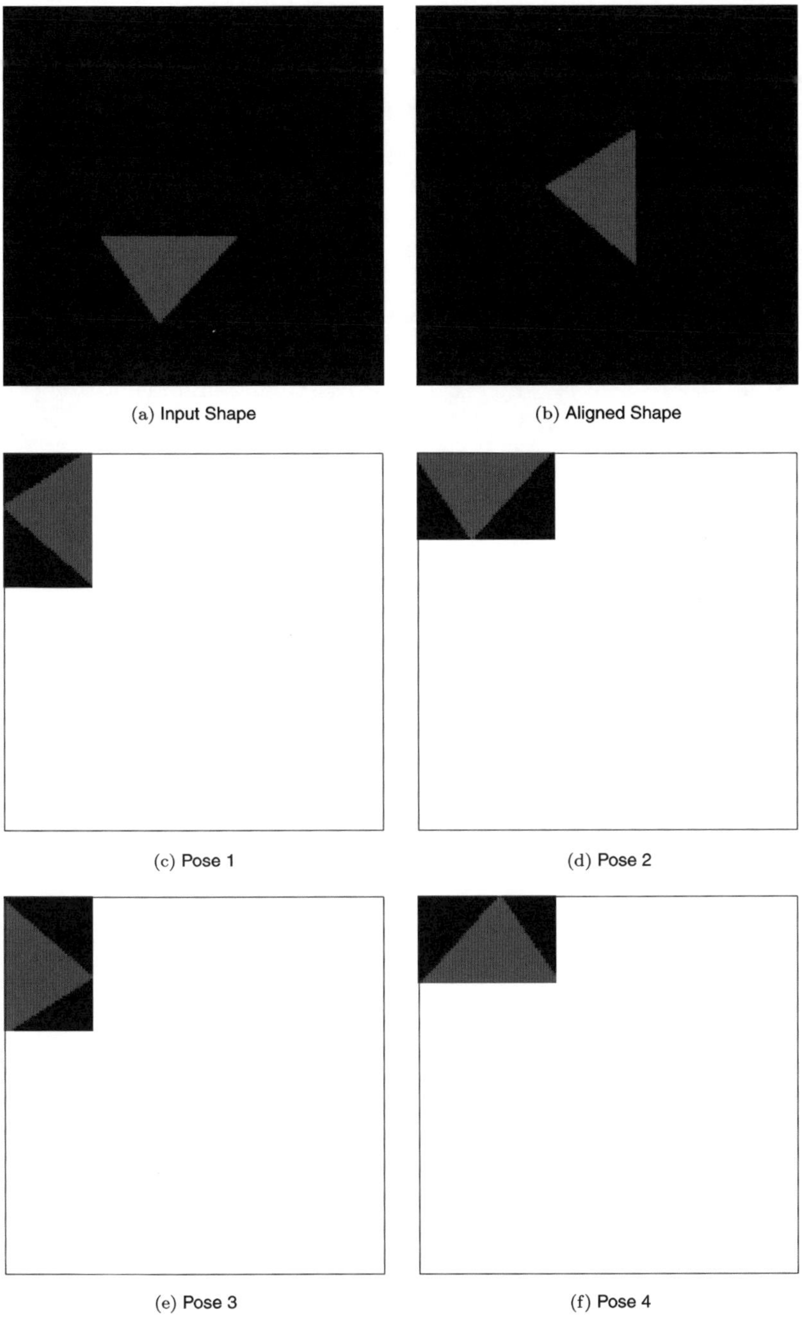

Fig. 8.31 Example of assigning a local coordinate system to a shape by minimum enclosing rectangle: (a) input image, (b) aligned shape, of which the minimum enclosing rectangle is parallel to the image, (c) the extracted sub-image, (d) a sub-image of the shape after 90 degrees of counterclockwise rotation, (e) a sub-image of the shape after 180 degrees of counterclockwise rotation, and (f) a sub-image of the shape after 270 degrees of counterclockwise rotation.

And, for each sub-image, the orthogonal edges at the upper-left corner of the minimum enclosing rectangle defines a local coordinate system for the shape.

<><><><><>

Rectangularity

By definition, the ratio, such as

$$R = \frac{A}{A_{mer}} \tag{8.8}$$

where A is the area of a shape, and A_{mer} the area of its minimum enclosing rectangle, is called the *rectangularity* measure. For example, if a shape is a rectangle, then the rectangularity measure will be equal to 1.

Circularity

Similarly, by definition, the ratio, such as

$$C = \frac{4\pi A}{P^2} \tag{8.9}$$

where P is the perimeter of a shape, is called the *circularity* measure. For example, when a shape is a circle, this measure is equal to 1.

Principal Axes

In data analysis and pattern recognition, we often read about the approach of principal component analysis, or PCA in short. In theory, PCA simply computes the eigenvalues, and eigenvectors of a co-variance matrix.

As a common practice, the eigenvalues, or eigenvectors, are treated as features of a set of data (or images). However, we think that PCA is an effective solution to determine the principal axes of a cluster of data in a multi-dimensional space, which is defined by a data's components. For instance, a pixel in an image has two coordinates, which can be treated as two components. These two components define a two-dimensional space (i.e. the image space). As a result, we can apply PCA to determine two principal axes of a cluster of proximal pixels from a shape in an image plane.

For instance, let $\{(r_i, c_i),\ i = 1, 2, 3, \ldots, N\}$ be the locations of a shape's pixels in an image plane. And, we first compute the mean values of coordinates r and c as follows:

$$\mu_r = \frac{1}{N}\sum_{i=1}^{N} r_i \tag{8.10}$$

and

$$\mu_c = \frac{1}{N}\sum_{i=1}^{N} c_i. \tag{8.11}$$

Knowing the mean values of coordinates r and c, we then compute the variances and co-variances of coordinates r and c as follows:

$$\begin{array}{ll} \sigma_{rr}^2 = \dfrac{\sum_{i=1}^{N}(r_i - \mu_r)^2}{N-1} & \sigma_{rc}^2 = \dfrac{\sum_{i=1}^{N}(r_i - \mu_r)(c_i - \mu_c)}{N-1} \\ \sigma_{cr}^2 = \dfrac{\sum_{i=1}^{N}(r_i - \mu_r)(c_i - \mu_c)}{N-1} & \sigma_{cc}^2 = \dfrac{\sum_{i=1}^{N}(c_i - \mu_c)^2}{N-1} \end{array} \tag{8.12}$$

Finally, the two orthogonal eigenvectors of the following covariance matrix

$$C = \begin{bmatrix} \sigma_{rr} & \sigma_{rc} \\ \sigma_{cr} & \sigma_{cc} \end{bmatrix} \tag{8.13}$$

indicate the directions of the two principal axes of the shape. Geometrically, these two principal axes of a shape define a local coordinate system, which can be assigned to the shape itself. Clearly, PCA is a solution to assign a local coordinate system to a shape, or a cluster of proximal points in any-dimensional space.

In mathematics, PCA can be interpreted as follows: A shape is rotated until it reaches an orientation, in which the variance of coordinates c_i of the shape's pixels is minimum. In this orientation, the vertical and horizontal axes are the two principal axes of the shape.

Example 8.9. We have a hand-shape in an arbitrary pose as shown in Fig. 8.32(a). We apply the PCA method to determine the eigenvector corresponding to the largest eigenvalue. And, we use this vector to define the vertical axis of the local coordinate system of the shape. Then, we align this vertical axis with the vertical axis of the image coordinate system. The result is shown in Fig. 8.32(b). And, the enclosing rectangle after the alignment of the shape delimits a nominal sub-image of the shape. This sub-image is shown in Fig. 8.32(c).

Similarly, as the shape may be in any arbitrary pose in the image plane, there are three other possible sub-images as follows:

- the nominal sub-image after 90 degrees of counterclockwise rotation,
- the nominal sub-image after 180 degrees of counterclockwise rotation,
- the nominal sub-image after 270 degrees of counterclockwise rotation.

Alternatively, we can also apply the solution of rotating the shape until an orientation, in which the variance of coordinates c_i of the shape's pixels is minimum. Not surprisingly, the results are exactly the same as the ones shown in Fig. 8.32.

<><><><><>

(a) Original Shape

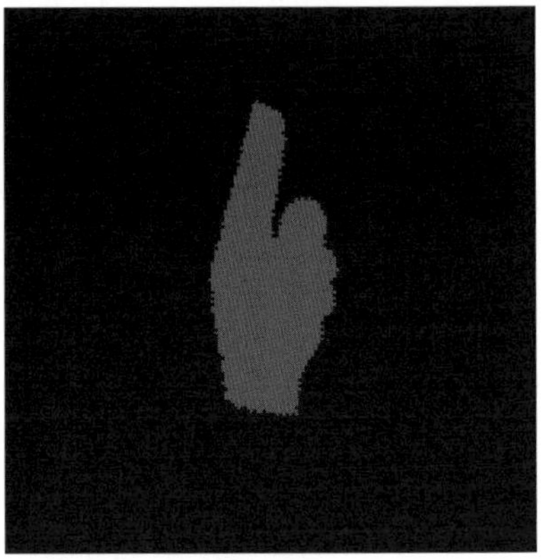

(b) Aligned Shape

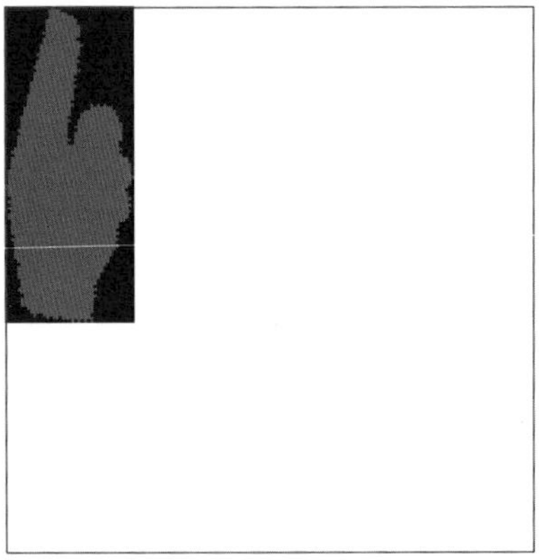

(c) Pose 1

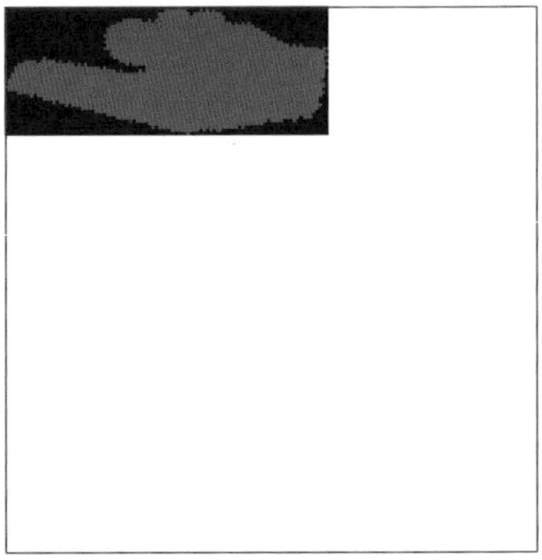

(d) Pose 2

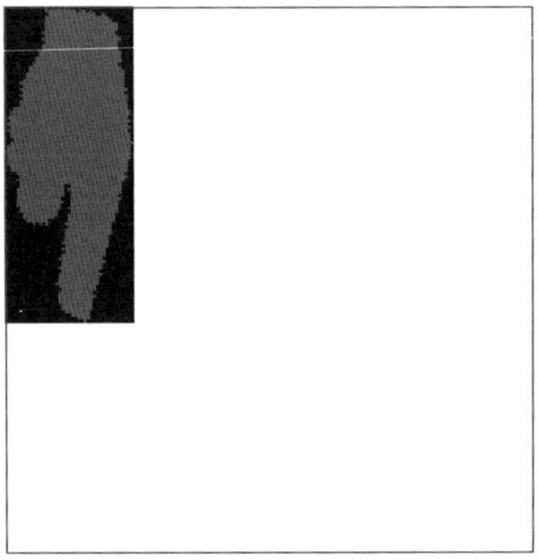

(e) Pose 3

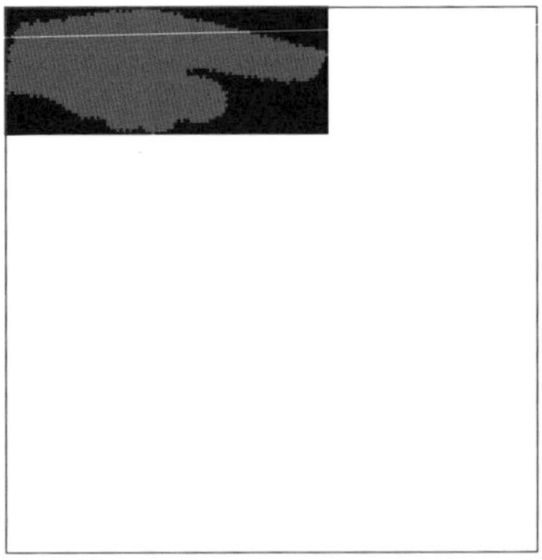

(f) Pose 4

Fig. 8.32 Example of assigning a local coordinate system to a shape by PCA: (a) input image, (b) aligned shape, (c) a nominal sub-image, (d) a sub-image of the shape after 90 degrees of counterclockwise rotation, (e) a sub-image of the shape after 180 degrees of counterclockwise rotation, and (f) a sub-image of the shape after 270 degrees of counterclockwise rotation.

Horizontal Profile of Pixel Distribution

A shape corresponds to a cluster of proximal pixels in an image. And, the pixels of a shape will be distributed spatially. Therefore, we can count the number of pixels at each column, which belong to a shape. If we put these numbers into a vector, the plot of this vector is called the *horizontal profile of pixel distribution.*

Vertical Profile of Pixel Distribution

Similarly, we can first count the number of pixels at each row, which belong to a shape. Then, we put these numbers into a vector. And, the plot of this vector is called the *vertical profile of pixel distribution.*

Example 8.10. We have a shape of letter "A". One of its nominal sub-images is shown in Fig. 8.33(a). We first scale the sub-image up to a standard size, which is chosen to be 128 × 128 in this example. Then, we count the number of pixels at each row, which belong to the shape. The plot of such count-numbers is shown in

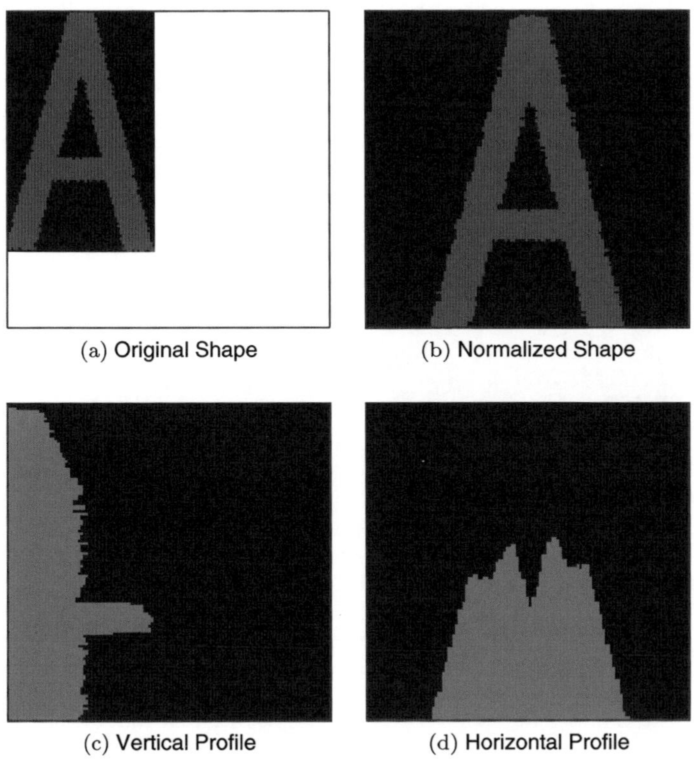

Fig. 8.33 Example of profiles of pixel distribution in both horizontal and vertical directions: (a) a nominal sub-image of a shape, (b) the sub-image scaled up to the size of 128 × 128, (c) vertical profile of the shape's pixel distribution, and (d) horizontal profile of the shape's pixel distribution.

Fig. 8.33(c). Similarly, we count the number of pixels at each column, which belong to the shape. Figure 8.33(d) shows the plot of such count-numbers.

<><><><><>

Moments

Let $f(u, v)$ denote the intensity at pixel location (u, v). If R denotes the set of locations of a shape's pixels, then

$$M_{ij} = \int_R u^i v^j f(u, v) du dv \tag{8.14}$$

is called the moment of order $i + j$ of a shape. Here, i and j are non-negative integers.

Symmetry

The properties listed above are interesting and useful. For instance, the minimum enclosing rectangle, and principal axes, are two possible solutions to the problem of assigning a local coordinate system to a shape. And, the measures of rectangularity and circularity allow us to categorize shapes based on the reference shapes of rectangle and circle.

Interestingly enough, for some special cases, we can consider that a shape is formed by a set of parallel lines of expansion, which are centered about an axis. And, such an axis is called an *axis of symmetry.*

Refer to Fig. 8.34. If a shape's parallel lines of expansion are perpendicular to the axis of symmetry, this shape is called a *symmetric shape.* Otherwise, it is called a *skew-symmetric shape.*

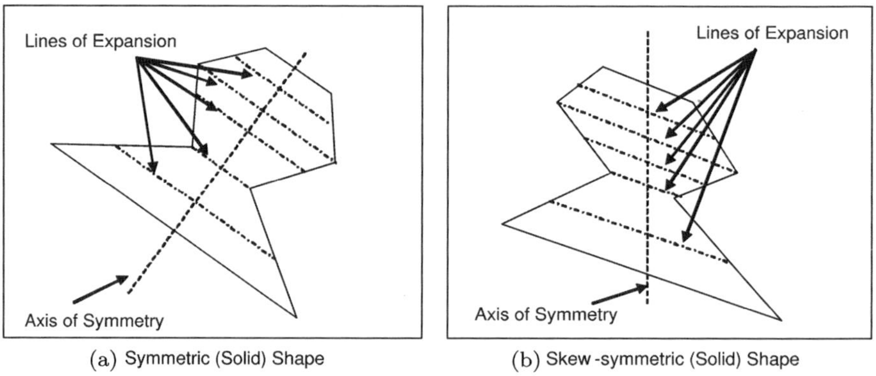

Fig. 8.34 Illustration of symmetric and skew-symmetric shapes: (a) a symmetric shape is formed by a set of parallel lines of expansion, which are perpendicular to, and centered at, an axis of symmetry, and (b) a skew-symmetric shape is formed by a set of parallel lines of expansion, which are not perpendicular to, but centered at, an axis of symmetry.

Generalized Symmetry

It is clear that the axis of symmetry and the parallel lines of expansion fully characterize the geometrical properties of a symmetric shape. However, this statement does not hold for other types of shapes.

In practice, we can extend the concept of symmetry to generally characterize an arbitrary shape. Here, we call the generalized concept of symmetry as the *generalized symmetry*. And, the idea behind the generalized symmetry is as follows:

- Generalized Axes of Symmetry:
  Refer to Example 8.10. When a shape is aligned in a nominal pose, and when it is scaled up to a standard size, we can identify the pixels at column c, which belong to the shape. Then, we can compute the mean value, denoted by $\mu_r(c)$, of row indices of these pixels at column c. If we put the mean values of row indices at all the columns into a vector, the plot of such a vector will give rise to a curved line in general. By definition, the curved line formed by the mean values (i.e. $\{\mu_r(c),\ c = 1, 2, 3, \ldots\}$) of row indices at all the columns is called a *generalized horizontal axis of symmetry*. For instance, the generalized horizontal axis of symmetry will be a straight line if the shape is a circle. Similarly, by definition, the curved line formed by the mean values (i.e. $\{\mu_c(r),\ r = 1, 2, 3, \ldots\}$) of column indices at all the rows is called a *generalized vertical axis of symmetry.*
- Generalized Lines of Expansion:
  When we know the pixels at column c, which belong to a shape, we can also compute the standard deviation, denoted by $\sigma_r(c)$, of row indices of these pixels. Then, by definition, the vertical line at column c within row interval $[\mu_r(c) - \sigma_r(c), \mu_r(c) + \sigma_r(c)]$ is called a generalized vertical line of expansion at column c. Similarly, by definition, the horizontal line at row r within column interval $[\mu_c(r) - \sigma_c(r), \mu_c(r) + \sigma_c(r)]$ is called a generalized horizontal line of expansion at row c.

Example 8.11. Figure 8.35(a) shows a shape of letter "A" in a nominal sub-image. We first scale the sub-image up to the size of 128×128. Then, we compute the generalized vertical axis of symmetry, and the generalized horizontal lines of expansion at every row. The results are shown in Fig. 8.35(c). Similarly, we can also compute the generalized horizontal axis of symmetry, and the generalized vertical lines of expansion at every column. The results are shown in Fig. 8.35(d).

<><><><><>

As we have mentioned earlier, the appearance model of a shape allows us to stipulate a sufficient condition, under which two shapes could be judged to be the same or not. However, with a feature model or vector, we could only state a necessary condition for two shapes to be the same (or very much similar). In other words, when two shapes are the same, it is necessary that their feature vectors will be the same. But, when some features of two shapes are the same (or very much

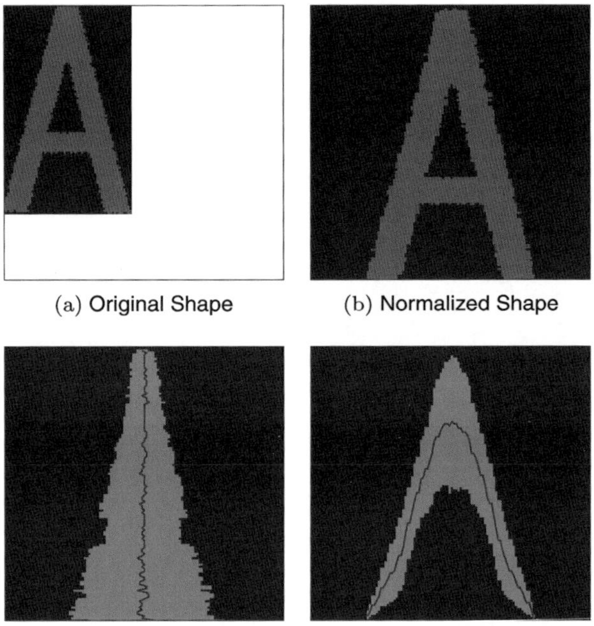

(a) Original Shape (b) Normalized Shape

(c) V–Axis of Symmetry (d) H–Axis of Symmetry

Fig. 8.35 Example of generalized symmetry: (a) a nominal sub-image of a shape, (b) the sub-image scaled up to the size of 128 × 128, (c) the generalized vertical axis of symmetry (in blue) together with the generalized horizontal lines of expansion (in green), and (d) the generalized horizontal axis of symmetry (in blue) together with the generalized vertical lines of expansion (in green).

similar), it is not necessary that the two shapes will be the same (or very much similar). However, in practice, we could state the following necessary and sufficient condition:

Necessary and Sufficient Condition: Two shapes will be the same, if all the possible features of these two shapes are the same.

8.3.2.3 *Composition Model of Shape*

Here, we may ask this question: Is there a shape model, which is suitable for shape synthesis? In order to provide an answer to this question, we must study composition models of shape.

Refer to Fig. 8.36. The question about shape synthesis arises because of three good reasons:

- To predict possible variations in visual learning:
  A shape corresponds to a spatial layout of proximal pixels in an image plane. For a specific shape such the shape of letter "A", it is possible that there are a large number of variations (e.g. different fonts). If a machine has learnt a composition

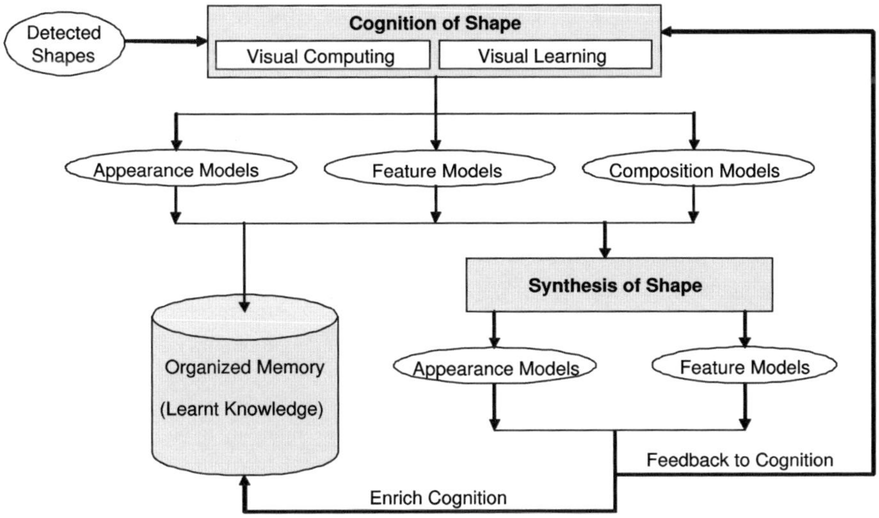

Fig. 8.36 Shape synthesis, in which the composition models play a very important role. A learnt, or cognised, composition model permits the synthesis of both appearance and features models, which in return enrich, or reinforce, shape cognition.

model of a shape, it should be able to synthesize the various versions of the same shape. Otherwise, a machine has to learn all the possible appearances of a shape. Therefore, it is clear that shape synthesis will make visual learning of shapes efficient.

- To achieve feedback-based cognition and recognition:
  We know that in control engineering, a feedback mechanism allows us to achieve the desired performance in controlling a system's dynamic responses. Also, we know that a fundamental problem in vision is the robustness in the presence of noises in visual signals. But, the issue of robustness in vision is partly due to the lack of an explicit feedback mechanism in a visual perception process. Interestingly, shape synthesis do provide a feedback to the visual perception process of cognition (or recognition). For instance, incomplete input to shape cognition will only result in an uncertain output. However, the level of uncertainty will be reduced if the feedback by shape synthesis tends to reinforce the output of shape cognition.
- To synthesize complex, and novel, shapes as part of the mental process of imagination:
  The mind of a human being is able to synthesize not only mental images, but also mental scenes. This mental ability is very useful for us to plan and validate actions, behaviors, decisions, and strategies in a virtual space. And, it is also a very useful mental skill for an autonomous machine, in which all actions, or behaviors, must first be validated in a virtual scene before they are to be executed in a real world.

In terms of physical properties and constraints, a complex shape could be considered as a composition of shape-elements such as lines, curves, squares, and circles, etc. Similarly, a composition model could also consider a shape's conceptual properties and constraints. We will study this aspect of shape composition in the next section under the title of generative model.

From the literature about computer-aided design and computer graphics, we could find many geometrical models, which emphasize the compositional properties of an object, or shape. And, we highlight three of them, because these three composition models are relevant to visual learning of shapes.

Structural Model of Composition

A shape's geometrical properties depend on its spatial layout in an image plane. According to the complexity of spatial layout, a shape may fall into one of these two categories: (a) solid shapes (e.g. a circle), and (b) structured shapes (e.g. letters, characters, and symbols).

For a structured shape, we can consider that it consists of a composition of some shape-elements in a specific manner. For instance, in Fig. 8.37, we show that a structured shape (i.e. a Chinese word) is composed of four shape-elements such as a horizontal line, a vertical line and two L-shape lines.

In general, a structural model of composition will consist of: (a) bar-like shape-elements, and (b) the linkage within a common coordinate system.

Model of Solid Composition

A complex structured shape looks like a linkage of bar-like shape-elements. But, a complex solid shape will be the result of a composition of some primitive solid-shapes such as circles, squares, rectangles, and bars, etc.

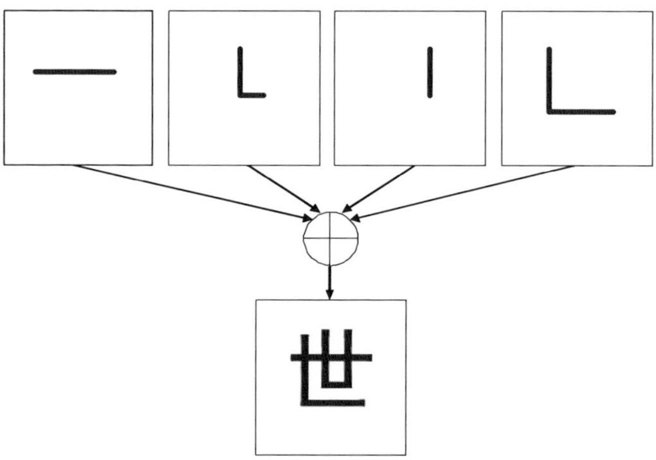

Fig. 8.37 An illustration of structural composition, in which a structured shape is formed by a composition of four shape-elements such as a horizontal line, a vertical line, and two L-shape lines.

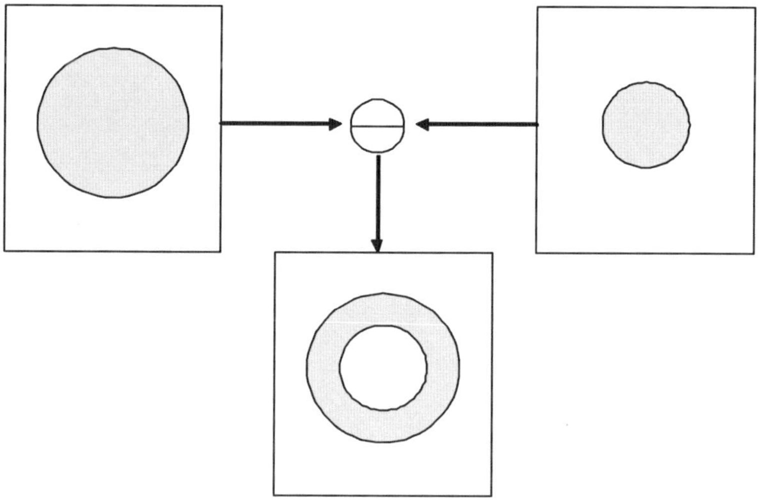

Fig. 8.38 An illustration of solid composition, in which a solid shape is formed by a composition of two circles of different sizes.

For instance, Fig. 8.38 shows that a ring shape is considered as a composition of two circles of different sizes, which are centered at the same point.

In general, a model of solid composition will consist of: (a) primitive solid-shapes, and (b) rules of composition such as union, difference, and intersection.

Model of Boundary Composition

As we discussed earlier, the appearance model of a shape could be simplified into a representation by its boundary. Then, it is interesting to consider the composition of such boundaries.

For instance, in Fig. 8.39, we show that the boundary of a face can be composed of the boundaries of eyes, nose, mouse and head. Here, eyes, nose, mouse and head are called the *constituent shapes* of a face shape.

In general, a model of boundary composition will consist of: (a) the boundaries of constituent shapes, and (b) the layout (or arrangement) of these boundaries.

8.3.3 *Representation of Shape's Conceptual Properties*

In an image, a shape refers to a solid shape, a structured symbol, a character, a pattern, or an object. Also, a shape do have conceptual meanings. For instance, the words such as rectangle, circle, triangle, and ellipse in English are specifically designated to name shapes. On the other hand, all conceptual entities (e.g. lexicons, concepts) do have a written form. For instance, a letter, or a character, has a physical shape. And, a sentence can be considered as a string of shapes. As a result, the modeling of a shape should include the representation in a conceptual world.

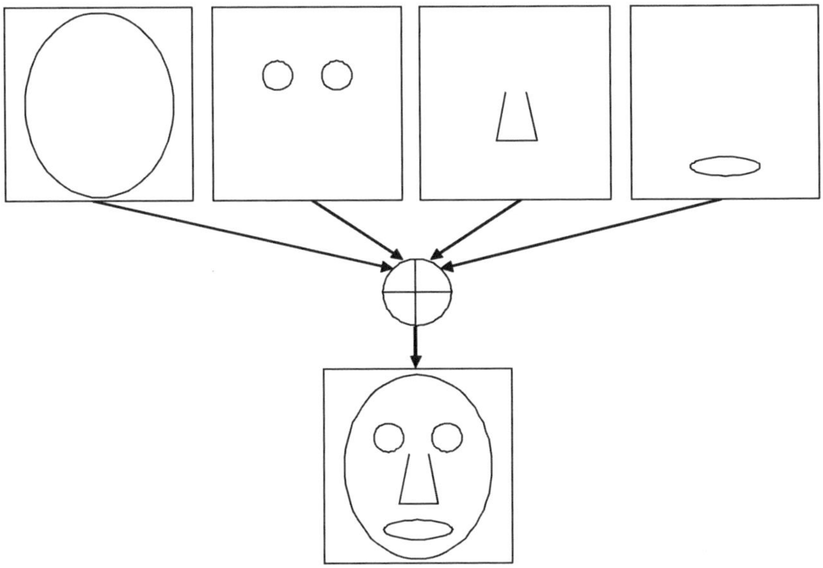

Fig. 8.39 An illustration of boundary composition, in which the boundary of a complex shape is formed by a composition of boundaries of its constituent shapes. In this example, the boundary of a face is composed of the boundaries of eyes, nose, mouse and head.

8.3.3.1 *Conceptual Label*

In an image plane, a shape corresponds to a spatial layout of clustered proximal pixels having similar identities, or categories. When a shape changes its physical size, it will not affect its conceptual meanings. For instance, a small circle is a circle, and a big circle is also a circle. Therefore, we can assign a conceptual label (or word) to designate a category, in which the shapes only differ in size.

However, if we consider language-related symbols or shapes, we will notice that most of human languages do have both the phonetic patterns and written forms. Interestingly enough, the written forms of a human language constitute a natural association between conceptual labels and shapes. By default, a letter, or word, in a human language is a conceptual label assigned to a category of shapes, which are represented by the letter, or word, itself. For instance, Fig. 8.40 shows three appearances which are not exactly the same. But, these three shapes belong to the same category labeled by the letter "w".

8.3.3.2 *Generative Model*

A conceptual label represents a conceptual property of a shape. For instance, the string of characters "circle" labels a circle-like shape. As a shape is manifested by a spatial layout of clustered proximal pixels in an image plane, we can examine the conceptual constraint imposed to the coordinates of a shape's pixels. In this way, a conceptual representation could also be based on a shape's conceptual constraint.

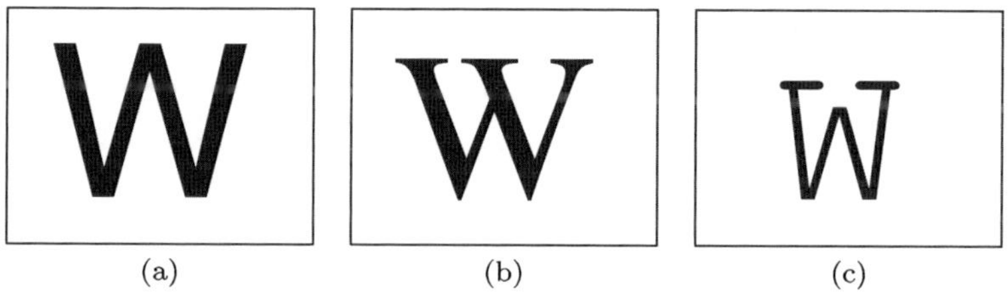

Fig. 8.40 Three shapes which belong to the same category labeled by the letter "w".

By definition, a representation derived from the conceptual constraint is called a *generative model.* And, a shape may have two types of generative models, namely: (a) model in the form of generative equation, and (b) model in the form of generative program (or behavioral process).

Equation of Generative Model

For example, let $f(r, c)$ be a binary function. When it equals 1, it means that the pixel at location (r, c) belongs to a shape. If a shape is a rectangle centered at (r_0, c_0), and is $w \times h$ for its width and height, then the generative equation of this shape could be described as follows

$$f(r,c) = \begin{cases} 1 & \text{if } -h/2 \leq r - r_0 \leq h/2 \text{ and } -w/2 \leq c - c_0 \leq w/2; \\ 0 & \text{otherwise.} \end{cases} \tag{8.15}$$

Now, assume that we have a circle which is centered at location (r_0, c_0) and its radius is denoted by R. Then, the generative equation of this circle will be

$$f(r,c) = \begin{cases} 1 & \text{if } (r - r_0)^2 + (c - c_0)^2 \leq R^2; \\ 0 & \text{otherwise.} \end{cases} \tag{8.16}$$

In general, the conceptual constraint imposed to the coordinates of a shape's pixels could take the form of a polynomial equation. For instance, the following equation

$$a_1 r^2 + a_2 c^2 + a_3 r \bullet c + a_4 r + a_5 c + a_6 = 0 \tag{8.17}$$

is a polynomial equation of order 2, in which $\{a_i, \ i = 1, 2, \ldots, 6\}$ are coefficients.

In computer graphics, other types of generative equations include cubic curves, Bezier curves and B-splines.

Program of Generative Model

It is easy to understand that not all shapes could be simply generated by an equation. For instance, the shape (i.e. a Chinese character) shown in Fig. 8.37 could not be generated by an equation. Instead, it can be written by a pen under the control of a hand, which is guided by a mental program (or behavioral process). In

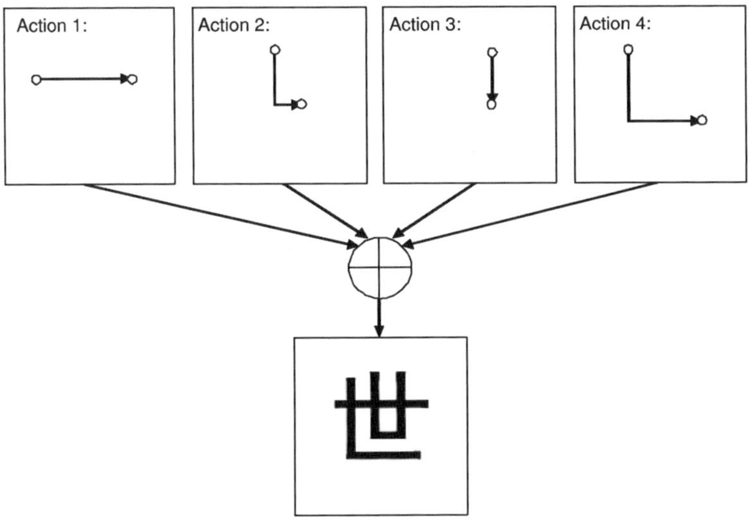

Fig. 8.41 Illustration of a program of generative model, which consists of four orderly actions. Each action has a starting point, and an ending point. And, the completion of these pre-programmed actions by a hand/pen will result in the final appearance of the shape.

this example, the mental program will guide the hand to undertake four actions, as shown in Fig. 8.41. In other words, the completion of these pre-programmed actions will result in the final appearance of the shape.

By definition, a program which controls a hand's actions for the sake of generating a shape is called a *program of generative model.* A program of generative model is a very powerful software, which permits the creation of very complex and novel shapes. Possibly, the creativity of human beings (and machines in future) in creating complex, and novel, shapes could be explained with programs of generative models. For instance, a textured shape, or an impressionist's painting, could be better modeled by a program of generative model, than any other feature model.

8.3.4 *Data Structure of Organized Meanings*

The astonishing power of a human being's mind in undertaking cognition and recognition not only depends on the dedicated mental processes but also an appropriately organized memory. Undoubtedly, a machine's mental ability of cognising and recognizing complex, and novel, shapes must be supported by an organized memory.

In the previous sections, we have studied the different aspects of a shape's representation in both the physical world and conceptual worlds. As shown in Fig. 8.42(a), an organized memory dedicated to the visual perception of shape must include

- Appearance model
- Structural model

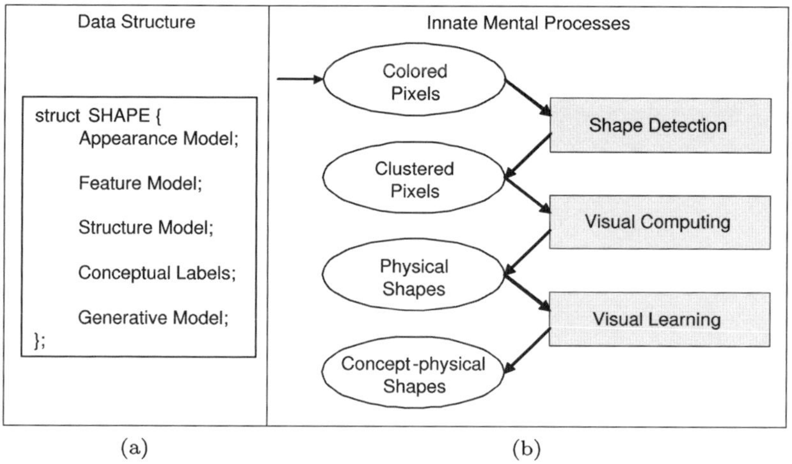

Fig. 8.42 Visual perception of shape must be supported by an appropriately organized memory, which consists of appearance model, structural model, composition model, conceptual labels, and generative model. With such a memory in place, the mental processes behind shape understanding will consist of shape detection, visual computing, and visual learning.

- Composition model
- Conceptual labels
- Generative model

With an appropriately organized memory in place, the mental processes for shape understanding will consist of (see Fig. 8.42(b))

- Shape detection, which transforms colored pixels into clusters of proximal pixels, each of which corresponds to one shape.
- Visual computing, which computes the physical meanings of a shape in terms of appearance model, structural model, and composition model.
- Visual learning, which autonomously learns the association between conceptual labels and physical shapes. Also, visual learning is the only effective way of apprehending a shape's generative model.

In the following sections, we will focus our attention on the study of two specific mental abilities, namely: (a) cognition of shape, and (b) recognition of shape.

8.4 Cognition of Shape

Refer to Fig. 8.43. The task of shape cognition is to understand the physical and conceptual meanings of shapes, which a machine is not familiar with, or does not know, beforehand. In the previous sections, we have studied the top-down design of various models for shapes. Now, the question is: What should be the innate principles, or solutions, which will make a machine to experience the achievement of shape cognition?

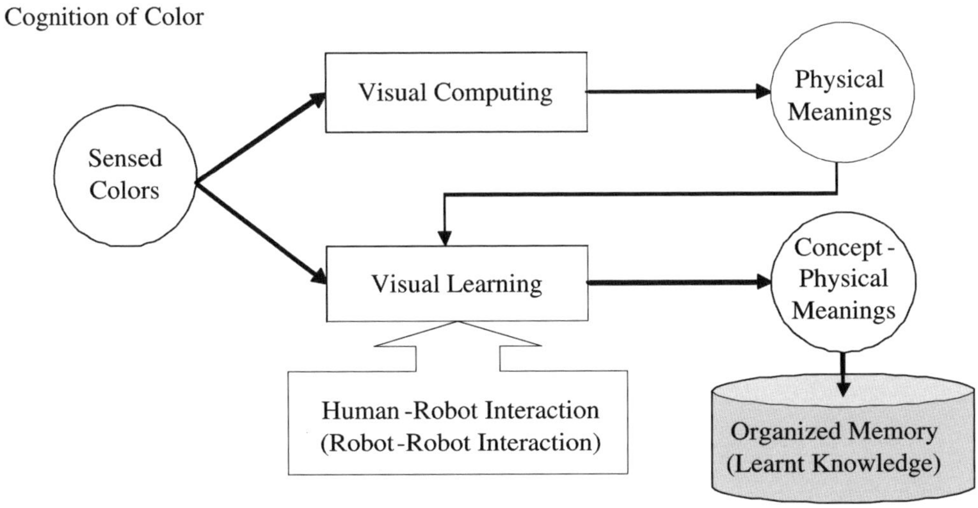

Fig. 8.43 Shape cognition is a bottom-up process, which consists of visual computing and visual learning. The former is for a machine to understand a shape's physical meanings, while the latter is for a machine to apprehend a shape's conceptual meanings through human-machine interaction.

We know that a shape is a cognitive entity which has the meanings in both the physical world and conceptual worlds. Therefore, in general, the mental process for shape cognition will include visual computing and visual learning. As shown in Fig. 8.43, the purpose of visual computing is to derive a shape's physical meanings from signals sensed at image plane (i.e. clustered pixels). And, it is necessary to invoke visual learning in order to apprehend a shape's conceptual meanings.

8.4.1 *Visual Computing*

In an image plane, a shape corresponds to a cluster of proximal pixels having similar identities, or categories. As a result of shape detection, a cluster of proximal pixels will be the input to shape cognition. And, the first concern in shape cognition is to compute a shape's physical properties and constraints in terms of appearance, feature and composition.

8.4.1.1 *Computing Appearance of Shape*

The presence of a shape in an image is physically manifested by a spatial layout of the shape's pixels. And, such a spatial layout of a shape's pixels in an image plane gives rise to the sensation of appearance in the mind of an observer. Mathematically, a spatial layout of a shape's pixels could be represented by a two-dimensional matrix, in which each element could take a binary number, an intensity value, or a trichromatic value. And, by definition, such a matrix is called an appearance model of a shape.

As we discussed earlier, an appearance model in an image plane can be treated as an object. In the coordinate system assigned to an image plane, an object can undergo translation, rotation, and scaling. Obviously, it is not desirable, nor effective, if a machine has to memorize all possible appearances of a shape, which are the results of all possible translations, rotations, and scalings. Instead, it will be interesting to know whether or not an arbitrary appearance of a shape could be transformed into one, or more, nominal appearance, which is translation-, rotation-, and scaling-invariant.

Interestingly enough, a simple solution of computing the nominal appearances of a shape is to first assign a local coordinate system to the shape, and then to scale the shape up to (or down to) a pre-defined standard size. In summary, the generic algorithm for a machine to compute a shape's appearance model will consist of the following steps:

- Step 1: Consistently assign a local coordinate system to a detected shape. In other words, every detected shape must have a unique way of assigning a local coordinate system to it.
- Step 2: Align the shape such that the local coordinate system of the shape coincides with the image coordinate system.
- Step 3: Extract the aligned shape, which is the sub-image within the enclosing rectangle defined by the ranges along the vertical and horizontal directions, respectively.
- Step 4: Scale the aligned shape up to, or down to, a standard size.
- Step 5: Rotate the normalized appearance with a multiple of 90^0 in order to obtain other nominal appearances.

Example 8.12. Figure 8.44(a) shows a detected hand-shape. The first task of shape cognition is to assign a local coordinate system to the detected shape. As we discussed earlier, one way of assigning a local coordinate system to a shape is to determine a principal axis of the shape. A simple solution of computing the principal axes is to employ the PCA method. Alternatively, one can rotate the detected shape until an orientation, in which the variance of the row (or column) coordinates of the rotated-shape's pixels is the minimum.

In this example, we rotate the shape until an orientation, in which the variance of the column coordinates of the rotated-shape's pixels is minimum. And, this orientation serves as the shape's vertical axis. Figure 8.44(b) shows the result of the aligned shape. In this orientation, the shape's local coordinate system coincides with the image coordinate system.

After the alignment of the shape's local coordinate system, we choose a standard size of 128×128, and scale the aligned shape up to this standard size. Here, the

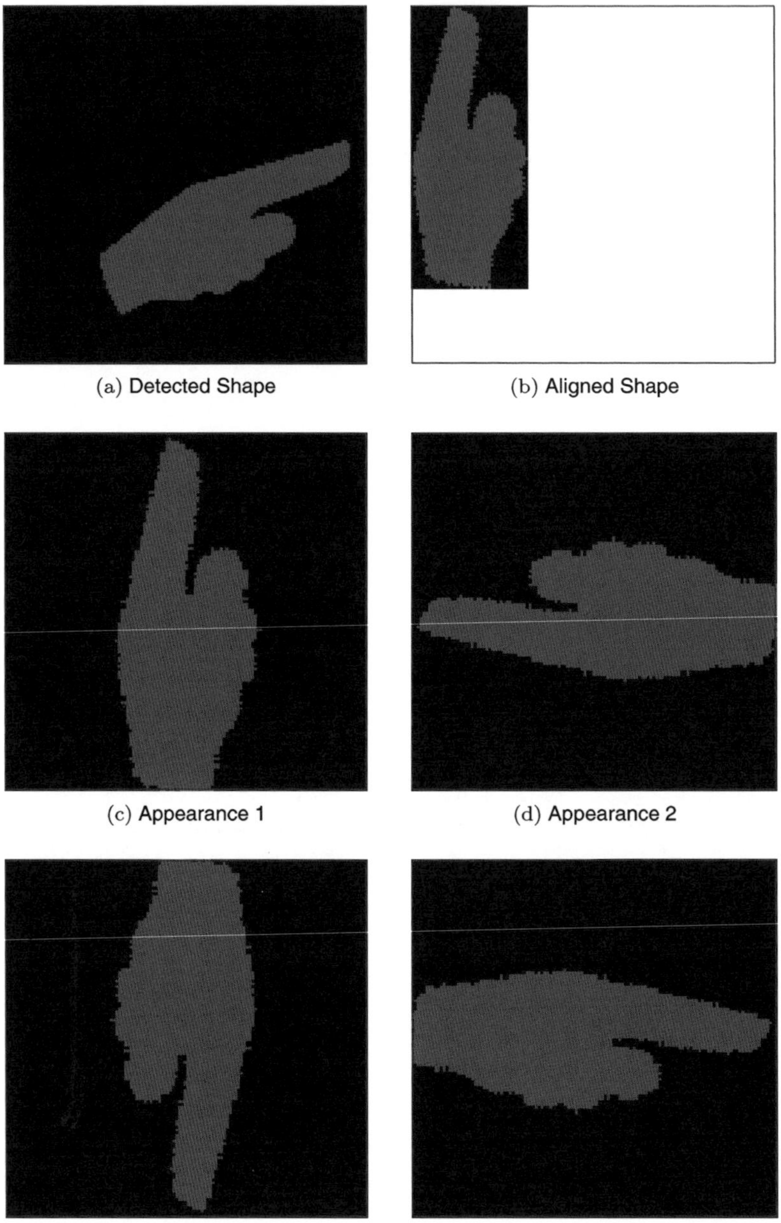

(a) Detected Shape (b) Aligned Shape

(c) Appearance 1 (d) Appearance 2

(e) Appearance 3 (f) Appearance 4

Fig. 8.44 Example of cognising the appearances of a detected hand-shape. A principal axis of the detected shape is fist determined. Then, a local coordinate system is assigned to the shape. After the alignment of the local coordinate system with the image coordinate system, the shape is scaled up to the size of 128×128. And, the rotations of the scaled shape with a multiple of 90^0 result in the four cognised appearances.

scaling factor, denoted by s, is determined as follows:

$$s = \begin{cases} \dfrac{128}{h} & \text{if } h \geq w \\[2ex] \dfrac{128}{w} & \text{if } h \leq w \end{cases} \tag{8.18}$$

where h and w are the height and width of the aligned shape, respectively. After the scaling, the shape is centered by the following offsets in row and column:

$$\begin{cases} r_0 = (128 - h * s)/2 \\ c_0 = (128 - w * s)/2 \end{cases} \tag{8.19}$$

Finally, the rotations of the scaled shape with a multiple of 90^0 will result in four cognised appearances, which are shown in Fig. 8.44(c)–(f).

<><><><><>

8.4.1.2 *Computing Feature Model of Shape*

When a detected shape is transformed into one nominal appearance, it is quite straightforward to compute the possible features of the shape. Mathematically, the features of a shape could be put together to form a vector, which is usually called a *feature vector*. And, the elements in a feature vector defines a space, which is called a *feature space*. Because of a large number of possible features manifested by a shape, a feature space is normally a high-dimensional space.

Geometrically, a feature vector can be treated as a position vector of a point in a feature space. Now, one may think of the possibility of partitioning a feature space into a set of sub-volumes, each of which represents a class of similar shapes in terms of identities, or categories. However, in practice, it is not so intuitive to partition a feature space because of the high-dimensionality of a feature space. And, further research is needed in order to devise a practical solution to the meaningful partition of a feature space.

Example 8.13. Figure 8.45(b) shows a nominal appearance of a hand-shape. First, we compute the generalized vertical axis of symmetry, and the generalized horizontal lines of expansion at every row. The results are shown in Fig. 8.45(c). Then, we compute the generalized horizontal axis of symmetry, and the generalized vertical lines of expansion at every column. The results are shown in Fig. 8.45(d).

<><><><><>

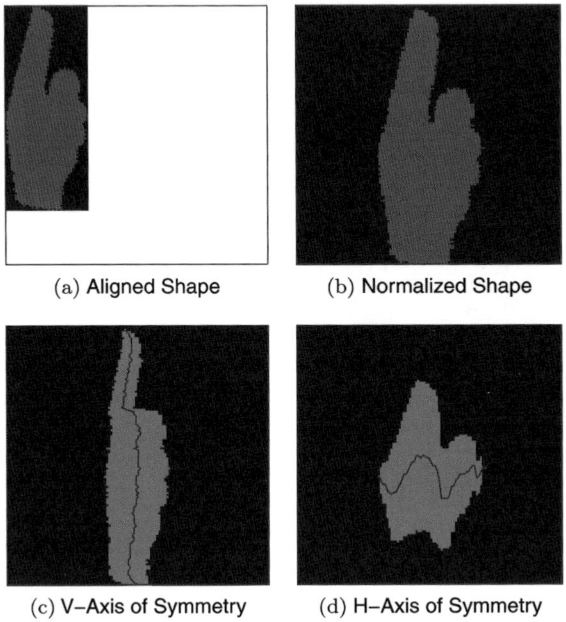

Fig. 8.45 Example of cognising features of a shape. A nominal appearance is given as input. Here, we compute the generalized vertical axis of symmetry (in blue) together with the generalized horizontal lines of expansion (in green), and the generalized horizontal axis of symmetry (in blue) together with the generalized vertical lines of expansion (in green).

8.4.1.3 *Computing Model of Shape's Composition*

If a nominal appearance of a shape is given as input, it is straightforward to compute all its possible features. However, it is still an open issue about whether or not a shape's composition model could directly be computed from the shape's appearance.

For instance, Fig. 8.46(a) shows a detected shape of letter "8". And, the background color is black. Here, we can compute two types of boundary:

- Inner Boundary:
  The inner boundary of a shape consists of the shape's pixels, each of which has at least one neighbor belonging to the background. And, the algorithm for computing the inner boundary of a shape can be as simple as follows:
  - Scan the appearance of the shape row by row and column by column.
  - When the scan reaches a pixel which belongs to the shape, count the number of its neighbors which belong to the background. If the number is greater than zero, mark the pixel as part of the inner boundary.
  - Continue until all the rows and columns have been scanned.
- Outer Boundary:
  The outer boundary of a shape consists of the background's pixels, each of which has at least one neighbor belonging to the shape. Similarly, the algorithm for computing the outer boundary of a shape can be as simple as follows:

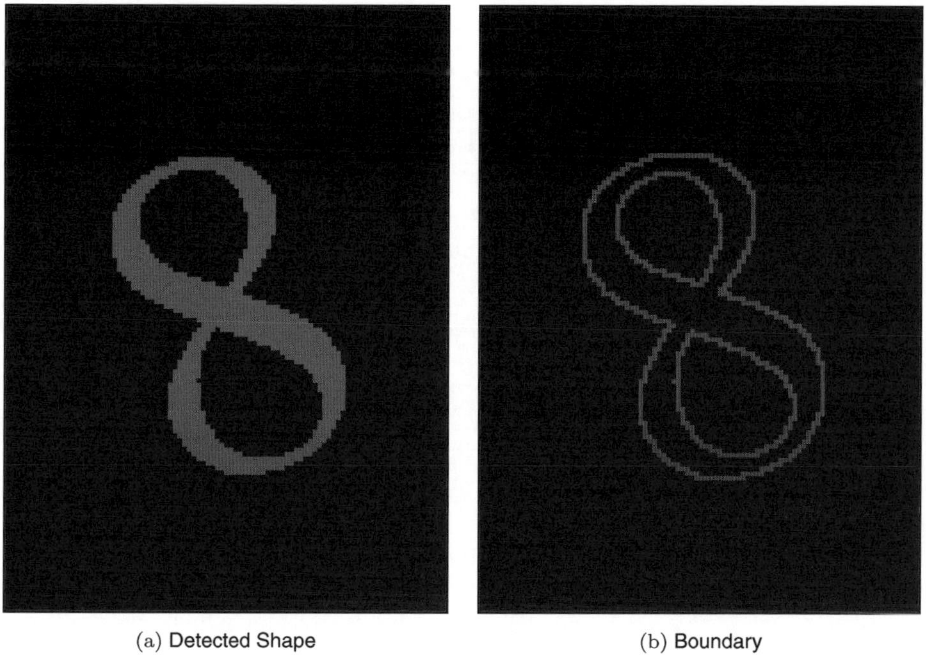

Fig. 8.46 Computation of the boundary of a detected shape: (a) the detected shape, and (b) the outer-boundary of the shape.

- Scan the appearance of the shape row by row and column by column.
- When the scan reaches a pixel which belongs to the background, count the number of its neighbors which belong to the shape. If the number is greater than zero, mark the pixel as part of the outer boundary.
- Continue until all the rows and columns have been scanned.

In Fig. 8.46(b), we show the result of outer boundary of the shape. Clearly, we can see that it is composed of three closed curves, each of which corresponds to the boundary of a constituent shape.

Now, if we ask this question of what the structural, or solid, composition of letter "8" should be, the answer will not be so straightforward. However, if a machine has learnt a generative model of letter "8" (i.e. a machine knows the way of writing letter "8"), then the answer should be easy to obtain. This observation indicates that in general, a shape's composition model should be derived from the shape's generative model, which depends on the visual learning.

8.4.2 *Visual Learning*

A shape is a cognitive entity. However, it is not possible to derive the conceptual meanings of a shape from visual signals directly. For instance, if a shape of a circle is given as input, it is impossible for any machine to derive its conceptual meanings in Chinese, or English, if the machine has no experience of visual learning.

In general, the visual learning of a shape's conceptual meanings plays two important roles, namely:

- Association of Physical Meanings to Conceptual Meanings:
  When a machine perceives a new, and novel, shape, the machine can make use of the innate programs to compute the shape's physical meanings. However, the machine must learn the shape's conceptual label(s) from a human master (or another experienced machine). And, the mechanism for a machine to associate a shape's physical meanings with the shape's conceptual label(s) can simply be a symbol grounding network.
- Acquisition of Generative Model:
  If a machine is cognitively intelligent in the sense that it has mastered a human language, then it is straightforward for the machine to learn a shape's generative models in terms of equations, or behavioral programs (e.g. the sequence of hand's movements in writing, or painting, a shape).

Now, we may ask this question of whether or not today's machines have been designed in such a way so as to make visual learning easy. As shown in Fig. 8.47, a learning machine must have an appropriately-designed body, which is apt to undertake active learning of physical and conceptual meanings through the autonomous interaction in a scene. And, the autonomous interaction will consist of behaviors, which are planned according to the decisions made by the machine itself.

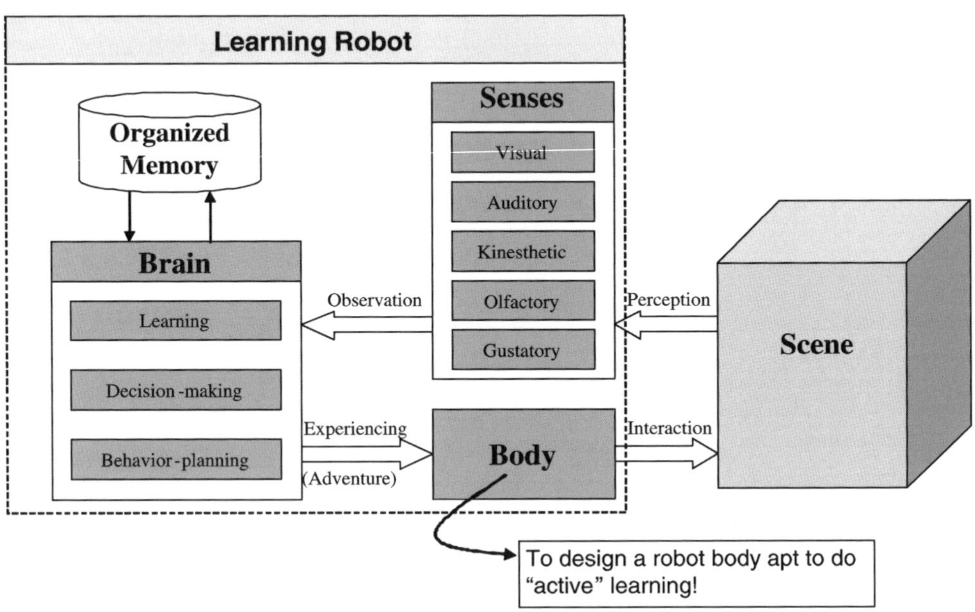

Fig. 8.47 A learning machine must possess a body which enables the machine to undertake active learning of physical and conceptual meanings through the autonomous interaction in a scene.

In addition, the importance of designing an appropriate body for active learning could also be understood by these two interesting observations:

- Domestic Animals Cannot Master Human Languages:
  For those who like domestic animals, it is interesting to take note that no matter how long an animal lives with us, it will never have the mental ability to speak a human language. This deficiency is not due to the lack of human-animal interaction, but to the lack of an appropriate body design which enables an animal to speak a human language.
- An Ape Can Master a Very Little Amount of a Human Language:
  Genetically, an ape is not very much different from a human being. Interestingly enough, experiments show that a well-trained ape could master about 500 to 600 words of a human language, in maximum. Beyond this limit, it is impossible for an ape to learn more regardless of any extra amount of human-ape interaction. This observation clearly indicates that the difference in the bodies (including the brains) will result in the difference in the results of learning.

As machine design is beyond the scope of this book, we will not study the design of human-machine interface suitable for autonomous human-machine interaction for the sake of visual learning.

8.5 Recognition of Shape

When a machine has accumulated enough visual experience of shape cognition, the machine should have built up an organized memory, in which the physical and conceptual meanings of learnt shapes are stored. Then, a relevant question will be: Will the machine be ready to undertake the visual task of shape recognition? In other words, we want to know the answer to the question of what the innate principles, or solutions, for a machine to recognize familiar shapes should be.

8.5.1 *Open-loop Approach to Shape Recognition*

In the literature, the general principle behind pattern recognition consists of these steps:

- Feature Extraction:
  For the types of patterns in consideration, we must define the features to be extracted from input data. Most importantly, we must design the algorithms, or solutions, which are effective for feature extraction.
- Classifier Training:
  Geometrically, the features of patterns in consideration define a feature space. In the feature space, a class of patterns will correspond to a cluster of points within a volume. Then, it is necessary to design a classifier, which will consistently partition the feature space into a set of sub-volumes, each of which will enclose

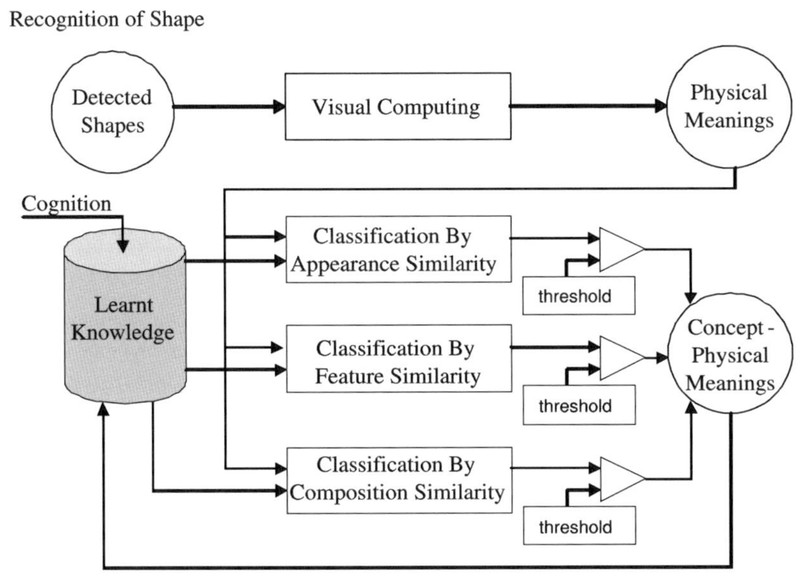

Fig. 8.48 An open-loop approach to shape recognition, in which the computed physical models of a detected shape are matched against the physical models of learnt shapes. And, the best match will be the output of recognition, which clearly depends on some thresholds.

one class of patterns. In general, a classifier has its own working parameters. And, the values of a classifier's working parameters have to be tuned according to a set of training samples of known patterns.

- Pattern Classification:
  When a classifier has been trained, then a feature space can be visualized as a set of sub-volumes, each of which has one, or more, prototype pattern. And, pattern classification can be interpreted as a process of matching an input pattern to one of the memorized prototype patterns. The best match will be the output of pattern recognition.

Clearly, the general principle of pattern recognition does not explicitly consider the feedback mechanism. If we apply the above principle to the case of shape recognition, we can devise an open-loop solution as shown in Fig. 8.48.

Refer to Fig. 8.48. When a detected shape is given as input, the first step is to compute the physical models of the shape. As we discussed earlier, the physical representation of a shape may include the appearance model, feature model, and composition model. Therefore, the open-loop solution for shape recognition can be implemented in one, or a combination, of the followings ways:

Appearance-based Classification

In this approach, the match between the appearance of an input shape and the appearance of a learnt shape is measured by a similarity function. And, the simplest way of implementing such a function is to compute the percent of overlap between two appearances. For instance, let S_a denote a similarity function of appearance, then

$$S_a = \frac{\text{Number of Overlapped Pixels}}{\text{Total Number of Pixels}} \tag{8.20}$$

where the numerator is the number of overlapped pixels between two shapes' appearance models, and the denominator is the total number of an input shape's pixels (or a learnt shape's pixels).

Equation 8.20 allows to select the best match, which maximizes S_a. However, the final decision of whether or not to accept the best match as the output of recognition depends on a threshold. In other words, if S_a of the best match is greater than a threshold, the best match will be accepted as the output of recognition. Then, a critical question arises: What should be the threshold value? Or, how to tune the threshold value?

Feature-based Classification

If we consider the feature vector of a detected shape as input, then we can define a distance function to measure the similarity between an input shape and a learnt shape. Let S_f denote a similarity function of feature. If $\|.\|$ denotes the Euclidean distance between two vectors, then

$$S_f = \frac{1}{\|\text{Input Feature Vector to Learnt Feature Vector}\|}. \tag{8.21}$$

Based on Eq. 8.21, we can implement a simple algorithm to identify the best match. For instance, among the learnt shapes in the organized memory, we can select the one which maximizes S_f in Eq. 8.21. And, if S_f of the best match is greater than a threshold, the best match will be accepted as the output of recognition.

Again, it is important to take note that the setting and tuning of the threshold value still remain an open issue in practice.

Composition-based Classification

In general, a shape's model of composition characterizes the shape's properties and constraints in terms of: (a) the constituent shapes, and (b) the rules of composition (including the rules imposed to the linkage of constituent shapes). Therefore, the match of two composition models depends on: (a) the match between two sets of constituent shapes, and (b) the match between two sets of composition rules. As the constituent shapes and composition rules could be quantified into numeric values similar to features, the basic mechanism underlying the matching between two composition models will be similar to Eq. 8.21.

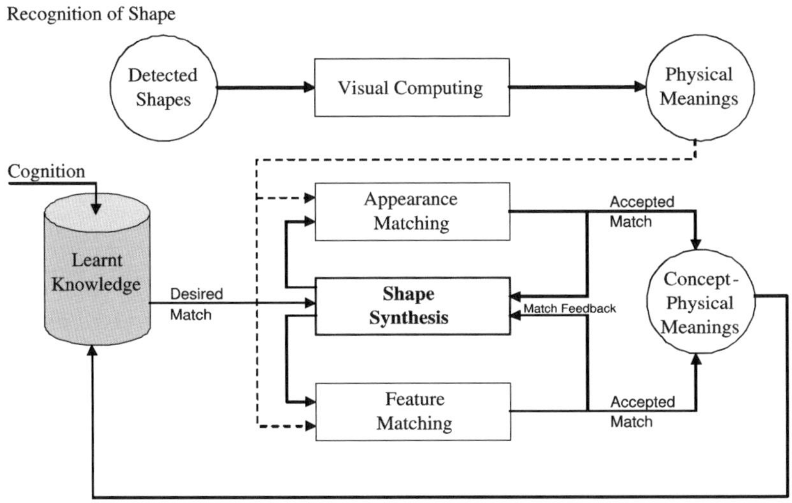

Fig. 8.49 A closed-loop approach to shape recognition, in which a learnt shape is chosen as a desired match. And, the generative model of the desired match is used to synthesize either the appearance model, or the feature model. If the generative model allows a machine to synthesize a series of appearance (or feature) models which converge to the appearance (or feature) model of a detected shape. The desired match will be accepted as the output of recognition.

8.5.2 *Closed-loop Approach to Shape Recognition*

The conventional principle underlying patter recognition is an open-loop framework. And, the critical issue in an open-loop framework is how to set, and tune, the threshold values. Hence, it is clear that the robustness of pattern recognition depends on the appropriate setting and tuning of threshold values (or other internal parameters). Now, we may raise this question: Can we adopt a closed-loop framework in order to avoid internal thresholds (or parameters) as much as possible?

Under the context of cognitive vision, a machine could learn the conceptual meanings of shapes through autonomous human-machine (or machine-machine) interaction in a scene. In addition, the conceptual representation of a shape includes the generative model, which enables a machine to synthesize the variations of a learnt shape. Interestingly enough, the ability for a machine to synthesize shapes do provide a mechanism to close the loop of shape recognition as shown in Fig. 8.49.

Refer to Fig. 8.49. If we adopt the closed-loop approach, the generic algorithm for shape recognition could be summarized as follows:

- Step 1: Choose a learnt shape as the desired match.
- Step 2: Synthesize an appearance (or feature) model.
- Step 3: Match the synthesized appearance (or feature) model against the appearance (or feature) model of a detected shape.
- Step 4: Feedback the discrepancy of match to shape synthesis module, and use the discrepancy to modify the output of shape synthesis, as long as the modification

is consistent with the generative model of the desired match (e.g. the attempt of modifying the appearance of letter "A" to be similar to the appearance of letter "B" is not consistent with the generative model of letter "A").

- Step 5: Match again the synthesized appearance (or feature) model against the appearance (or feature) model of the detected shape. If the discrepancy diminishes, continue Steps 4 and 5 until the discrepancy converges to zero. Otherwise, go to Step 1 and choose another learnt shape.
- Step 6: If the discrepancy of match converges to zero, the desired match is accepted as the output of recognition.

It is clear that there is no threshold in the above algorithm. And, the above algorithm is simple enough for us to make it an innate solution inside a machine's visual perception system.

8.6 Summary

A shape is a spatial layout of a cluster of proximal pixels having the similar identity, or category, in an image plane. Also, a shape is a cognitive entity in an image plane. In addition, the visual perception of shape is an indispensable ability that an autonomous machine must possess. And, we have learnt that the visual perception of shape consists of: (a) shape detection, (b) shape modelling, (c) shape cognition and (d) shape recognition.

For shape detection, one effective way is to make use of the information about the identities, or categories, of colors at the pixels in an image plane. Also, we have learnt that by dynamic programming, we can group the pixels having the similar colors together to form a cluster, which may correspond to a shape. Alternatively, one can first detect the color edges, and then to identify the areas enclosed by color edges. An additional advantage of using color edges is the ability to detect regions formed by color textures.

As a shape is a cognitive entity, it's important to design a concept-physical model in order to represent the physical and conceptual meanings of shapes. We have learnt that the physical representation of a shape should include: (a) appearance model, (b) feature model, and (c) composition model. As for the conceptual representation of a shape, it is important to consider both the conceptual labels and the generative model. A notable advantage of incorporating generative models is to be able to synthesize complex, or novel, shapes, which could enforce shape cognition, and make shape recognition robust (i.e. a closed-loop approach).

With the top-down design of a shape's concept-physical representation, the next step is to implement an appropriately organized memory, with which the mental processes behind shape cognition and recognition will interact. The purpose of shape cognition is for a machine to learn new, or unknown, shapes. We have learnt that the two important processes behind shape cognition are: (a) visual computing, and (b) visual learning. The former aims at computing the physical meanings of a shape, while the latter is to autonomously learn a shape's conceptual meanings. In particular, we understand that it is important for a machine to learn generative models of shapes.

When a machine has gained enough experience in visual perception of shape, another important visual task is shape recognition, which aims at matching an input shape against those learnt shapes. And, we have learnt that there are two working principles for a machine to recognize shapes. The first principle consists of open-loop approaches, in which the setting and tuning of thresholds still remain an open issue. However, with the ability of synthesizing shapes, shape recognition could be implemented in a closed-loop manner. And, the advantage of the closed-loop approach is the robustness because of the absence of thresholds.

8.7 Exercises

(1) What is shape in an image?
(2) Explain the usefulness of color histogram.
(3) What is color edge?
(4) Search in the literature articles which discuss about intensity edge detection. And, explain the difference in complexity between intensity edge detection and color edge detection.
(5) What are the major difficulties in defining the term *texture*?
(6) When a texture appears at a certain spatial scale in an image plane, will it remain at a much smaller scale?
(7) What are the physical properties of a shape in an image?
(8) What is a shape's appearance model?
(9) What is a shape's feature model?
(10) What is a shape's composition model?
(11) What is a shape's generative model?
(12) Explain the difference between shape cognition and shape recognition.
(13) Comment on the importance for an autonomous machine to master a human language.
(14) Project 1: Develop a program which computes color histogram, when a color image is given as input.
(15) Project 2: Develop a program, which groups proximal pixels of similar colors together.
(16) Project 3: Develop a program, which takes a color image as input and produces color edges as output.
(17) Project 4: Develop a program, which a color image as input and produces textured color edges as output.
(18) Project 5: Develop a program, which computes principal axes of a shape.
(19) Project 6: Develop a program, which computes the generalized axes of symmetry of a shape.

(20) Project 7: Develop a program, which determines the appearance models of a shape.
(21) Project 8: Develop a program, which computes the percent of overlapped pixels between the appearance models of two similar shapes.
(22) Project 9: Develop a human-machine interface, which allows a machine to learn the conceptual labels of shapes.

Chapter 9

Knowledge Acquisition from Depth

9.1 Introduction

The environment, in which we are living, is a three-dimensional space. Due to our sense of vision (or sight), we are able to survive, act and interact in a complex and three-dimensional environment. Most importantly, the mental ability of perceiving depths as well as understanding dynamic scenes enables human beings to discover new knowledge.

Indeed, vision allows us to assess the distances and relationships among the objects in a scene. Also, vision enables a human being to gain the awareness of the pose (i.e. position and orientation) of an object in a scene. Now, we may ask this curious question: What is the visual information, which enables our brain to sense the poses of objects in a dynamic scene? Not surprisingly, the answer is the depth of objects perceived from images (i.e. image understanding) and scenes (i.e. scene understanding).

So far, we have learnt that vision could enable machines to perceive colors. Also, through the use of vision, a machine can sense the shapes in an image plane. However, a necessary step toward making a machine to gain the visual awareness of a physical scene is to achieve the visual perception of depth from images or scenes.

Literally, depth in vision refers to the third dimension in the direction which is perpendicular to the image plane of a camera. Loosely speaking, the visual perception of depth is to recover the third dimension for every pixel in the image plane of a camera. But, in order to understand a scene, it is important to determine the positions from which lights emanate towards a machine's cameras. This is because such positions are actually on the surfaces of the objects, or terrains, in a scene. And, these positions are sufficient enough for a machine to discern the objects, and also to compute their poses in a scene. Therefore, the visual perception of depth actually means the visual perception of the positions, from which lights emanate towards eyes, or cameras. However, as the depth (i.e. the third dimension) associated to a pixel is sufficient enough for a machine to determine the position (i.e. the coordinates), at which a light emanates and illuminates the pixel, we will continue to use the term *depth* in this chapter.

Broadly speaking, the perception of depth from a scene could be achieved in two different ways. The first category of approaches is the so-called *scanning*, or *echolocation*. When a machine (or animal) actively emits (electro-magnetic or mechanical) waves toward a scene and detects the echo, the machine can then compute the depth from the signals of echo and time. And, the computation is quite simple, because the distance traveled by an echo is a function of both the time-of-flight and the speed of the emitted waves. Since the design of human vision does not follow the principle of scanning, or echolocation, for depth perception, we will not study this category of approaches in this book.

The second category of approaches is the so-called *depth from disparity*, or *depth from apparent motion*. Here, the apparent motion refers to the displacement of a pixel within two, or more, consecutive images. And, the disparity refers to the displacement of a pixel between a left-image and a right-image in a binocular vision. Without loss of generality, disparity could be considered as being equivalent to apparent motion. This is because a left-image and a right-image in a binocular vision could be mathematically treated as a pair of consecutive images.

As shown in Fig. 9.1, the visual perception of depth is equally important, compared to the visual perception of color as well as the visual perception of shape. In

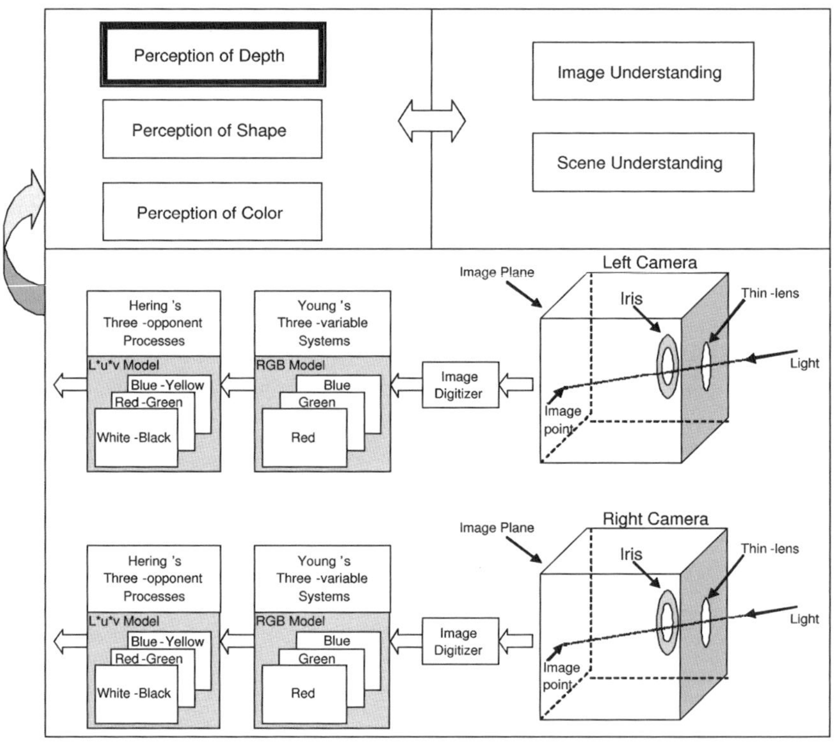

Fig. 9.1 Visual perception of depth under the cognitive vision paradigm, which emphasizes the importance of achieving the cognitive state of knowing color, shape, depth, and behavior of objects in a scene through the mental processes of cognition and recognition.

fact, these three modules are the pillars underlying the visual tasks of image understanding (i.e. interpretation of images or photos of monocular views) and scene understanding (i.e. interpretation of images or photos of binocular views). In addition, the shapes in an image plane are the manifestation of the shapes in a scene. Hence, the detection of shapes in a scene is related to the detection of shapes in an image. Without loss of generality, we will call the shapes in a scene the *objects*. In this way, we will avoid any confusion between 2D shapes (in an image) and 3D shapes (in a scene).

In this chapter, we will study four important topics underlying the visual perception of depth, namely:

- From Depth to Image;
- From Image to Apparent Motion;
- Cognition of Depth (i.e. to derive depth from apparent motion of binocular views);
- Recognition of Depth (i.e. to infer depth from a photo or a video of monocular views).

9.2 Transformation From Depth to Image

Human vision is designed in a specific manner in order to sense colors. Also, the specific design of human vision makes it easier to achieve the astonishing power of perceiving depth from apparent motion, or photo. In addition, we know that human vision consists of two eyes, which form a binocular vision. Most importantly, a human being's binocular vision should be treated as two cooperative monocular visions. And, the reasons behind this claim are:

- First of all, a human being's brain processes the visual signals coming from both eyes, independently. And, the sudden close, or obstruction, of one eye will not affect human vision, because of the inflow of visual signals from the other eye. Moreover, the visual awareness constructed by human vision is the sum of the outputs from both eyes, instead of the common portion of the contents from both eyes.
- Second, a human being's brain is able to coherently match a pair of left-image and right-image in order to infer the geometry of a scene from disparity. And, the visual awareness constructed by human vision is represented in an unified manner.

After the birth, a human being is able to autonomously perceive colors, shapes, and depth of the objects in a scene. Such a performance in undertaking visual perception is only possible with a top-down design, which pre-builds the innate principles and solutions underlying visual perception in advance.

Clearly, the importance of the top-down design behind the success of human vision should inspire human designers of machine vision to answer these two fundamental questions:

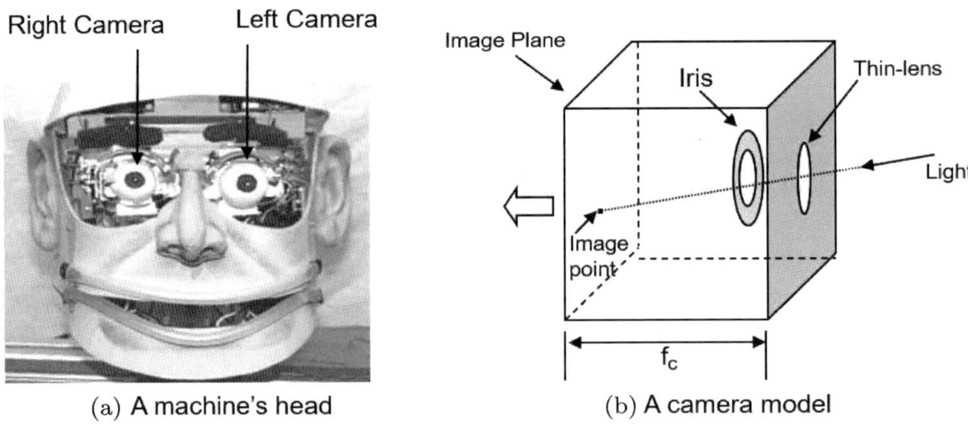

Fig. 9.2 A machine's head consisting of a binocular vision, in which a camera is modeled as a combination of optic lens, iris, and image plane.

- What are the innate principles and solutions, which enable the success of machine vision?
- How to accurately incorporate such principles and solutions into the physical system of machine vision?

9.2.1 *Properties of Image Plane*

Human vision is binocular. If we arrange two cameras in a configuration similar to that of human eyes, we can build a binocular vision for machine. In Fig. 9.2, we show a robot head, in which the left-camera and the right-camera are part of the binocular vision. And, inside a machine's vision system, there are many geometrical parameters, which a human designer of machine vision must specify.

9.2.1.1 *Pixel*

A camera will mainly consist of: (a) a thin-lens, (b) an iris, and (c) an image plane (i.e. an array of photo-detectors). As shown in Fig. 9.2b, when an incoming light passes through the center of the thin-lens of a camera, it will go straight inside the camera, and intersect with the image plane. And, the point of intersection will form a spot, at which the color (including intensity) of the incoming light will be sensed by the photo-detectors placed behind this spot. By definition, the locations where the photo-detectors are placed are called the *picture elements*, or *pixels* in short.

In the literature of image processing, there is no concern about the design of camera, or vision. Therefore, a pixel in the context of image processing generally refers to an element of a digital image, or a digitized photo.

9.2.1.2 *Resolution*

When the image plane of a camera is being designed, it is desirable to arrange the photo-detectors into a two-dimensional array, in which the rows are perpendicular to the columns. In this way, the image plane of a camera will consist of an image-sensor array.

For the design of an image-sensor array, two important design parameters will be: (a) the number of pixels in a row, and (b) the number of pixels in a column. By definition, these two parameters are called the *resolution* of image, or *image size* in short. In other words, the resolution of image equals the size of image-sensor array.

It is worth noting that today's cameras are not designed specifically for machine vision. In fact, the resolution of images from today's cameras is the result of the sampling during the process of image digitization. Therefore, it does not depend on the size of image-sensor array. For instance, when the resolution of image is greater than the size of image-sensor array, the quality of images will be poor because of over-sampling.

Vision is an important system for a machine to gain the visual awareness about a scene. Therefore, the design of machine vision should not unnecessarily compromise the quality of images. If it is possible, the resolution of image should equal the size of image-sensor array.

9.2.1.3 *Image Coordinate System*

For the sake of geometrical modeling, it is also necessary to assign a coordinate system to a camera's image plane. Refer to Fig. 9.4. The common practice of assigning a coordinate system to an image plane is as follows:

- The origin of an image plane's coordinate system is chosen to be at the intersection between the optical axis and the image plane (i.e. the optical center).
- The x axis of an image plane's coordinate system is chosen to be parallel to the rows of the image sensor array.
- The y axis of an image plane's coordinate system is chosen to be parallel to the columns of the image sensor array.

9.2.1.4 *Index Coordinate System*

Inside a digital camera, the signals from the photo-detectors of an image plane will be digitized into a set of numeric values. And, these numeric values will be stored into a two-dimensional matrix in the processing unit of machine vision. Therefore, the location of a pixel with respect to the array of photo-detectors can be measured by two indices, namely: (a) row index r, and (b) column index c.

As a result of accurate design, the rows and columns of an image-sensor array are perpendicular to one another. In this case, the row and column indices actually define an orthogonal coordinate system, which is called an *index coordinate system*.

It goes without saying that the row axis of the index coordinate system is perpendicular to rows, and that the column axis of the index coordinate system is perpendicular to the columns. Then, one may ask this question: Where is the origin of the index coordinate system in an image plane?

We know that an image plane consists of an array of photo-detectors. And, the signals from these photo-detectors will be digitized into numeric values. When the numeric values are collected and arranged into a two-dimensional matrix, the collecting process must first start from one of the four corners in an image plane, and then scan the image-sensor array row by row and column by column. As there are four corners in an image plane, there will be four possible ways of positioning the origin of the index coordinate system in the image plane.

Refer to Fig. 9.3. The four possible assignments of the index coordinate system to an image plane are as follows:

- Case 1: Axes x and c are pointing toward the same direction. Also, axes y and r are pointing toward the same direction.
- Case 2: Axes x and c are pointing toward two opposite directions. But, axes y and r are pointing toward the same direction.
- Case 3: Axes x and c are pointing toward the same direction. But, axes y and r are pointing toward two opposite directions.

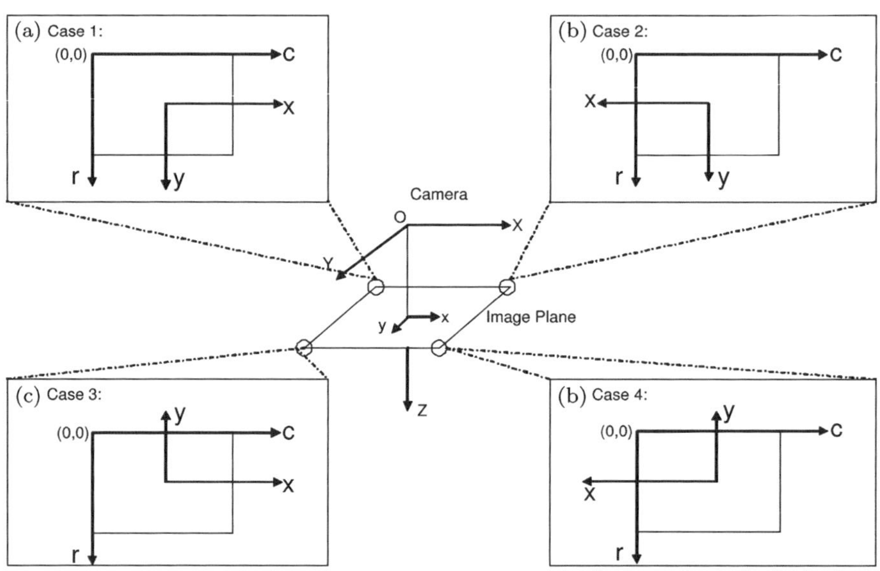

Fig. 9.3 Four possible ways of positioning the index coordinate system in an image plane.

- Case 4: Axes x and c are pointing toward two opposite directions. Also, axes y and r are pointing toward two opposite directions.

9.2.1.5 *Image Center*

By definition, the intersection point between an optical axis and an image plane is called an *image center.* In terms of the row and column indices, the location of the image center in an image plane is represented by (r_0, c_0).

9.2.1.6 *Coordinates of Pixel*

By definition, the coordinates of a pixel with respect to an image coordinate system are called *image coordinates.* And, the row-column indices of a pixel with respect to an index coordinate system are called *index coordinates.*

If we denote (x, y) the image coordinates of a pixel and (r, c) the index coordinates of the same pixel, then the relationship between index coordinates (r, c) and image coordinates (x, y) will be uniquely determined. For instance, let w and h denote the width and height of a pixel, respectively. Then, it is easy to verify the following results:

- If the index coordinate system is the one shown in Fig. 9.3a, we have

$$\begin{cases} x = (c - c_0) \bullet w \\ y = (r - r_0) \bullet h \end{cases} \tag{9.1}$$

- If the index coordinate system is the one shown in Fig. 9.3b, we have

$$\begin{cases} x = -(c - c_0) \bullet w \\ y = (r - r_0) \bullet h \end{cases} \tag{9.2}$$

- If the index coordinate system is the one shown in Fig. 9.3c, we have

$$\begin{cases} x = (c - c_0) \bullet w \\ y = -(r - r_0) \bullet h \end{cases} \tag{9.3}$$

- If the index coordinate system is the one shown in Fig. 9.3d, we have

$$\begin{cases} x = -(c - c_0) \bullet w \\ y = -(r - r_0) \bullet h \end{cases} \tag{9.4}$$

Obviously, parameters w and h should be treated as design parameters. And, the fabrication of cameras should accurately follow the design specifications for w and h. It is important to take note that parameters w and h in today's cameras have to be estimated by a calibration process.

Now, if we consider (r_0, c_0) and (w, h) as design parameters, it is clear that Eq. 9.1, Eq. 9.2, Eq. 9.3 or Eq. 9.4 will uniquely determine (x, y) if (r, c) are known, or vice versa.

9.2.2 *Properties of Camera*

An image plane is an important component inside a camera. In addition, a camera has other components such as optical lens, diaphragm and support casing, etc.

9.2.2.1 *Optical Axis*

Geometrically, the process for straight light rays (i.e. light beams) to pass through a fixed point (e.g. the center of the thin-lens of a camera) and to intersect with an image plane is called a *perspective projection.* In particular, a straight light ray, which passes through the center of a thin-lens and is perpendicular to an image plane, is called an *optical axis.*

9.2.2.2 *Focal Length*

By definition, the distance between the center of a thin-lens and an image plane is commonly called the *focal length* of a camera, which is denoted by f_c, as shown in Fig. 9.2b.

Under the context of cognitive vision, it makes sense to believe that f_c should be a designed variable, and that f_c could be precisely controlled by a machine. However, it is worth noting that parameter f_c in today's cameras has to be estimated in one way, or another, by a calibration process. And, any change in f_c must re-invoke the calibration process. Obviously, this is one drawback, which undermines the success of machine vision.

9.2.2.3 *Image Plane of Equivalence*

Refer Fig. 9.2b. We can see that the image-plane is physically located behind the thin-lens at the distance of f_c. However, for the sake of convenience, we can imagine that there is a plane, which is in front of the thin-lens at the distance of f_c and is parallel to the actual image-plane. By definition, such a plane is called an *virtual image-plane* or *image-plane of equivalence.*

Mathematically, the actual and virtual image-planes will be equivalent, if the content of the virtual image-plane equals the content of the actual image-plane. By default, an image-plane will implicitly refer to the virtual image-plane of a camera in the following discussions.

9.2.2.4 *Camera's Coordinate System*

In order to study the geometrical relationship between a camera and a scene, it is necessary to assign a coordinate system to a camera, which is a rigid body in a scene.

Refer to Fig. 9.4a, or Fig. 9.4b. Each camera in a machine's binocular vision is assigned with a coordinate system. And, in a camera's coordinate system, a light

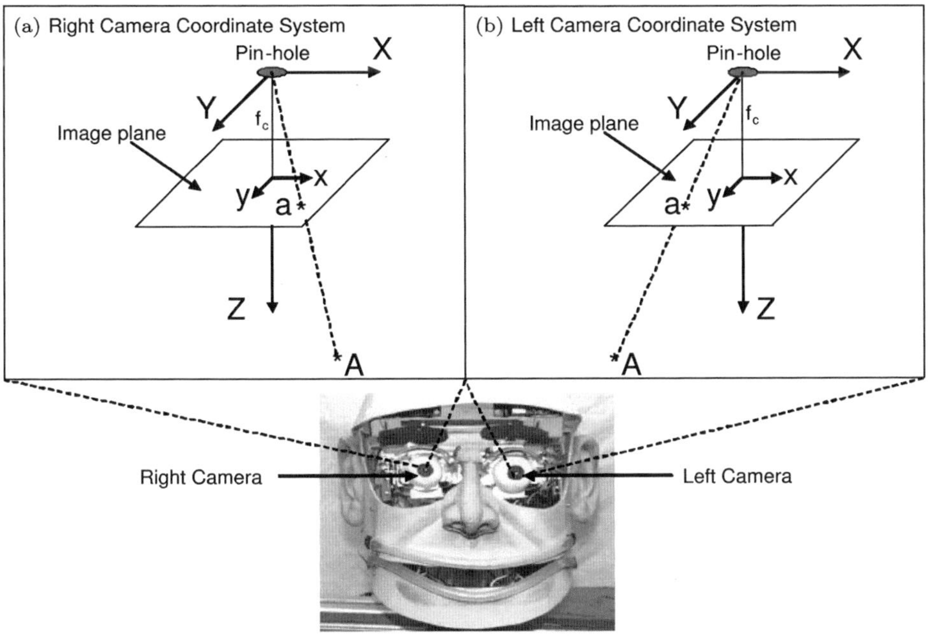

Fig. 9.4 A machine's head with a binocular vision, in which each camera is assigned with a coordinate system. And, a light coming from point A will pass through the center of a camera's thin-lens, and intersect with the camera's image plane at image point a. In an image plane, a coordinate system could also be defined, which is called the image coordinate system.

coming from point A will pass through the center of the camera's thin-lens, and intersect with the camera's image plane at image point a. Because the pixels in an image plane are arranged as a two-dimensional array, we can expect that by design, the rows and columns in the array of photo-detectors will be perfectly perpendicular.

Refer to Fig. 9.4 again. The convention of assigning a coordinate system to a camera is as follows:

- The origin of a camera's coordinate system is chosen to be the center of the camera's thin-lens.
- The Z axis of a camera's coordinate system is chosen to coincide with the camera's optical lens. And, the Z axis points outward from the center of the camera's optical lens.
- The X axis of a camera's coordinate system is chosen to be in a direction parallel to the rows of the array of photo-detectors on the camera's image plane.
- The Y axis of a camera's coordinate system is chosen to be in a direction parallel to the columns of the array of photo-detectors on the camera's image plane. And, the X, Y and Z axes form a right-hand coordinate system.

Experienced readers may notice that researchers in computer vision, or machine vision, have to use cameras which are not specifically designed for vision. In this case, the assumption about the orthogonality of rows and columns is not valid. And, one has to do careful calibration in order to rectify the non-orthogonality of a camera's image plane. However, under the context of cognitive vision which emphasizes the importance of top-down design, it makes sense to consider the effort of accurate design and fabrication of cameras, instead of repeatedly calibrating the cameras.

9.2.2.5 *Viewer's Coordinate System*

Our daily visual experiences tell us that vision's three-dimensional impression about a scene is referenced to a common coordinate system. Such a coordinate system is called a *viewer's frame.*

In robotics, a coordinate system attached to a rigid body is called a *frame.* A camera is usually housed inside a head which is also a rigid body in a scene. When we assign a coordinate system to a head, such a coordinate system could be called as a *viewer's frame.*

As we have mentioned earlier, the sudden close, or obstruction, of one of our eyes will not make human vision fail. This evidence suggests that a viewer's frame cannot coincide with neither the left-eye's coordinate system nor the right-eye's coordinate system. In human vision, it is likely that a viewer's frame is fixed somewhere between the left-eye and the right-eye. Therefore, the specification of a viewer's frame should be an important design parameter for machine vision.

Refer to Fig. 9.5. The usefulness of a viewer's frame in the design of machine vision is two-fold:

- First of all, a viewer's frame actually serves as a common reference for the control (or intentional change) of the orientation of the two cameras. For instance, each

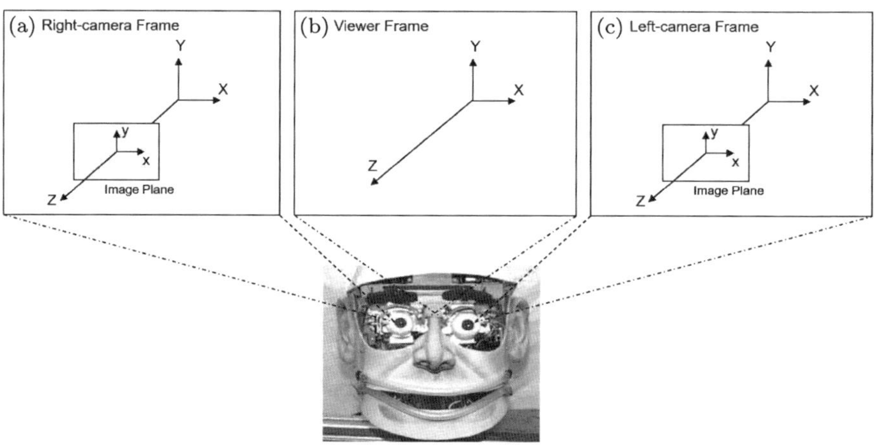

Fig. 9.5 Viewer's frame as an important design parameter, to which both the left-camera's coordinate system and the right-camera's coordinate system are referenced.

eye in human vision could rotate about two axes for motions such as left-right panning and up-down tilting. Similarly, when we design a machine's vision, the two cameras should not be fixed. Instead, each camera should be allowed to pan, and to tilt, about two orthogonal axes, which must be referenced to the viewer's frame. And, the viewer's frame is better to be fixed at the midway between the left camera and the right camera. In this way, the position and orientation of either the left-camera's or the right-camera's coordinate system with respect to the viewer's frame will be precisely known at any time instance. Interestingly enough, through the transformation of coordinate systems, the relative pose between the left-camera's and right-camera's coordinate systems will also be precisely known.

- Second, a viewer's frame serves as the common reference for the three-dimensional representation of a scene perceived by a machine. In human vision, the final output of visual perception is the sum of the perceptual outputs from both the left eye and the right eye. However, it is possible to add the perceptual outputs from both the left eye and the right eye, if and only if these perceptual outputs are expressed with respect to a common coordinate system. Clearly, this property should inspire the design of machine vision for the sake of achieving a high level of perceptual performance similar to human vision.

9.2.3 *Relative Pose Between Camera's Coordinate Systems*

The two eyes in human vision are separated by a distance. And, the optical axes of the two eyes are always converging at a changeable angle. Therefore, it is important to model the relative pose of one eye's coordinate system with respect to another eye's coordinate system. In machine vision, we call it the *relative pose* between two cameras.

Without loss of generality, we consider that the left-camera's coordinate system serves as a reference coordinate system. In this way, the relative pose between two cameras in a binocular vision will be fully described by the position and orientation of the right-camera's coordinate system, with respect to the left-camera's coordinate system.

For instance, let (t_x, t_y, t_z) be the coordinates of the origin of the right-camera's coordinate system, with respect to the left-camera's coordinate system. Then, the translation from the left-camera's origin to the right-camera's origin can be represented by vector T, as follows:

$$T = (t_x, t_y, t_z)^t \tag{9.5}$$

where the superscript t denotes the transpose of a vector. And, it is easy to verify that

$$\begin{pmatrix} t_x \\ t_y \\ t_z \end{pmatrix} = T + \begin{pmatrix} 0 \\ 0 \\ 0 \end{pmatrix} \tag{9.6}$$

or

$$^{l}O_r = T +^{r} O_r \tag{9.7}$$

with

$$^{l}O_r = \begin{pmatrix} t_x \\ t_y \\ t_z \end{pmatrix} \quad \text{and} \quad ^{r}O_r = \begin{pmatrix} 0 \\ 0 \\ 0 \end{pmatrix}. \tag{9.8}$$

In Eq. 9.8, $^{l}O_r$ means the coordinates of the origin of the right-camera's coordinate system, with respect to the left-camera's coordinate system. And, $^{r}O_r$ means the coordinates of the origin of the right-camera's coordinate system, with respect to the right-camera's coordinate system itself. In other words, the superscript l, or r, before symbol O, simply denotes the reference coordinate system.

On the other hand, if

- (r_{11}, r_{21}, r_{31}) is a unit vector representing the X axis of the right-camera's coordinate system, with respect to the left-camera's coordinate system,
- (r_{12}, r_{22}, r_{32}) is a unit vector representing the Y axis of the right-camera's coordinate system, with respect to the left-camera's coordinate system, and
- (r_{13}, r_{23}, r_{33}) is a unit vector representing the Z axis of the right-camera's coordinate system, with respect to the left-camera's coordinate system,

then the rotation which makes the left-camera's coordinate system to coincide with the right-camera's coordinate system can be represented by matrix R, which will be

$$R = \begin{pmatrix} r_{11} & r_{12} & r_{13} \\ r_{21} & r_{22} & r_{23} \\ r_{31} & r_{32} & r_{33} \end{pmatrix}. \tag{9.9}$$

Similarly, it is easy to verify that

$$\begin{pmatrix} r_{11} \\ r_{21} \\ r_{31} \end{pmatrix} = R \bullet \begin{pmatrix} 1 \\ 0 \\ 0 \end{pmatrix} \tag{9.10}$$

or

$$^{l}e_X = R \bullet {}^{r}e_X \tag{9.11}$$

with

$$^{l}e_X = \begin{pmatrix} r_{11} \\ r_{21} \\ r_{31} \end{pmatrix} \quad \text{and} \quad ^{r}e_X = \begin{pmatrix} 1 \\ 0 \\ 0 \end{pmatrix}. \tag{9.12}$$

In Eq. 9.12, $^{l}e_X$ is a unit vector, which represents the X axis of the right-camera's coordinate system, with respect to the left-camera's coordinate system. And, $^{r}e_X$ is a unit vector, which represents the X axis of the right-camera's coordinate system, with respect to the right-camera's coordinate system itself.

In general, the combination of Eq. 9.7 and Eq. 9.11 can be expressed as follows:

$$^{l}P = T + R \bullet {}^{r}P \tag{9.13}$$

where ${}^{r}P = ({}^{r}X,\ {}^{r}Y,\ {}^{r}Z)^t$ is the position vector of point P with respect to the right-camera's coordinate system, and ${}^{l}P = ({}^{l}X,\ {}^{l}Y,\ {}^{l}Z)^t$ is the position vector of the same point P with respect to the left-camera's coordinate system. In other words, Eq. 9.13 transforms the coordinates expressed in the right-camera's coordinate system into the coordinates expressed in the left-camera's coordinate system.

Mathematically, Eq. 9.13 can also be rewritten in the following matrix form:

$$\begin{pmatrix} {}^{l}X \\ {}^{l}Y \\ {}^{l}Z \\ 1 \end{pmatrix} = M \bullet \begin{pmatrix} {}^{r}X \\ {}^{r}Y \\ {}^{r}Z \\ 1 \end{pmatrix} \tag{9.14}$$

with

$$M = \begin{pmatrix} r_{11} & r_{12} & r_{13} & t_x \\ r_{21} & r_{22} & r_{23} & t_y \\ r_{31} & r_{32} & r_{33} & t_z \\ 0 & 0 & 0 & 1 \end{pmatrix}. \tag{9.15}$$

By definition, $({}^{l}X,\ {}^{l}Y,\ {}^{l}Z,\ 1)^t$ is called the *projective coordinates* of point P expressed in the left-camera's coordinate system, and $({}^{r}X,\ {}^{r}Y,\ {}^{r}Z,\ 1)^t$ the *projective coordinates* of the same point P expressed in the right-camera's coordinate system. And, matrix M is called a *homogeneous transformation matrix*, which describes the *relative pose* of the right-camera's coordinate system, with respect to the left-camera's coordinate system.

In the above discussion, we have considered a set of two camera coordinate systems. One is the left-camera's coordinate system. And, the other one is the right-camera's coordinate system. In theory, a camera's coordinate systems at two different time instances may also constitute a set of two camera coordinate systems. For instance, when a machine moves within a time interval $[t_1, t_2]$, there is a relative pose of the left (or right) camera's coordinate system at time t_2, with respect to the left (or right) camera's coordinate system at time t_1. Therefore, Eq. 9.15 is general enough to describe the relative pose between any pair of camera coordinate systems.

Example 9.1. Suppose that we are in a process of designing a machine's head, in which there is a binocular vision. And, we choose the conceptual design solution shown in Fig. 9.6.

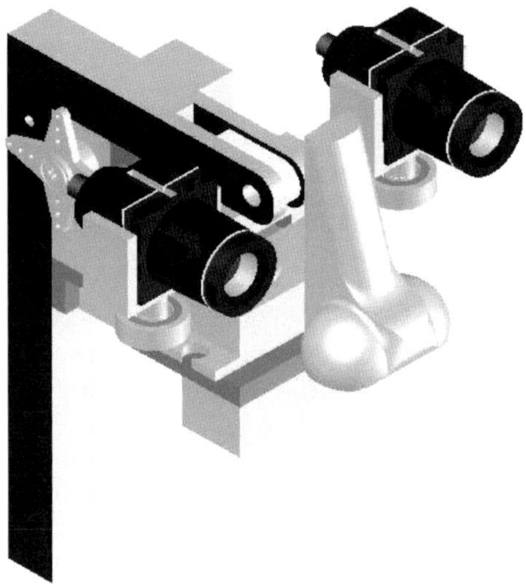

Fig. 9.6 A conceptual design solution of a machine's head having a binocular vision.

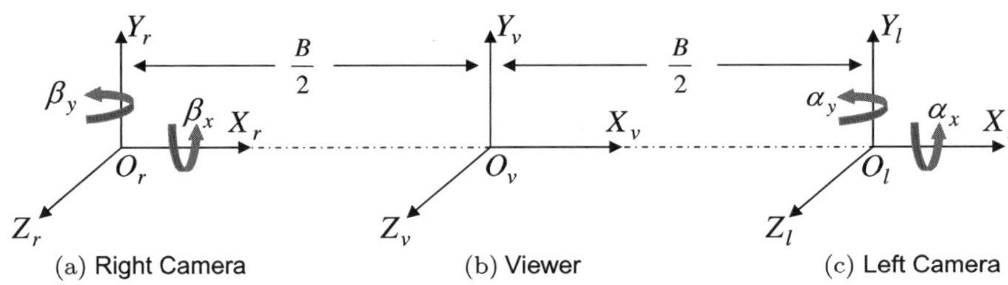

Fig. 9.7 Coordinate systems assigned to the right-camera, left-camera and viewer.

For this conceptual design solution, some design specifications are as follows:

- The origins of the left-camera, right-camera and view coordinate systems are on a straight line.
- The distance between the right-camera and view coordinate systems equals the distance between the left-camera and view coordinate systems.
- Each camera has two degrees of freedom, of which one is for pan motion, and the other for tilt motion.

Based on the above specifications, the coordinate systems assigned to the right-camera, left-camera and viewer should be the ones as shown in Fi.g. 9.7. In other words, coordinate system $O_vX_vY_vZ_v$ is assigned to the viewer, coordinate system $O_rX_rY_rZ_r$ to the right-camera and coordinate system $O_lX_lY_lZ_l$ to the left-camera.

Now, let variables α_x and α_y denote the left-camera's angles of pan and tilt rotations, respectively. Then, the orientation of the left-camera's coordinate system

$O_lX_lY_lZ_l$ with respect to the viewer's coordinate system $O_vX_vY_vZ_v$ will be

$$^vR_l = \begin{pmatrix} 1 & 0 & 0 \\ 0 & \cos\alpha_x & -\sin\alpha_x \\ 0 & \sin\alpha_x & \cos\alpha_x \end{pmatrix} \bullet \begin{pmatrix} \cos\alpha_y & 0 & \sin\alpha_y \\ 0 & 1 & 0 \\ -\sin\alpha_y & 0 & \cos\alpha_y \end{pmatrix}. \tag{9.16}$$

If the distance between the origin of the viewer's coordinate system and the origin of the left-camera's coordinate system is $\frac{B}{2}$, the relative pose of the left-camera's coordinate system with respect to the viewer's coordinate system will be

$$^vM_l = \begin{pmatrix} \cos\alpha_y & 0 & \sin\alpha_y & \dfrac{B}{2} \\ \sin\alpha_x\sin\alpha_y & \cos\alpha_x & -\sin\alpha_x\cos\alpha_y & 0 \\ -\sin\alpha_y\cos\alpha_x & \sin\alpha_x & \cos\alpha_x\cos\alpha_y & 0 \\ 0 & 0 & 0 & 1 \end{pmatrix}. \tag{9.17}$$

Similarly, if variables β_x and β_y denote the right-camera's angles of pan and tilt rotation respectively, the relative pose of the right-camera's coordinate system with respect to the viewer's coordinate system will be

$$^vM_r = \begin{pmatrix} \cos\beta_y & 0 & \sin\beta_y & -\dfrac{B}{2} \\ \sin\beta_x\sin\beta_y & \cos\beta_x & -\sin\beta_x\cos\beta_y & 0 \\ -\sin\beta_y\cos\beta_x & \sin\beta_x & \cos\beta_x\cos\beta_y & 0 \\ 0 & 0 & 0 & 1 \end{pmatrix}. \tag{9.18}$$

In practice, if we control angles α_x and β_x in such a way that $\alpha_x = \beta_x$, axes Z_l and Z_r will always be coplanar. Alternatively, we can design the binocular vision in such a way that angle α_x is always coupled with angle β_x. In other words, this property of coplanarity can be achieved by a proper design.

When axes Z_l and Z_r are coplanar, it is easy to verify that they will intersect as long as $\alpha_y \neq \beta_y$. In vision, when the optical axes of two eyes intersect, we say that the two eyes converge at a common location in a scene. And, this property of convergence is a necessary condition for achieving visual attention.

<><><><><>

9.2.4 *Relative Motion Between Camera's Coordinate Systems*

In Eq. 9.15, matrix M describes the position and orientation of the right-camera's coordinate system, with respect to the left-camera's coordinate system. Mathematically, we can imagine that the right camera initially coincides with the left camera, and that the right camera actually undergoes a general motion, within a unit time-interval, in order to reach the relative pose described by matrix M. In this imagined scenario, it is clear that there is a relative motion of the right-camera's coordinate system, with respect to the left-camera's coordinate system. Now, the question will be: How do we determine this relative motion if matrix M is known in advance?

We know that a coordinate system in a three-dimensional space consists of three mutually orthogonal vectors, which intersect at the coordinate system's origin. Hence, a coordinate system must be physically treated as a rigid body in a three-dimensional space. In physics, we know that the general motion of a rigid body in a space will consist of two parts: (a) the linear motion, and (b) the angular motion. And, these two motions are described by linear and angular velocities, respectively.

It is a fact that a rigid body's linear velocity is equal to the linear velocity of any point on the rigid body. Therefore, we can consider that the linear velocity of a coordinate system is measured by the linear velocity of its origin. In other words, the linear velocity of a coordinate system simply refers to the rate of change of its origin's position vector, with respect to a reference coordinate system. For instance, if matrix M in Eq. 9.15 describes the relative pose reached by the right-camera's coordinate system within a unit time-interval, then the linear velocity of the right-camera's coordinate system, with respect to the left-camera's coordinate system, will be

$$v_c = (t_x,\ t_y,\ t_z)^t \tag{9.19}$$

where the superscript t denotes the transpose of a vector.

Also, it is a fact that a rigid body's angular velocity consists of an equivalent axis of rotation and a rate of change of the rotated angle. For instance, if matrix R in Eq. 9.9 describes the relative orientation reached by the right-camera's coordinate system within a unit time-interval, the angular velocity of the right-camera's coordinate system, with respect to the left-camera's coordinate system, will be

$$\omega_c = \frac{\dot{\theta}}{\sqrt{(r_{32}-r_{23})^2+(r_{13}-r_{31})^2+(r_{21}-r_{12})^2}} \bullet \begin{pmatrix} r_{32}-r_{23} \\ r_{13}-r_{31} \\ r_{21}-r_{12} \end{pmatrix}_{3\times 1} \tag{9.20}$$

with

$$\dot{\theta} = \arccos\left(\frac{r_{11}+r_{22}+r_{33}-1}{3}\right). \tag{9.21}$$

Example 9.2. In binocular vision, we imagine that the right-camera's coordinate system initially coincides with the left-camera's coordinate system, and that the relative pose of the right-camera's coordinate system is reached after a rotation followed by a translation within a unit time-interval. Assume that the orientation of the right-camera's coordinate system, with respect to the left-camera's coordinate system, is

$$R = \begin{pmatrix} 0.7071 & 0.0 & -0.7071 \\ 0.0 & 1.0 & 0.0 \\ 0.7071 & 0.0 & 0.7071 \end{pmatrix}$$

By applying Eq. 9.20 and Eq. 9.21, we obtain

$$\omega_c = \dot{\theta} \bullet \begin{pmatrix} 0.0 \\ -1.0 \\ 0.0 \end{pmatrix}$$

with

$$\dot{\theta} = \arccos(0.7071) = 45^0.$$

The above result means that the right-camera's coordinate system has rotated 45^0 about the negative Y axis of the left-camera's coordinate system.

<><><><><>

In summary, if matrix M describes the relative pose reached by the right-camera's coordinate system within a unit time-interval, the relative motion of the right-camera's coordinate system, with respect to the left-camera's coordinate system, will consist of linear velocity v_c and angular velocity ω_c. Most importantly, this result of relative motion is valid for any pair of camera coordinate systems.

9.2.5 *Baseline Between Camera's Coordinate Systems*

In binocular vision, the distance between the origin of the left-camera's coordinate system and the origin of the right-camera's coordinate system is commonly called the *baseline* of binocular vision (or stereovision).

For instance, if matrix M in Eq. 9.15 describes the relative pose of the right-camera's coordinate system with respect to the left-camera's coordinate system, then the baseline of the binocular vision will be

$$B = \sqrt{t_x^2 + t_y^2 + t_z^2}. \tag{9.22}$$

And, baseline B is an important design parameter for machine vision.

Example 9.3. Human mind can not only process visual signals coming from the eyes, but also synthesize mental images. The mental ability of synthesizing images is crucial to the visual feedback in the process of photo interpretation, or depth recognition from photos. Therefore, when we build a machine's vision, we must also consider the machine's mental ability of synthesizing images.

In order to construct a synthetical binocular vision, it is necessary to consider two aspects. The first aspect is the configuration of the virtual cameras in a synthetic

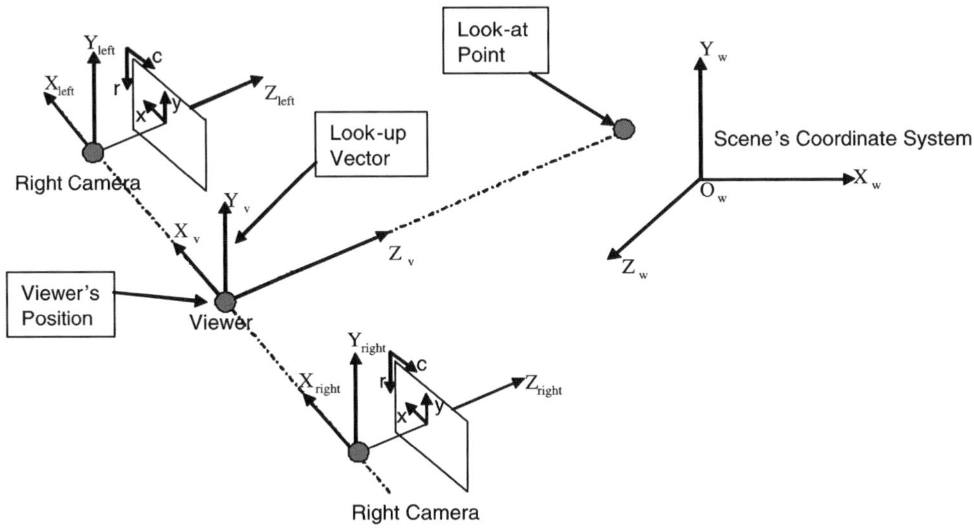

Fig. 9.8 Configuration of virtual-viewer and virtual cameras for a synthetic binocular vision.

binocular vision. In other words, one must specify the pose of the virtual-viewer's coordinate system, and the relative poses of the virtual-cameras with respect to the virtual-viewer. And, the second aspect is to specify the important parameters of the virtual-cameras before synthetic images could be generated.

Now, suppose that we are using the popular graphic library, called OpenGL, to implement a synthetic binocular vision. In order to configure the virtual-cameras, we first specify the virtual-viewer's coordinate system, denoted by $O_vX_vY_vZ_v$, with respect to a virtual world's coordinate system, denoted by $O_wX_wY_wZ_w$.

Refer to Fig. 9.8. In OpenGL, the pose of $O_vX_vY_vZ_v$ with respect to $O_wX_wY_wZ_w$ is defined by

- The look-at point, which specifies the point pointed by axis Z_v;
- The look-up vector, which specifies axis Y_v with respect to $O_wX_wY_wZ_w$;
- The viewer's position, which specifies origin O_v with respect to $O_wX_wY_wZ_w$.

Here, it is worth noting that OpenGL is designed for computer graphics. Actually, the loop-at point is the point pointed by the negative Z axis of a viewer's coordinate system. However, if we invert both axes Z and X, we will make OpenGL follow the convention of coordinate systems for vision. In Fig. 9.8, axes Z_v and X_v have already been inverted.

With the above specifications, OpenGL will automatically calculate the pose of the virtual-world's coordinate system with respect to the virtual-viewer's coordinate system. Alternatively, OpenGL allows an user to directly load transformation matrix ${}^vM_w$, which represents the pose of the virtual-world's coordinate system with respect to the virtual-viewer's coordinate system.

Before OpenGL is ready to synthesize stereoscopic images, it is necessary to further specify where the left and right virtual-cameras are. For instance, we can position the two virtual cameras along axis X_v. And, let parameter B be the distance between the two cameras (i.e. the baseline). Assume that the left virtual-camera has two degrees of freedom for rotation: one for panning and the other for tilting. Then, the relative pose of the left virtual-camera's coordinate system with respect to the virtual-viewer's coordinate system will be ${}^vM_l$, as described by Eq. 9.17. In this case, the relative pose of the virtual-world's coordinate system with respect to the left virtual-camera's coordinate system will be

$$ {}^lM_w = ({}^vM_l)^{-1} \bullet {}^vM_w. \tag{9.23}$$

Similarly, the relative pose of the virtual-world's coordinate system with respect to the right virtual-camera's coordinate system will be

$$ {}^rM_w = ({}^vM_r)^{-1} \bullet {}^vM_w \tag{9.24}$$

where ${}^vM_r$ is the relative pose of the right virtual-camera's coordinate system with respect to the virtual-viewer's coordinate system, as described by Eq. 9.18.

Now, the loading of matrix ${}^lM_w$ into OpenGL will correctly place the left virtual-camera. In order to correctly place the right virtual-camera, one must load matrix ${}^rM_w$ into OpenGL.

Finally, in order to capture synthetic images through OpenGL, we must specify

- Image size (i.e. the maximum number of pixels per row, and the maximum number of pixels per column)
- Angle of the field of view in plane XZ of a virtual-camera's coordinate system
- Range $[Z_{min}, Z_{max}]$, which define two clipping planes parallel to plane XY: one at distance Z_{min} (i.e. the near plane) and the other at distance Z_{max} (i.e. the far plane)

for each virtual camera.

In OpenGL, a pixel corresponds to a small square-area in an image plane. If we want to simulate an actual camera with OpenGL, the image size in OpenGL must be set to the image size of the actual camera. And, the angle of the field of view in OpenGL can be determined by solving the following simple equation:

$$ \tan\frac{\theta}{2} = \frac{c_{max} w}{2 f_c} \tag{9.25}$$

where θ is the angle of the field of view, c_{max} the maximum number of pixels per row, w the width of pixel in the actual camera, and f_c the focal length of the actual camera.

Now, let us apply the above detail to create a synthetical binocular vision with the following specifications:

- The origin of the virtual viewer's coordinate system is at (40 cm, −90 cm, 60 cm) in $O_wX_wY_wZ_w$.

- The Z axis of the virtual viewer's coordinate system is pointed at $(0, 0, 0)$ in $O_w X_w Y_w Z_w$.
- The Y axis of the virtual viewer's coordinate system is parallel to vector $(0, 0, 1)$ in $O_w X_w Y_w Z_w$.
- The pan and tilt angles of both virtual cameras are set to zero.
- The baseline between the left and right cameras is 2 cm.
- The image size of each virtual camera is 400 columns (i.e. image width) by 400 rows (i.e. image height).
- The size of pixel is 0.001 cm (i.e. width) by 0.001 cm (i.e. height).
- The focal length (i.e. f_c) of each camera is 0.5 cm.

By applying Eq. 9.25, we can easily determine the angle of the filed of view, which is actually 43.603^0. After loading a virtual office scene, OpenGL will automatically synthesize a pair of images as shown in Fig. 9.9.

<><><><><>

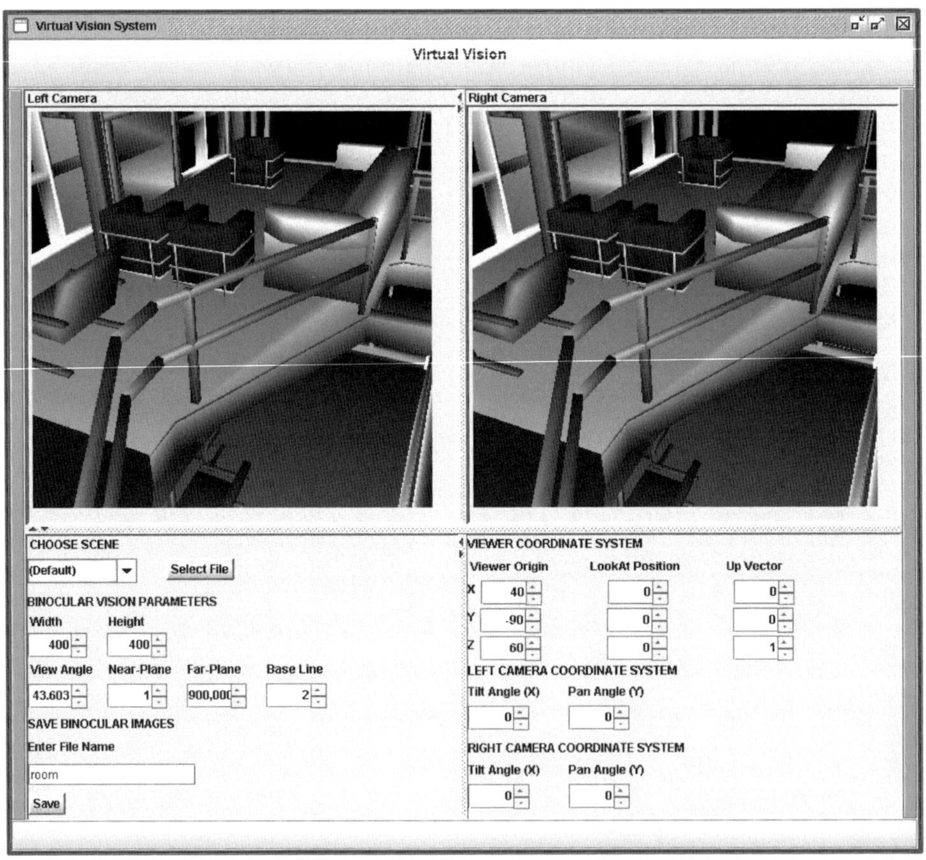

Fig. 9.9 A synthesized binocular vision, in which the binocular images are synthesized by OpenGL.

9.2.6 *Projection of Points from Scene to Image*

Now, we examine the relationship between the coordinates of a point in a scene and the coordinates of its image-point in the image plane of a camera.

Refer to Fig. 9.4c, or Fig. 9.4d. Let (x, y, f_c) denote the coordinates of point a in a camera's coordinate systems, where f_c is the focal length of the camera (i.e. the distance from the center of the camera's thin-lens to the image plane). If the coordinates of point A in the camera's coordinate system are (X, Y, Z), then we have

$$\frac{x}{X} = \frac{y}{Y} = \frac{f_c}{Z} = S \tag{9.26}$$

where variable S is a scale. Eq. 9.26 explains that under perspective projection, the coordinates of an image point (e.g. a) are linearly proportional to the coordinates of an object point (e.g. A) by a scale.

Because

$$S = \frac{f_c}{Z} \tag{9.27}$$

and f_c is a camera's parameter, scale S will be unique if object coordinates (X, Y, Z) are known. And, image coordinates (x, y, f_c) will be uniquely determined by

$$\begin{cases} x = S \bullet X = f_c \bullet \dfrac{X}{Z} \\ y = S \bullet Y = f_c \bullet \dfrac{Y}{Z}. \end{cases} \tag{9.28}$$

However, because

$$S = \frac{x}{X} \quad \text{or} \quad S = \frac{y}{Y}, \tag{9.29}$$

scale S cannot be determined from image coordinates (x, y, f_c) alone. If we rewrite Eq. 9.26 into the following form:

$$\begin{cases} X = \dfrac{x}{S} \\ Y = \dfrac{y}{S} \\ Z = \dfrac{f_c}{S}, \end{cases} \tag{9.30}$$

it is clear that object coordinates (X, Y, Z) are not unique, if scale S is unknown.

Now, let $a = (x, y)^t$ and $A = (X, Y, Z)^t$ where t indicates the transpose of a vector. Then, it is easy to verify that Eq. 9.28 can be rewritten into the following matrix form

$$a = \frac{1}{Z} \begin{pmatrix} f_c & 0 & 0 \\ 0 & f_c & 0 \end{pmatrix} A \tag{9.31}$$

which projects the coordinates of point A in a scene into the coordinates of point a in an image plane.

As Eq. 9.31 fully describes the geometrical aspect of image formation by a pin-hole camera, we call it the *geometrical model* of a pin-hole camera.

9.2.7 *Projection of Velocities from Scene to Image*

From Eq. 9.28, we have

$$x = f_c \frac{X}{Z}. \tag{9.32}$$

The differentiation of Eq. 9.32 with respect to time yields

$$\dot{x} = f_c \frac{\dot{X}}{Z} - f_c \frac{X}{Z^2} \dot{Z} \tag{9.33}$$

where $\dot{x} = \frac{dx}{dt}$, $\dot{X} = \frac{dX}{dt}$, and $\dot{Z} = \frac{dZ}{dt}$. Also, from From Eq. 9.28, we have

$$y = f_c \frac{Y}{Z}. \tag{9.34}$$

If we compute the first-order derivative of y with respect to time, we obtain

$$\dot{y} = f_c \frac{\dot{Y}}{Z} - f_c \frac{Y}{Z^2} \dot{Z} \tag{9.35}$$

where $\dot{y} = \frac{dy}{dt}$.

If we denote $\dot{a} = (\dot{x}, \ \dot{y})^t$ and $\dot{A} = (\dot{X}, \ \dot{Y}, \ \dot{Z})^t$, Eq. 9.33 and Eq. 9.35 can be combined into a matrix form, as follows:

$$\dot{a} = \begin{pmatrix} \frac{f_c}{Z} & 0 & -f_c \frac{X}{Z^2} \\ 0 & \frac{f_c}{Z} & -f_c \frac{Y}{Z^2} \end{pmatrix} \dot{A}. \tag{9.36}$$

In Eq. 9.36, $\dot{a}$ is the velocity of image point a within the image plane of a camera. And, $\dot{A}$ is the velocity of object point A with respect to a camera's coordinate system. In fact, Eq. 9.36 describes the mapping from an object point's velocity into its image point's velocity. As a result, Eq. 9.36 is called the *differential model* of a pin-hole camera.

It is worth mentioning that the velocity of an object point in a camera's coordinate system is composed of two components: (a) the linear velocity, and (b) the circular velocity. For instance, when a point is undergoing a pure translation in a camera's coordinate system, its circular velocity is zero. Let v denote the linear velocity of point A in a camera's coordinate system. Then, we have

$$\dot{A}_{linear} = v. \tag{9.37}$$

On the other hand, if a point is rotating about an axis in a camera's coordinate system, its linear velocity is zero. Moreover, when a point is rotating about an axis, the rotation itself is described by the so-called *angular velocity.* If vector ω denotes the angular velocity of a rotation by point A about an axis passing through the origin of a camera's coordinate system, then

$$\dot{A}_{circular} = \omega \times A \tag{9.38}$$

where "×" means the cross-product of two vectors, and $A = (X,\ Y,\ Z)^t$.

Therefore, in general, the velocity of point A in a camera's coordinate system will be

$$\dot{A} = \dot{A}_{linear} + \dot{A}_{circular} = v + \omega \times A. \tag{9.39}$$

And, the substitution of Eq. 9.39 into Eq. 9.36 will yield

$$\dot{a} = \begin{pmatrix} \frac{f_c}{Z} & 0 & -f_c\frac{X}{Z^2} \\ 0 & \frac{f_c}{Z} & -f_c\frac{Y}{Z^2} \end{pmatrix} (v + \omega \times A). \tag{9.40}$$

In Eq. 9.40, $\dot{a}$ is the perceived velocity at an image location. By definition, $\dot{a}$ is also called the *apparent motion*, or *optic flow* of a pixel in an image plane. And, Eq. 9.40 relates the general motion of an object point to the apparent motion of its image point.

Now, we consider the following scenario: A machine is moving within time interval $[t_1, t_2]$, which is short enough. And, the robot's scene is actually static within this short time interval. In this case, the relative motion between object point A and one of the machine's cameras uniquely depends on the camera's linear velocity v_c and angular velocity ω_c. And, we have

$$\begin{cases} v = -v_c \\ \omega = -\omega_c. \end{cases} \tag{9.41}$$

On the other hand, we can also consider this imagined scenario: In binocular vision, the right-camera's coordinate system initially coincides with the left-camera's coordinate system. Then, the right-camera's coordinate system reaches its actual pose, with respect to the left-camera's coordinate system, after the simultaneous rotation and translation within a unit time-interval. Assume that the relative motion of the right-camera's coordinate system is described by linear velocity v_c and angular velocity ω_c. In this case, Eq. 9.40 and Eq. 9.41 are still valid, if apparent motion $\dot{a}$ actually refers to the disparity between the left and right images.

Under the context of above two scenarios, Eq. 9.40 can be rewritten as follows:

$$\dot{a} = \begin{pmatrix} \frac{f_c}{Z} & 0 & -f_c\frac{X}{Z^2} \\ 0 & \frac{f_c}{Z} & -f_c\frac{Y}{Z^2} \end{pmatrix} (-v_c - \omega_c \times A). \tag{9.42}$$

Finally, Eq. 9.31 and Eq. 9.42 allows us to define three different problems, such as

- How to determine the apparent motion $\dot{a}$, if A and $(v_c,\ \omega_c)$ are known?
- How to determine A, if $(a,\ \dot{a})$ and $(v_c,\ \omega_c)$ are known?
- How to determine $(v_c,\ \omega_c)$, if $(a,\ \dot{a})$ and A are known?

However, under the context of visual perception of depth from apparent motion, what we are interesting in is the question of how A could be determined, if $(a,\ \dot{a})$ and $(v_c,\ \omega_c)$ are known. Obviously, the solution to this problem depends on the result of $(a,\ \dot{a})$. In the next section, we will study the way of estimating apparent motion $\dot{a}$ at image point a.

9.3 Transformation from Image to Apparent Motion

A machine such as a robot will perceive changing images when it moves its body, neck, or head in a scene. Also, a machine will perceive changing images if it is within a scene in which objects are changing their positions and orientations in various manners. Therefore, when apparent motion occurs in the image plane of a camera, it is important for a machine to autonomously detect and track the apparent motion. And, the detection of apparent motion is a necessary condition, which has to be satisfied before depth estimation by Eq. 9.31 and Eq. 9.42.

9.3.1 *Formation of Apparent Motion*

If a machine such as a robot continuously observes a static scene, the motion from the machine's body, neck or head will make the cameras of the robot's vision system change their position and/or orientation. As a result, the image inside a camera will not be still, but be constantly changing. In this case, we say that there is an apparent motion in the image plane of a camera.

On the other hand, if a machine stands still, its cameras will maintain a stable pose with respect to the ground. Then, the image inside a camera will remain unchanged if nothing is moving inside the scene seen by the camera. However, if at least one object is moving, the images seen by the camera will not remain the same. In this case, apparent motion also appears in the image plane of a camera.

Fig. 9.10 illustrates the above-mentioned two scenarios. For instance, the trajectory of a point in a dynamic scene will be projected into a trajectory of a pixel in the image plane of a camera. Similarly, when a machine's camera moves around a scene, the projection of a fixed point in a scene onto the camera's image plane will also form a trajectory. In order to record a trajectory in an image plane, images must be captured at a certain rate. In this way, the snapshots of an apparent motion will be stored in the form of a sequence of recorded images as shown in Fig. 9.10c.

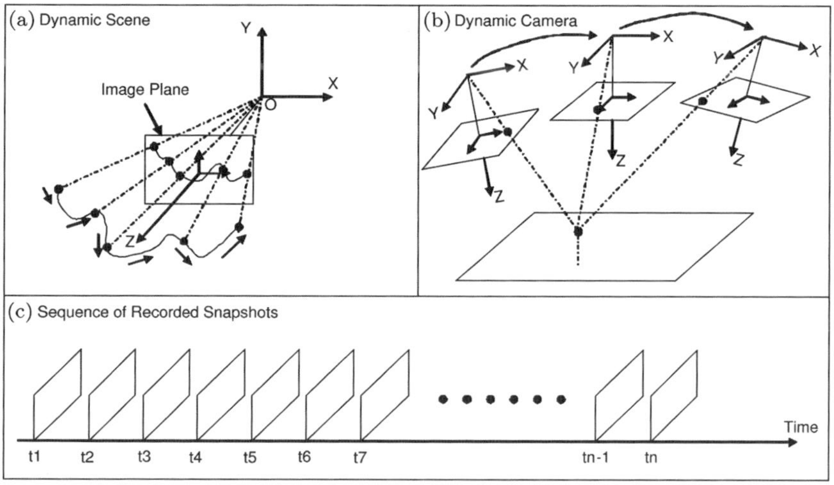

Fig. 9.10 Apparent motion occurs as a result of a dynamic scene, a dynamic camera, or the combination of both. As the signal processing is not done at sensor level, images must be recorded at a certain rate in order to capture the snapshots of apparent motion in the image plane of a camera.

9.3.2 *Registration of Image Sequence*

According to the study in neuroscience, the signals inside the neural systems of human brain are in pulsed-waveforms. This implies that the visual signals in human vision are discrete in nature. Similarly, a modern computer is a digital system. Therefore, if a machine uses computers to process input signals, these signals must be converted into digital signals before they can be processed by digital computers for various purposes.

As a matter of fact, the design of machine vision must consider the module called *image digitizer*, which plays the important role of generating a sequence of digital (color) images from a camera at a fixed rate such as 25 images per second. For instance, in Fig. 9.1, we can see that an image digitizer takes input signals from the photo-detectors, or photo-receptors, placed at an image plane, and that the output from an image digitizer will be a sequence of digital (color) images referenced at discrete time instances.

When a machine is equipped with a binocular vision, the output from the two cameras in the machine's binocular vision will be two sequences of images, namely: (a) the left image-sequence and (b) the right image-sequence, as shown in Fig. 9.11. And, within each image sequence, apparent motion may appears, which will give rise to this challenging problem of how to robustly detect apparent motion within an image-sequence.

Refer to Fig. 9.11. The two image-sequences in a machine's binocular vision are related to one another because of the binocular configuration. This constraint implies that there is a correspondence between the left-image and the right-image

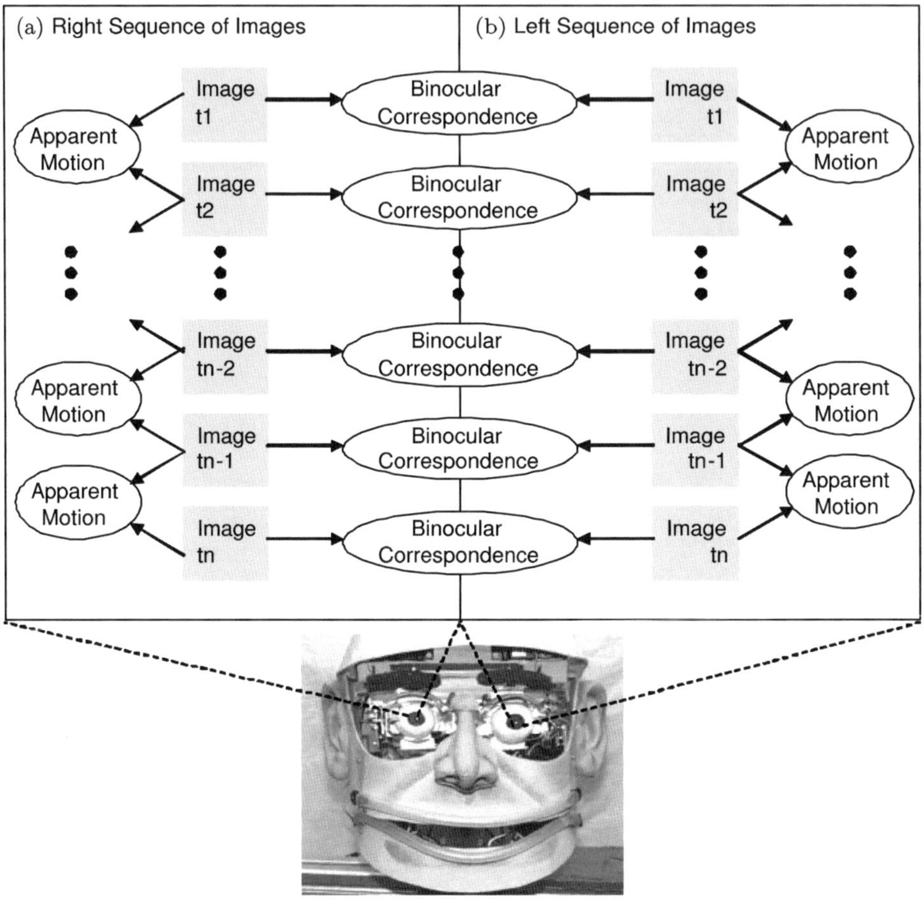

Fig. 9.11 Two image-sequences from a machine's binocular vision. These two sequences of images pose two challenges to machine vision. The first challenge is how to detect apparent motions within an image-sequence. And, the second challenge is how to detect disparities, or establish binocularcorrespondence, between two image-sequences.

at each discrete time. In other words, an object in a scene may be visible to both the left-camera and the right-camera. And, the binocular images of an object will correspond to one another. Therefore, how to establish the binocular correspondence between two image-sequences poses another challenge to machine vision.

9.3.3 *Use of Conglomerated Appearance*

Now, we consider the case, in which there is apparent motion in the image plane of a camera. Without loss of generality, we assume that an image point moves inside the image plane within two consecutive time instances, as shown in Fig. 9.12.

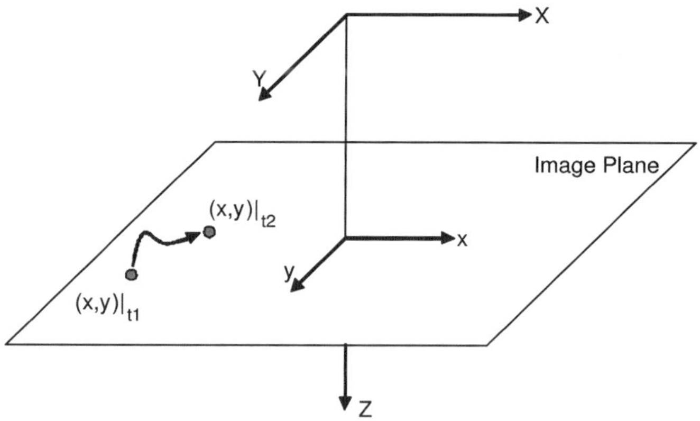

Fig. 9.12 Within two consecutive time instances, an image point moves from one location to another location inside the image plane of a camera.

In an image plane, a pixel is characterized by its image coordinates, denoted by (x, y), and also by its color. When a pixel moves inside an image plane, we can further assume that the color of a moving pixel remains the same within two consecutive time instances. If in practice, the color of a moving pixel changes within a short period of time, such change could be ignored as long as we focus on the time intervals in which the color of a moving pixel remains relatively stable (e.g, the time interval between two consecutive images is extremely short).

Now, assume that a pixel moves within two consecutive time instances. Its image coordinates change, while its color remains stable. In this case, the detection of apparent motion is to identify the location change of a pixel's color. For instance, if a pixel has an unique color in an image plane, then it is relatively easy to detect the locations of this unique color at two consecutive time instances t_1 and t_2. And, the apparent motion of this pixel having an unique color will be

$$\begin{cases} \triangle x = x(t_2) - x(t_1) \\ \triangle y = y(t_2) - y(t_1) \end{cases} \tag{9.43}$$

where $(x(t_1), y(t_1))$ denote the pixel's coordinates at time t_1 of the pixel, and $(x(t_2), y(t_2))$ the pixel's coordinate at time t_2.

However, an image in general is quite complex because of the complexity of a scene. This fact implies that the assumption about the uniqueness of colors at pixels is not valid. For instance, it is generally true that the pixels within a neighborhood of an image plane will have an identical color at a certain color scale. As a result, the detection of apparent motion at a pixel will not work if we only consider the color at this pixel. In general, it is easy to conclude that the detection of apparent motion at a pixel must also consider the colors of this pixel's neighbors.

Here, for the sake of notation, we define the appearance given rise by a pixel and its neighbors the *conglomerated appearance* at this pixel. Now, a relevant question will be: How to determine the conglomerated appearance at a pixel?

(a) Left Image (b) Right Image

Fig. 9.13 When two images are taken at two different viewing positions or orientations, some portions of a scene will appear in one image, but not in the other image. In this case, occlusions occur.

Experienced readers will notice that many existing algorithms for detecting apparent motion (or binocular disparity) rely on the use of *template*, which is normally a rectangular region centered at a pixel. And, the image inside a template is called a *template-image*, which is treated as the conglomerated appearance of a pixel, by default. However, template-based approach to determine the conglomerated appearance of a pixel has two severe drawbacks:

- First of all, when a camera changes its viewing position or orientation, occlusion will occur within two consecutive images. Technically speaking, occlusion means that some portions of a scene will appear in one image, but not in the other image. For instance, Fig. 9.13 shows a pair of binocular images, in which some occlusions are highlighted in the left image. When a template overlaps with an occluded area, it is clear that the content inside the template will not remain stable within two consecutive images. In this case, the use of template-image at a pixel for the purpose of detecting its apparent motion (or disparity) is no more valid.
- Second, a template has a certain size. Normally, the size of a template is pre-specified. In this case, if an image has a large uniform area, the content inside a template will not be unique. For instance, the white-paper placed on the table in Fig. 9.13 is uniform. In this case, the templates placed inside the area of the white-paper will have similar contents.

Today, the success of depth perception in machine vision has been severely undermined by the above-mentioned two drawbacks. Clearly, we should not restrict ourselves to the use of template-based approach to determiner the conglomerated appearance at a pixel.

Under the context of cognitive vision, there is an effective way of determining the conglomerated appearance at a pixel. As the detection of apparent motion has to make use of conglomerated appearance at a pixel, let us answer the following question first: What should be the desirable properties of the conglomerated appearance at a pixel?

As we have mentioned earlier, it is easy to detect the apparent motion at a pixel if this pixel has a unique color in an image plane. Similarly, it will be easy to detect the apparent motion at a pixel if the conglomerated appearance at this pixel is unique in an image plane. Therefore, there are two desirable properties of the conglomerated appearance at a pixel, namely:

- Uniqueness:
  It means that the conglomerated appearances at two different pixels in a same image plane should be different.
- Constancy:
  If a pixel is not occluded within two consecutive images, the conglomerated appearance of this pixel should remain relatively the same within these two consecutive images.

Now, we can introduce the exact definition of the conglomerated appearance at a pixel, by answering the following specific question: What is the conglomerated appearance at a pixel, which satisfies the properties of uniqueness and (relative) constancy?

We know that the images from a camera are color-images in general. Also, we have learnt that the colors at the pixels in an image could be effectively represented by a set of multi-scale color indices. Under the context of cognitive vision, the color indices at a specific scale represent the color categories at this scale. Therefore, it is easy to identify the pixels in an image, which have the similar (or identical) color indices at a given scale. Interestingly enough, the appearance given rise by these pixels having the similar (or identical) color indices at a given scale will be unique in an image, and will be relatively stable within two consecutive images.

As a result, the use of the multi-scale color indices allows us to define the conglomerated appearance at a pixel, as follows:

Definition 9.1. The conglomerated appearance at a pixel in consideration is a sub-image, which is always centered at the pixel in consideration. And, the content of the sub-image consists of pixels having the color indices, which are similar (or identical), at a given color scale, to the color indices of the pixel in consideration.

Based on the above definition, it is easy to conclude:

- If the color indices at two pixels are different, the conglomerated appearances at these two pixels will be definitely different.
- If the color indices at two pixels are the same, the content of the conglomerated appearances at these two pixels will be the same. But, these two conglomerated appearances can still be different because the distributions of pixels inside these two conglomerated appearances may be different, as illustrated by Fig. 9.14.

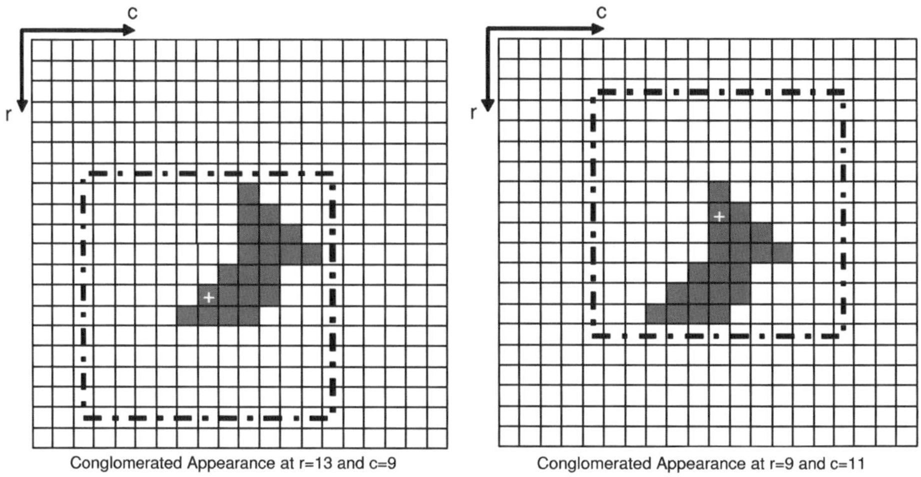

(a) (b)

Fig. 9.14 An conglomerated appearance is referenced to the location of a pixel in consideration. Two conglomerated appearances of the same content are still considered to be different if the distributions of their internal pixels are not the same.

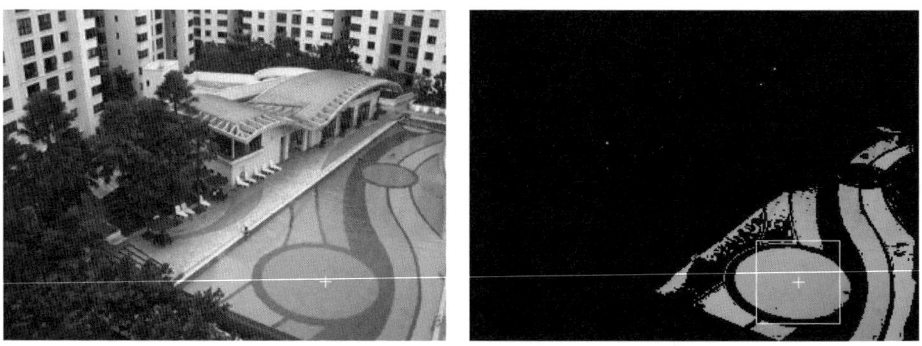

(a) Image and Selected Point (b) Conglomerated Appearance at Selected Point

Fig. 9.15 A real example of conglomerated appearance at a selected pixel, which is marked by "+" in (a). The original image is represented in the RGB color space. At scale 4, the pixels having the color indices similar (or identical) to the color indices of the selected pixel are identified. These pixels are shown in (b), in which the conglomerated appearance at the selected pixel is enclosed by the rectangle.

Example 9.4. Fig. 9.15a shows a color image taken by a digital camera from an outdoor scene. We first compute the multi-scale color indices for every pixel in the input image. Then, we randomly select a pixel, which is marked by "+" in Fig. 9.15a. At the selected pixel, we read out the RGB color indices at scale 4. Subsequently, we scan the original image row by row and column by column. At every scanned pixel, we compare the RGB color indices of the scanned pixel with the RGB color indices of the selected pixel. If the difference between the indices of any one of the red, green and blue colors is not more than one, we declare that the color at the scanned pixel is similar (or identical) to the color of the selected pixel. After the complete scan of the whole image, the result of pixels having the color indices similar (or identical) to the color indices of the selected pixel is shown in Fig. 9.15b. Finally, we take a sub-image of the size of 61×61, which is centered at the location of the selected pixel. And, this sub-image is the conglomerated appearance at the selected pixel, which is enclosed by the rectangle shown in Fig. 9.15b. It is worth noting that the size of the sub-image could be adaptively determined. For instance, one could incrementally increase the size until the sub-image become unique.

<><><><><>

9.3.4 *Matching of Conglomerated Appearances*

When the image of a point in a scene moves in the image plane of a camera, this will give rise to the so-called apparent motion. In order to detect the apparent motion of a moving pixel, it is necessary to record a sequence of digital images. Furthermore, it is reasonable to assume that the conglomerated appearances of a moving pixel are unique, and remain relatively stable within two consecutive images. In this way, the detection of apparent motion will consist of matching the conglomerated appearances of a moving pixel between two consecutive images. Now, the question will be: Is there any constraint which restricts the way in which a pixel moves in the image plane of a camera?

In fact, without any prior knowledge, there will be no specific constraint which conditions the way in which a moving pixel changes its location in the image plane of a camera. However, in the absence of any specific constraint, we can assume that within two consecutive images, a pixel will move in any direction, and that the magnitude of a pixel's displacement will be within a maximum radius, denoted by D_{max}.

Refer to Fig. 9.16. Let us consider the case in which two consecutive images are taken at two distinct time instances: t_1 and t_2. And, within these two consecutive images, a pixel moves from (r_1, c_1) at time t_1 to (r_2, c_2) at time t_2. Then, the apparent motion of this pixel will be $(\triangle r, \triangle c)$ with $\triangle r = r_2 - r_1$ and $\triangle c = c_2 - c_1$. Finally, the problem of detecting the apparent motion could be defined as follows:

> Given a pixel's location (r_1, c_1) in the image at time t_1, find a displacement vector, denoted by $(\triangle r, \triangle c)$, such that:
>
> (1) The magnitude of $(\triangle r, \triangle c)$ is less than D_{max}, and
> (2) The conglomerated appearance at location $(r_1 + \triangle r, c_1 + \triangle c)$ in the image at time t_2 has the maximum overlap with the conglomerated appearance at location (r_1, c_1) in the image at time t_1.

The solution to the above problem could be implemented by a simple algorithm, or function, called **detect-apparent-motion(r1, c1, D, s)**, where D denotes parameter D_{max} and s indicates the color scale at which the conglomerated appearances are computed.

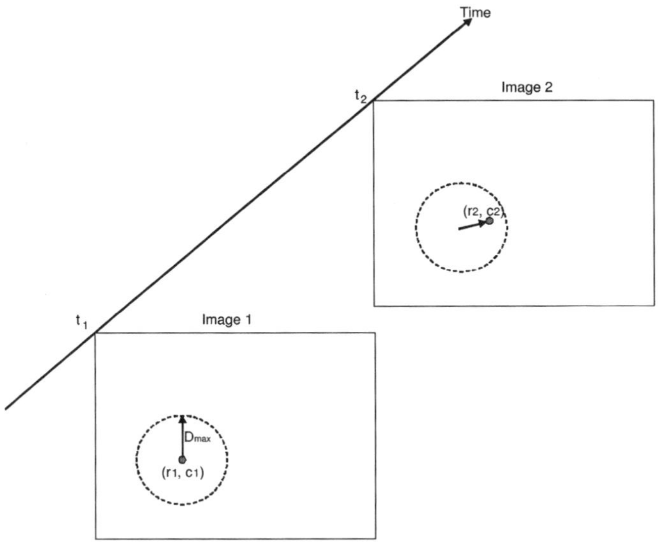

Fig. 9.16 Illustration of two consecutive images which are referenced to two consecutive time instances: t_1 and t_2. Within the two consecutive images, a pixel moves from (r_1, c_1) at time t_1 to (r_2, c_2) at time t_2. And, the magnitude of the displacement is within a circle of a maximum radius, denoted by D_{max}.

In a pseudo-form of program, function **detect-apparent-motion(r1, c1, D, s)** could be as follows:

```
detect-apparent-motion(r1, c1, D, s) {
    Read the color indices of scale s at pixel location (r1, c1)
    in the image at time t1;

    Scan the image at time t1 row by row and column by column,
    and identify the pixels having the color indices, at scale s,
    which are similar (or identical to) the color indices at
    pixel location (r1, c1) in the image at time t1;

    Get the conglomerated appearance at pixel location (r1, c1)
    in the image at time t1;

    Scan the image at time t2 row by row and column by column,
    and identify the pixels having the color indices, at scale s,
    which are similar (or identical to) the color indices at
    pixel location (r1, c1) in the image at time t1;

    At every possible location (r2, c2) in the image at
    time t2, get the conglomerated appearance, and
    compute the number of overlapped pixels with
```

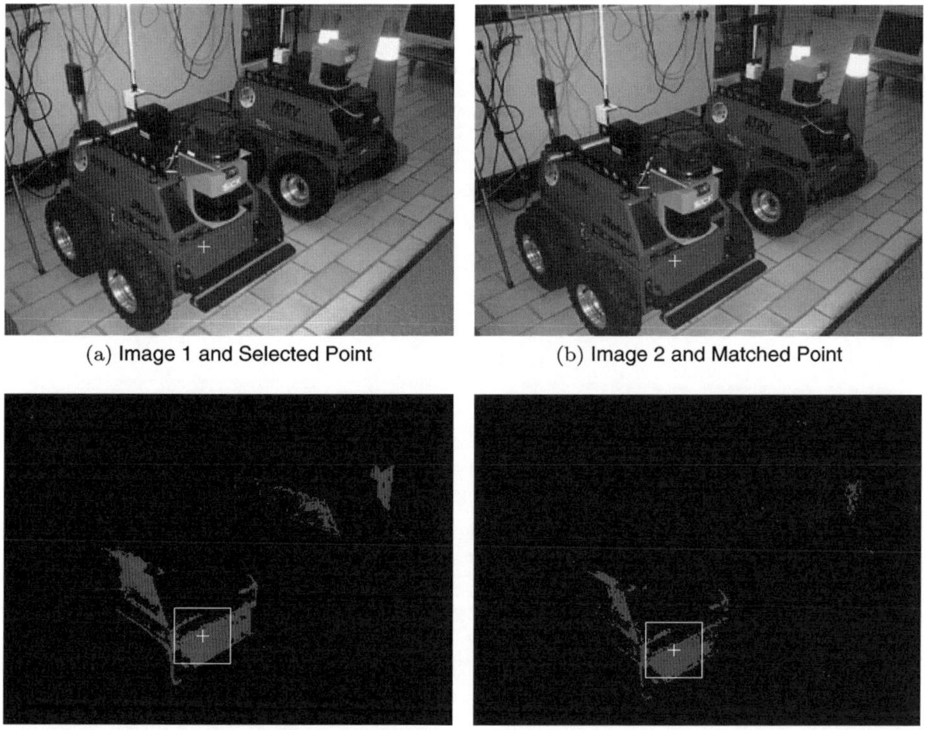

(a) Image 1 and Selected Point
(b) Image 2 and Matched Point
(c) Conglomerated Appearance at Selected Point
(d) Conglomerated Appearance at Matched Point

Fig. 9.17 A real example of matching conglomerated appearances. Two images are taken at two different positions and orientations. A pixel is selected from the first image, which is marked by "+" in (a). In the first image, the pixels having the color indices, at scale 4, similar (or identical) to the color indicates at the selected pixel are shown in (c). Similarly, in the second image, the pixels having the color indices, at scale 4, similar (or identical) to the color indices at the selected pixel are shown in (d). At the selected pixel, the conglomerated appearance is enclosed by the rectangle in (c). And, the matched point and the conglomerated appearance at the matched point are shown in (b) and (d), respectively.

```
    the conglomerated appearance at pixel location (r1, c1)
    in the image at time t1.

    Among the possible locations of (r2, c2), output the one
    in which the number of overlapped pixels is the maximum.
}
```

Example 9.5. We have taken two color images from an indoor scene at two different positions and orientations. These two images are shown in Fig. 9.17a and Fig. 9.17b, respectively.

First of all, we compute the multi-scale RGB color indices at every pixel of these two images. Then, we select a pixel from the first image, which is marked by "+" as shown in Fig. 9.17a. At the selected pixel, the color indices at scale 4 are read out.

Subsequently, we identify the pixels from both the images, which have the color indices at scale 4 that are similar (or identical) to the color indices at the selected pixel. The result of the identified pixels from the first image is shown in Fig. 9.17c. And, the result of the identified pixels from the second image is shown in Fig. 9.17d.

Now, we set maximum radius D_{max} to be 20, and the size of conglomerated appearances to be 21×21. Then, we invoke function **detect-apparent-motion(r1, c1, D, s)**. Finally, the matched point and the conglomerated appearance at the matched are highlighted in Fig. 9.17b and Fig. 9.17d, respectively.

<><><><><>

9.3.5 *Match Search Space*

In the previous discussion, we assume that we do not know any constraint which restricts the way in which a pixel moves in the image plane of a camera. However, if the apparent motion is due to the ego-motion of a camera, then there is a constraint which restricts the variation of (r_2, c_2) in the image at time t_2. Alternatively, if two consecutive images are taken by a pair of cameras, of which the relative pose is known in advance, such a constraint also exists.

Without loss of generality, we consider the case in which two consecutive images are taken by a pair of cameras, of which the relative pose is known in advance.

Refer to Fig. 9.18. A pair of cameras observes a static scene. Assume that point A in the scene is visible to both cameras, and that the images of point A in these two cameras are point B and point C, respectively. In other words, point C is the match of point B. Then, according to perspective projection, the extension of line AB will pass through origin O_{left} of the left camera's coordinate system. And, the extension of line AC will pass through origin O_{right} of the right camera's coordinate system. When we draw a line which connects origin O_{left} to origin O_{right}, it is clear that points A, O_{left} and O_{right} will form a triangle, which is planar.

Refer to Fig. 9.18 again. Let vector $\overrightarrow{O_{left}B}$ denote the direction of line $O_{left}A$, and vector $\overrightarrow{O_{right}C}$ denote the direction of line $O_{right}A$. If vector $\overrightarrow{O_{left}O_{right}}$ represents the direction of line $O_{left}O_{right}$, then the normal vector to the plane defined by triangle $AO_{left}O_{right}$ will be

$$\overrightarrow{O_{left}O_{right}} \times \overrightarrow{O_{left}B}.$$

As the normal vector to triangle $AO_{left}O_{right}$ is perpendicular to vector $\overrightarrow{O_{right}C}$, the product between normal vector $\overrightarrow{O_{left}O_{right}} \times \overrightarrow{O_{left}B}$ and vector $\overrightarrow{O_{right}C}$ will be equal to zero. In other words, we have the following constraint

$$\left(\overrightarrow{O_{left}O_{right}} \times \overrightarrow{O_{left}B}\right)^t \bullet \overrightarrow{O_{right}C} = 0. \tag{9.44}$$

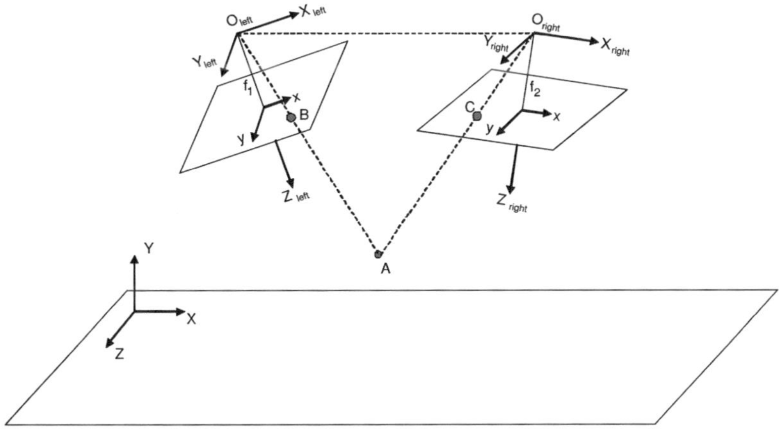

Fig. 9.18 A pair of cameras observes a static scene, in which point A is visible to both cameras. When point A projects onto the image plane of the left camera, its image is point B. And, in the right camera, the image of point A is point C.

In Eq. 9.44, all vectors must be expressed with respect to a common coordinate system. Without loss of generality, we choose the left camera's coordinate system as the common coordinate system for reference. In the left camera's coordinate system, vector $\overrightarrow{O_{left}B}$ will be

$$\overrightarrow{O_{left}B} = (x_1, y_1, f_1)^t \tag{9.45}$$

where (x_1, y_1) are the coordinates of point B in the image plane of the left camera, and f_1 the focus length of the left camera. However, if we factorize out f_1, vector $\overrightarrow{O_{left}B}$ could also be expressed as

$$\overrightarrow{O_{left}B} = \left(\frac{x_1}{f_1}, \frac{y_1}{f_1}, 1\right)^t. \tag{9.46}$$

Similarly, in the right camera's coordinate system, vector $\overrightarrow{O_{right}C}$ will be

$$\overrightarrow{O_{right}C} = \left(\frac{x_2}{f_2}, \frac{y_2}{f_2}, 1\right)^t \tag{9.47}$$

where (x_2, y_2) are the coordinates of point C in the image plane of the right camera, and f_2 the focus length of the right camera.

Now, assume that the relative pose between the left-camera and the right-camera are known in advance. In particular, let T denote the translation of origin O_{right} of the right camera's coordinate system with respect to the left camera's coordinate system, and R denote the rotation matrix which represents the orientation of the right camera's coordinate system with respect to the left camera's coordinate system. Then, vector $\overrightarrow{O_{left}O_{right}}$ with respect to the left camera's coordinate system will be

$$\overrightarrow{O_{left}O_{right}} = T \tag{9.48}$$

where translation vector T equals $(t_x, t_y, t_z)^t$.

Recall Eq. 9.47. When there is no relative rotation between the left-camera and the right-camera, vector $\overrightarrow{O_{right}C}$ with respect to the right camera's coordinate system will remain the same as vector $\overrightarrow{O_{right}C}$ with respect to the left camera's coordinate system. However, if there is a relative rotation, denoted by matrix R, between the left-camera and the right-camera, then vector $\overrightarrow{O_{right}C}$ with respect to the left camera's coordinate system will be

$$\overrightarrow{O_{right}C} = R \bullet \left(\frac{x_2}{f_2}, \frac{y_2}{f_2}, 1\right)^t \tag{9.49}$$

or

$$\overrightarrow{O_{right}C} = R \bullet \begin{pmatrix} \frac{x_2}{f_2} \\ \frac{y_2}{f_2} \\ 1 \end{pmatrix}. \tag{9.50}$$

Now, the substitution of Eq. 9.46, Eq. 9.48 and Eq. 9.49 into Eq. 9.44 will yield

$$\left(T \times \left(\frac{x_1}{f_1}, \frac{y_1}{f_1}, 1\right)^t\right)^t \bullet \left(R \bullet \left(\frac{x_2}{f_2}, \frac{y_2}{f_2}, 1\right)^t\right) = 0. \tag{9.51}$$

If we define skew-symmetrical matrix S_T as follows

$$S_T = \begin{pmatrix} 0 & -t_z & t_y \\ t_z & 0 & -t_x \\ -t_y & t_x & 0 \end{pmatrix}, \tag{9.52}$$

we will have the following equality

$$T \times \left(\frac{x_1}{f_1}, \frac{y_1}{f_1}, 1\right)^t = S_T \bullet \left(\frac{x_1}{f_1}, \frac{y_1}{f_1}, 1\right)^t. \tag{9.53}$$

And, the application of Eq. 9.53 to Eq. 9.51 will result in the following expression

$$\left(\frac{x_1}{f_1}, \frac{y_1}{f_1}, 1\right) \bullet (S_T^t \bullet R) \bullet \begin{pmatrix} \frac{x_2}{f_2} \\ \frac{y_2}{f_2} \\ 1 \end{pmatrix} = 0. \tag{9.54}$$

Eq. 9.54 simply means that given image coordinates (x_1, y_1) in the image plane of the left camera, image coordinates (x_2, y_2) of the match in the image plane of the right camera must satisfy the linear equation described by Eq. 9.54, in which T, R, f_1 and f_2 are known in advance. In other words, if point C is the match of point B, image coordinates of point C must satisfy Eq. 9.54, which describe a line in the image plane of the right camera.

Because Eq. 9.54 describes a line, it is also called an *epipolar line constraint* in computer vision. Refer to Fig. 9.18. When we search for match C of point B, we just need to examine the pixels which intersect with the epipolar line. Therefore, an epipolar line defines a search space for a match.

Example 9.6. We place four cameras at four different views, which are named as upper-left view, lower-left view, upper-right view and lower-right view. At these four views, four images are captured at the same time. These four image are shown in Fig. 9.19.

As the four cameras are not precisely designed for the purpose of achieving cognitive vision, we have to undertake an elaborated calibration in order to obtain

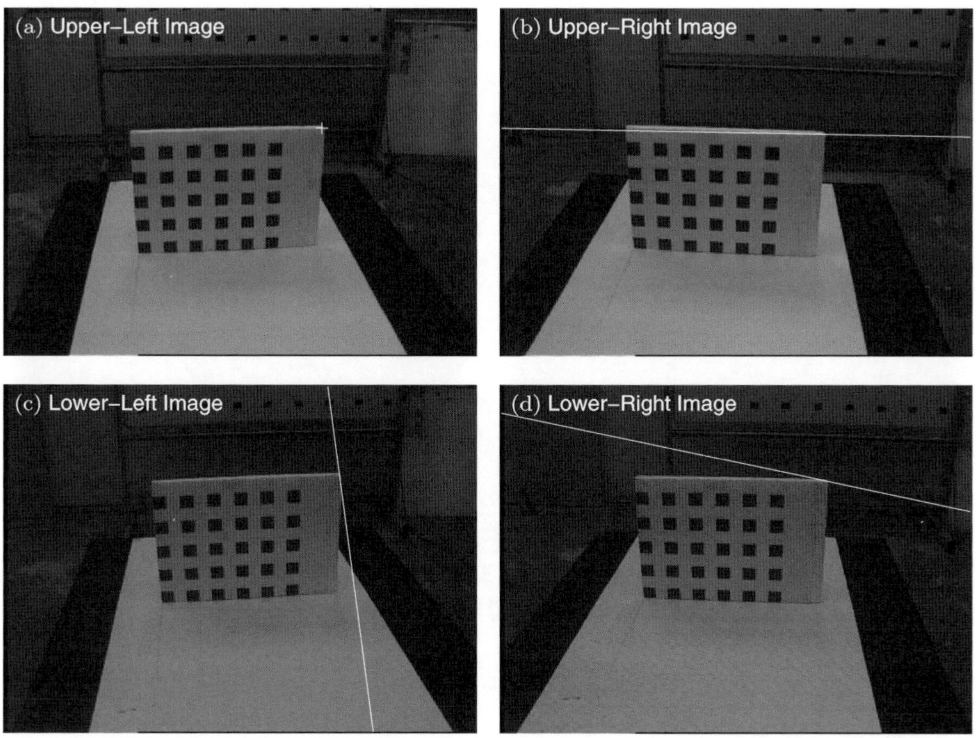

Fig. 9.19 Examples of epipolar lines between pairs of images. The four images are taken by four cameras placed at four views, namely: upper-left view, lower-left view, upper-right view and lower-right view. The parameters of, and the relative poses between, these cameras are obtained by a calibration process. When we select a pixel in the image of the upper-left view shown in (a), the matches in the other three images are constrained by three epipolar lines which can be computed from Eq. 9.54. And, the results are shown in (b), (c) and (d), respectively.

the parameters of, and the relative poses between, these cameras. As this book advocates precise design of camera and vision system, there will be no discussion on camera calibration here.

Now, we manually select a pixel from the upper-left image, which is marked by "+" in Fig. 9.19a. By applying Eq. 9.54, the three epipolar lines in the other three images can be computed. The results are shown in Fig. 9.19b, Fig. 9.19c, and Fig. 9.19d, respectively. And, we can see that the matches to the selected pixel are on these three epipolar lines. In other words, an epipolar line defines the search space for a match.

<><><><><>

9.4 Representation of Depth

An image is a two-dimensional matrix, in which each element is called a pixel. And, in terms of a two-dimensional matrix of pixels, an image is also called as a pixel-map, or pixmap in short. Therefore, a pixmap is the most elementary way of representing an image. As an image is the projection of a scene onto the image plane of a camera, now the question will be: What should be the most elementary way of representing a scene?

We know that every pixel in an image not only records the chrominance, intensity and direction of an incoming light, but also witnesses the existence of a position (or depth) at which the incoming light emanates. In other words, for every pixel in an image seen by a camera, there should be a corresponding position (or depth) in a scene. As the pixels of an image form a pixmap, it is natural to consider that the depths of a scene will form a so-called *depth-map*.

9.4.1 *Use of Depth-map*

When a light emanates from a position in a three-dimensional scene toward a camera, a pixel in the image plane of the camera will receive stimulus. Most importantly, the photo-receptors placed behind the pixel will output signals, which not only measure the color of the incoming light, but also witness the existence of a position at which the incoming light emanates. Therefore, in vision, depth refers to the third dimension (or coordinate), which a pixel witnesses at a time instance.

Refer to Fig. 9.20. Assume that lights emanate from positions A, B and C at a time instance. When these lights enter a camera, pixels a, b and c in the image plane of the camera will be illuminated. Clearly, the sensor outputs from pixels a, b and c will record the colors of the incoming lights from positions A, B and C. And, at this time instance, coordinates Z of positions A, B and C will be the depths of pixels a, b and c.

Now, let us denote (X_A, Y_A, Z_A) the coordinates of position A with respect to the camera's coordinate system, and (x_a, y_a) the image coordinate of pixel a.

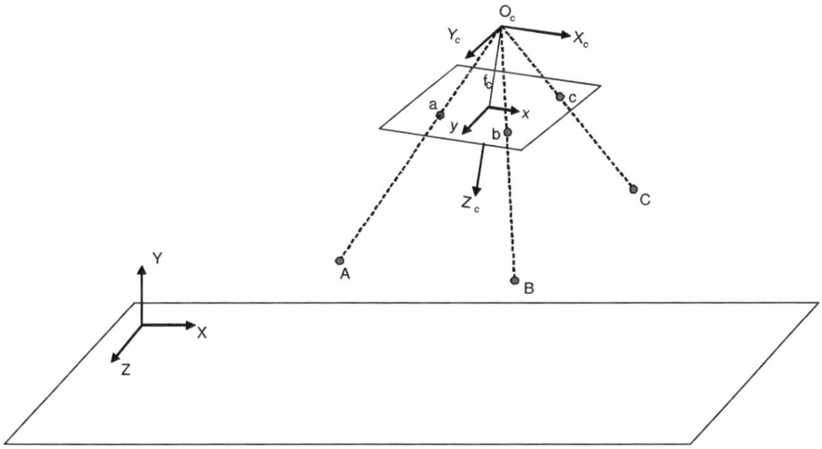

Fig. 9.20 Depth in vision refers to the third dimension which a pixel witnesses at a time instance. For example, pixels a, b and c record colors coming from positions A, B and C at a time instance. And, coordinates Z of positions A, B and C will be the depths of pixels a, b and c.

According to the perspective projection by the optical lens of the camera, we have $x_a = f_c \frac{X_A}{Z_A}$ and $y_a = f_c \frac{Y_A}{Z_A}$. Moreover, we could express coordinates X_A and Y_A as a function of coordinate Z_A (i.e. the depth of pixel a), as follows:

$$\begin{cases} X_A = \dfrac{x_a}{f_c} Z_A \\ Y_A = \dfrac{y_a}{f_c} Z_A. \end{cases} \tag{9.55}$$

And, the application of Eq. 9.55 will yield the following expression for the coordinates at position A:

$$\begin{pmatrix} X_A \\ Y_A \\ Z_A \end{pmatrix} = \begin{pmatrix} \dfrac{x_a}{f_c} \\ \dfrac{y_a}{f_c} \\ 1 \end{pmatrix} \bullet Z_A. \tag{9.56}$$

Eq. 9.56 indicates that position A will be fully determined if and only if the depth (i.e. Z_A) of pixel a could be computed. In other words, position A only depends on one unknown, which is depth Z_A, if pixel a is given.

Now, it is clear that at a time instance, there is an unknown at every pixel in the image plane of a camera, and that this unknown at a pixel is called a *depth*. As the pixels in the image plane of a camera form a pixmap, it is natural that the unknown depths of the pixels will also form a map, which is called a *depth-map*.

Mathematically, a pixmap is a two-dimensional matrix of pixels, in which each element records the color of an incoming light at a time instance. Now, if we substitute the color at a pixel by the depth of an incoming light at a time instance, a pixmap will become a depth-map. Hence, a formal definition of depth-map could be stated as follows:

Definition 9.2. A depth-map is a two-dimensional matrix of pixels, in which each element stores a value called a depth.

For instance, in Eq. 9.56, Z_A is the depth of pixel a which is at location (x_a, y_a).

We have learnt that a color image at a time instance, denoted by I, could be described by

$$I = \{(r, c, R_{r,c}, G_{r,c}, B_{r,c}) | r \in [1, n],\ c \in [1, m]\} \tag{9.57}$$

where n is the maximum value that row r can take, and m the maximum value that column c can take.

Similarly, a depth-map at a time instance, denoted by D, can be described as follows:

$$D = \{(r, c, Z_{r,c}) \,|\, r \in [1, n],\ c \in [1, m]\} \tag{9.58}$$

where $Z_{r,c}$ is the depth at pixel location (r, c) at a time instance.

9.4.2 *Continuous Depth-map in Camera Frame*

In Eq. 9.56, depth Z_A is a coordinate with respect to the coordinate system assigned to a camera. By default, a depth in a camera's coordinate system could take any real value, as long as this value is greater than focus length f_c. In other words, depth can be modeled by a continuous variable which takes continuous values from interval $(f_c, -\infty)$.

9.4.3 *Discrete Depth-map in Camera Frame*

An image from a camera is the projection of the positions in a scene, from which lights emanate toward the camera. Furthermore, an image inside a human being's mind, or a machine's brain, is represented by a two-dimensional matrix. This means that an image represents a discrete space. However, due to the high resolution of an image, we do not feel the discrete nature of an image (or a photo).

On the other hand, a pixel in an image could only take discrete locations in the image plane of a camera. This implies that when depths are determined by a binocular vision, all the possible outcomes will not occupy the entire interval $(f_c, -\infty)$. This property suggests that we could divide interval $(f_c, -\infty)$ into a set of discrete values, which are the permissible values for depth. When we divide interval $(f_c, -\infty)$ into a set of discrete values, it means that depth is modeled by a discrete variable, which takes discrete values from a predefined set.

Refer to Fig. 9.21. Assume that we divide interval $(f_c, -\infty)$ into a set of discrete values $\{z_i,\ i = 1, 2, 3, \ldots\}$. When depth Z takes discrete value z_i, the following

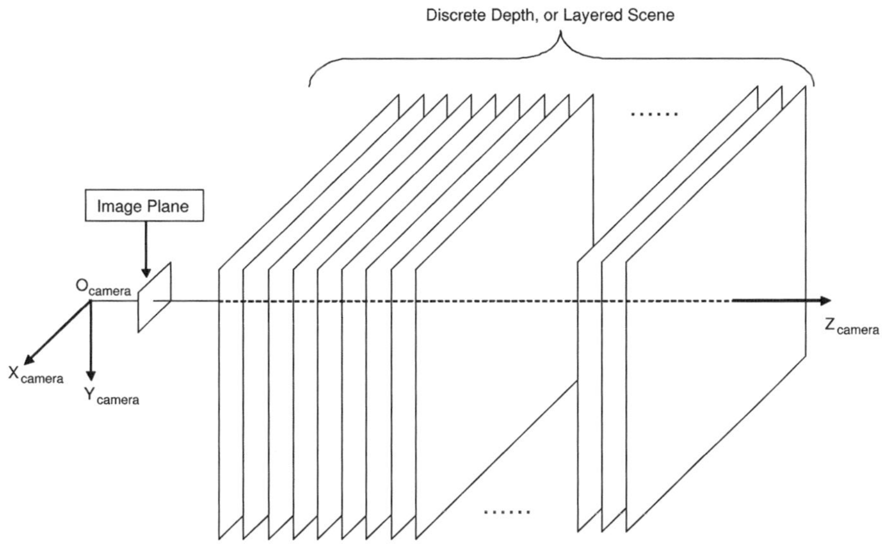

Fig. 9.21 Discrete depth, or layered scene. When depth Z takes discrete value z_i, equation $Z = z_i$ defines a plane which is perpendicular to axis Z_{camera}. When all such planes corresponding to $i = 1, 2, 3, \ldots$ are put together, the result will be a layered scene.

equation

$$Z = z_i$$

will define a plane, which is perpendicular to axis Z. We could put all the planes corresponding to $i = 1, 2, 3, \ldots$, together. In this way, the result will be a set of parallel planes in a camera's coordinate system. And, these parallel planes define a so-called *layered scene.* In other words, a discrete depth-map represents a layered scene.

Refer to Eq. 9.56. When image coordinates (x_a, y_a) are given, coordinates (X_A, Y_A, Z_A) could not be uniquely determined because there is a lack of one additional constraint imposed on Z_A. In fact, depth Z_A in Eq. 9.56 could take any value within interval $(f_c, -\infty)$. However, if we know the layer which contains coordinates (X_A, Y_A, Z_A), then depth Z_A will be constrained to take a specific value imposed by the layer. In other words, a layer in a scene imposes an useful constraint on depth.

In theory, the concept of layered scene could be extended to the case of parallel planes which are not perpendicular to axis Z. In this case, a general plane in a camera's coordinate system could be expressed as

$$Z = \alpha X + \beta Y + \gamma \tag{9.59}$$

where α, β and γ are coefficients, which are known for a given plane.

Now, if depth Z_A is constrained to a plane described by Eq. 9.59, we have

$$Z_A = \alpha X_A + \beta Y_A + \gamma. \tag{9.60}$$

And, the substitution of Eq. 9.60 into Eq. 9.56 will yield

$$\begin{pmatrix} X_A \\ Y_A \\ Z_A \end{pmatrix} = \begin{pmatrix} \frac{x_a}{f_c} \\ \frac{y_a}{f_c} \\ 1 \end{pmatrix} \bullet (\alpha X_A + \beta Y_A + \gamma). \tag{9.61}$$

In Eq. 9.61, the first two rows give two independent linear equations, which are sufficient enough to determine the unknown X_A and Y_A. Once we know X_A and Y_A, depth Z_A could be determined by Eq. 9.60. Hence, a layered scene could be modeled by a set of parallel planes which are not necessarily to be perpendicular to axis Z in a camera's coordinate system. This conclusion also inspires us to represent a discrete scene in a viewer's frame.

9.4.4 *Discrete Scene in Viewer Frame*

In the above discussion, depth is referenced to a camera's coordinate system. More precisely, depth actually refers to coordinate Z in a camera's coordinate system. However, according to the fact that the final output from human vision is expressed in a common coordinate system which is called a viewer's frame, the representation of depth perceived by machine vision should be referenced to a viewer's frame, instead of a camera's frame.

9.4.4.1 *Layered Scene in Viewer Frame*

In geometry, we know that the transformation between two coordinate systems will preserve the geometrical properties such as linearity and parallelism. In other words, two parallel planes in one coordinate system will remain parallel in another coordinate system. Therefore, in a viewer frame, we can define a set of parallel planes which are perpendicular to axis Z of the viewer frame. And, these planes will remain parallel in either the left-camera's coordinate system or the right-camera's coordinate system.

Now, assume that we divide internal $(0, -\infty)$ along axis Z of a viewer frame into a set of discrete values $\{z_i, \ i = 1, 2, 3, \ldots\}$. Then, equation ${}^vZ = z_i$ describes a plane which is perpendicular to axis Z of the viewer frame (NOTE: The superscript v indicates that coordinate Z is referenced to the viewer frame).

In a machine's head, the positions and orientations of the left-camera and the right-camera could be represented by translation vectors and rotation matrices, which are referenced to a viewer frame. Conversely, a viewer frame's position and orientation could be described by a translation vector and a robot matrix, which are referenced to either the left-camera's coordinate system, or the right-camera's coordinate system.

For instance, let $({}^{l}R_v,\ {}^{l}T_v)$ denote the orientation and position of a viewer frame with respect to the left-camera's coordinate system. If $({}^{v}X,\ {}^{v}Y,\ z_i)$ are the coordinates of a point on the plane ${}^{v}Z = z_i$ in the viewer frame, the coordinates of the same point in the left-camera's coordinate system will be

$$\begin{pmatrix} {}^{l}X \\ {}^{l}Y \\ {}^{l}Z \end{pmatrix} = {}^{l}T_v + {}^{l}R_v \bullet \begin{pmatrix} {}^{v}X \\ {}^{v}Y \\ z_i \end{pmatrix}. \tag{9.62}$$

In Eq. 9.62, the first two rows give two linear equations which allow us to solve for $({}^{v}X,\ {}^{v}Y)$. And, the substitution of $({}^{v}X,\ {}^{v}Y)$ into the third row in Eq. 9.62 will yield the equation which describes the plane ${}^{v}Z = z_i$ in the left camera's coordinate system.

Similarly, in the right camera's coordinate system, we have

$$\begin{pmatrix} {}^{r}X \\ {}^{r}Y \\ {}^{r}Z \end{pmatrix} = {}^{r}T_v + {}^{r}R_v \bullet \begin{pmatrix} {}^{v}X \\ {}^{v}Y \\ z_i \end{pmatrix} \tag{9.63}$$

where $({}^{r}X,\ {}^{r}Y,\ {}^{r}Z)$ are the coordinates of a point, in the right camera's coordinate system, and $({}^{r}R_v,\ {}^{r}T_v)$ the orientation and position of the view frame with respect to the right camera's coordinate system. By eliminating $({}^{v}X,\ {}^{v}Y)$ from Eq. 9.63, we could obtain the equation which describes the plane ${}^{v}Z = z_i$ in the right camera's coordinate system.

It is clear now that a scene could also be divided into a set of layers, which are parallel in a viewer frame. And, each layer implicitly indicates a possible depth of a scene. Here, we call such a layered representation a *layered scene* in a viewer frame.

9.4.4.2 *Discrete Scene in Viewer Frame*

If a machine perceives a scene with its binocular vision, the depths computed from the images of the binocular vision will be located on the parallel planes in a viewer frame, which divide the scene into a set of parallel layers.

On the other hand, we know that an image is a two-dimensional matrix of pixels, each of which records the color of an incoming light at a time instance. Therefore, coordinates (x, y) of a pixel in the image plane of a camera could only take discrete values. This implies that when we determine the positions from which lights emanate toward the cameras of a robot's binocular vision, these computed positions in a viewer frame should be discrete not only among the layers, but also within a layer. As a result, a scene perceived by a machine in a viewer frame could

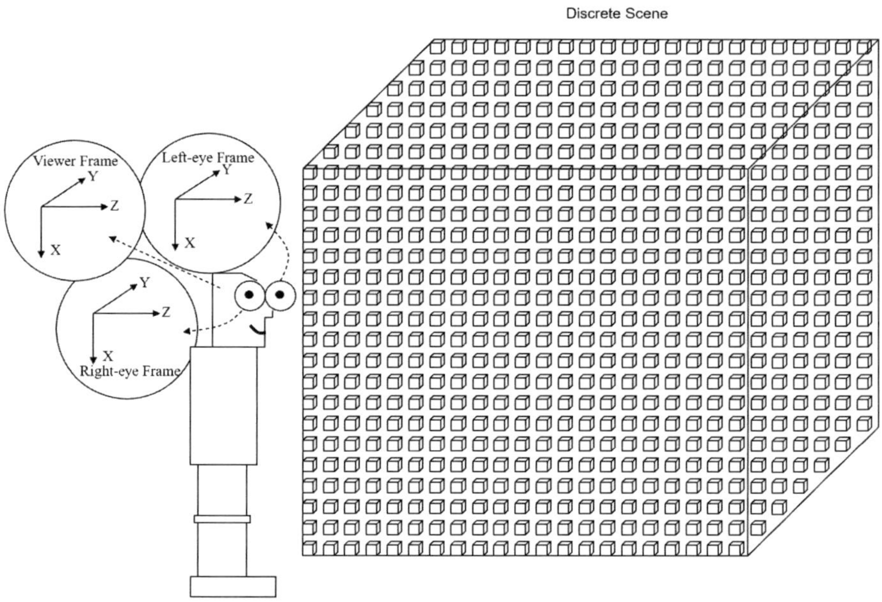

Fig. 9.22 A perceived scene by a machine is represented by a set of discrete positions, which are constrained by a set of parallel layers, in a viewer frame.

be represented by a set of discrete positions, which are constrained by a set of parallel layers, as illustrated in Fig. 9.22.

Here, a set of discrete positions, which represent a perceived scene by a robot's binocular vision, is called a *discrete scene* in a viewer frame.

9.4.5 *Generalized Depth-map in Viewer Frame*

By default, an image, or a pixmap, refers to a digital image, which is a two-dimensional matrix of pixels. Most importantly, a pixel in a pixmap records a set of three measured values, which are

- the output from the photo-receptor, which has a peak response to red
- the output from the photo-receptor, which has a peak response to green, and
- the output from the photo-receptor, which has a peak response to blue.

And, in literature, such a triplet is called *tristimulus values.*

Now, if we substitute the tristimulus values stored at a pixel by a value called *depth*, a pixmap becomes what we call a *depth-map*, which has been discussed earlier. Mathematically, a depth-map is a two-dimensional matrix of pixels, in which each element stores a coordinate at axis Z of a camera's coordinate system. However, what is actually more important is the depth-map in a viewer frame. This is because the perceptual outputs from both the left-camera and the right-camera must be fused (or combined) together with respect to a viewer frame. Therefore, it is necessary to extend the concept of depth-map, and to make it applicable to a viewer frame.

A simple way of extending the concept of depth-map is to substitute the tristimulus values at a pixel in a pixmap with the coordinates of a discrete position from which a light emanates toward the cameras of a robot's binocular vision. In this way, a pixmap becomes a two-dimensional matrix of pixels, in which each element records the coordinates of a position in a viewer frame. Here, we call such a two-dimensional matrix of pixels a *generalized depth-map*. And, a formal definition of generalized depth-map could be stated as follows:

Definition 9.3. A generalized depth-map is a two-dimensional matrix of pixels, in which each element stores the coordinates of a position in a viewer frame.

In the above definition, we could also call the coordinates of a position in a viewer frame a *generalized depth*. If we denote the generalized depth at pixel location (r, c) by $({}^vX_{r,c},\ {}^vY_{r,c},\ {}^vZ_{r,c})$, a generalized depth-map, denoted by G, could be described as follows:

$$G = \{(r,\ c,\ {}^vX_{r,c},\ {}^vY_{r,c},\ {}^vZ_{r,c})\,|\, r \in [1, n],\ c \in [1, m]\}. \tag{9.64}$$

9.4.6 *Concept of Look-through Scene*

Physically, a generalized depth-map depends on the image plane of a camera. For instance, the resolution of a generalized depth-map is the same as the resolution of the images from a camera. Moreover, the content of a generalized depth-map depends on the pose of a camera. For instance, if a machine's camera changes its position and orientation, the content of a generalized depth-map will also change accordingly.

Therefore, if we consider that the image plane of a machine's camera is a window through which a machine looks at an outside scene, a generalized depth-map actually represents a so-called *look-through scene* seen by a machine's camera (or eye) at a time instance. So, by definition, the look-through scene from image I at a time instance, denoted by S, could be described by

$$S = \{(r, c, R_{r,c},\ G_{r,c},\ G_{r,c},\ {}^vX_{r,c},\ {}^vY_{r,c},\ {}^vZ_{r,c})|r \in [1, n],\ c \in [1, m]\} \tag{9.65}$$

where $(R_{r,c},\ G_{r,c},\ G_{r,c})$ is the color at pixel location (r, c) at a time instance.

9.4.7 *Concept of Depth Redundancy*

If a machine has a binocular vision, there must be two cameras inside the machine's vision system. Therefore, a binocular vision could output two generalized depth-maps at a time instance, namely: (a) the look-through scene seen by the left camera, and (b) the look-through scene seen by the right camera. As these generalized depth-maps are referenced to a same viewer frame, there is a redundancy with

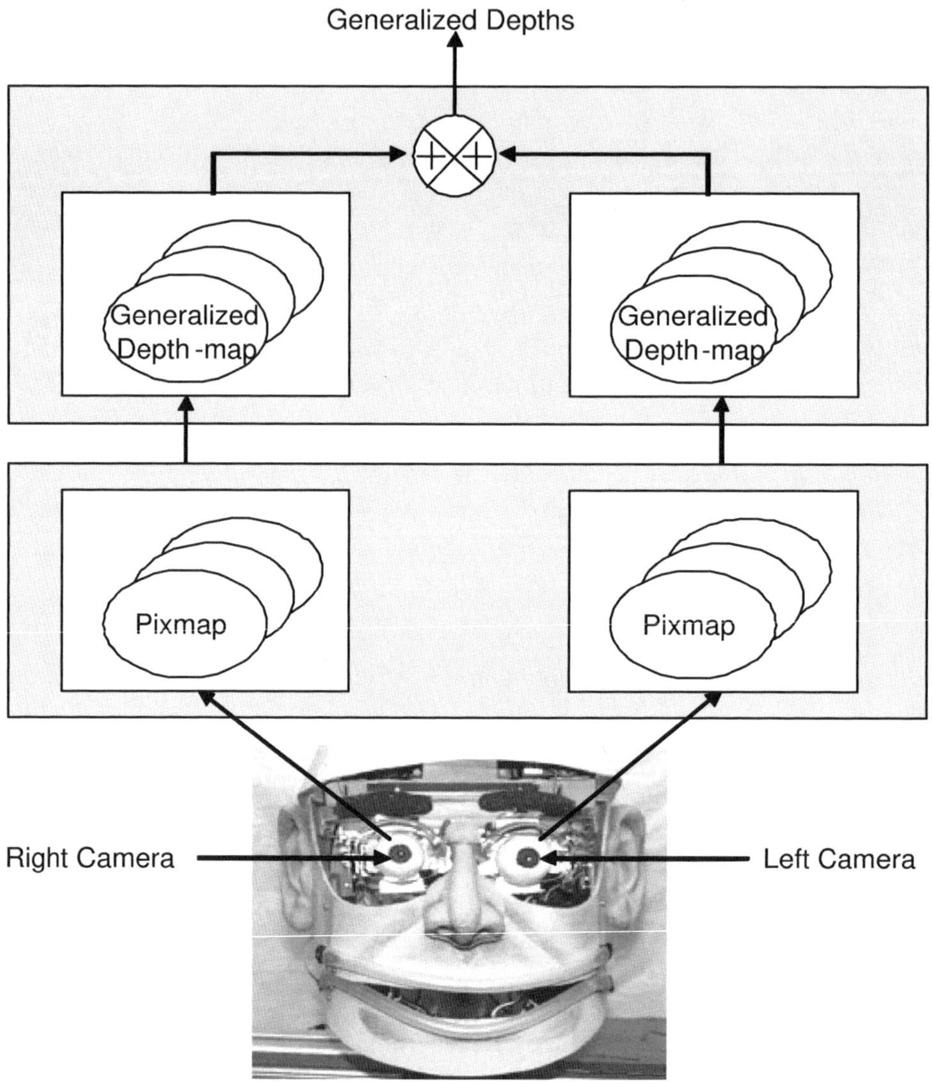

Fig. 9.23 Depth redundancy from a binocular vision, in which there are two series of generalized depth-maps. The final output is the sum of these two series. And, the loss of one series for whatever reason will not cause the binocular vision to fail.

regard to the generalized depths. Here, we call this property the *depth redundancy*, which is illustrated by Fig. 9.23.

In Fig. 9.23, we can see that the images from a machine's left camera are sufficient enough to enable the machine's mind to determine a series of generalized depth-maps. Also, the machine's mind should be able to determine a series of generalized depth-maps from the images of the right camera. Now, let G_1 denote the generalized depth-map seen by the left camera at a time instance, and G_2 the generalized depth-map seen by the right camera at the same time instance. It is clear that the final output, denoted by G, should be the sum of G_1 and G_2. In other words, we have

$$G = \{G_1, G_2\} \tag{9.66}$$

where

$$G_1 = \{(r1, c1, {}^vX_{r1,c1}, {}^vY_{r1,c1}, {}^vZ_{r1,c1}) | r1 \in [1, n],\ c1 \in [1, m]\} \tag{9.67}$$

and

$$G_2 = \{(r2, c2, {}^vX_{r2,c2}, {}^vY_{r2,c2}, {}^vZ_{r2,c2}) | r2 \in [1, n],\ c2 \in [1, m]\}. \tag{9.68}$$

(NOTE: Here, we assume that the images from both the cameras have the same resolution)

With such a redundancy described by Eq. 9.66, it is easy for us to explain why the sudden close, or obstruction, of one eye (or camera) will not cause human (or eventually robot) vision to fail.

Also, it should be clear to us now that during the dynamic process of visual perception, each camera inside a machine's vision system could output a series of look-through scenes. In addition, several look-through scenes seen by a camera could be fused, or combined, together in order to form a more comprehensive scene which surrounds a machine.

9.4.8 *Conceptual Meaning of Depth*

Under the context of cognitive vision, we consider that the depth of a scene is a concept-physical entity. This is because the depth of a point, or an object, in a scene could be conceptually described by conceptual labels such as far, near, in-front, behind, left, right, above, and below, etc.

As the depth of a scene must be referenced to a coordinate system, the conceptual meanings of depth should be referenced to a viewer frame. In this way, the conceptual meanings of depth will be viewer-centric, but not scene-centric.

9.4.9 *Data Structure of Organized Meanings*

When we open our eyes, we instantly feel that we see a three-dimensional scene. At the same time, we do have the sensation that we are within a three-dimensional scene, despite the fact that our mental processes underlying the visual perception of depth actually occur inside our brain.

Certainly, human mind is capable of creating a generalized depth-map from the images of human vision. Most importantly, such a generalized depth-map is organized and stored inside the memory of our mind. Hence, when we design a machine's vision system, we must also consider the organized storage of depths perceived by a machine itself.

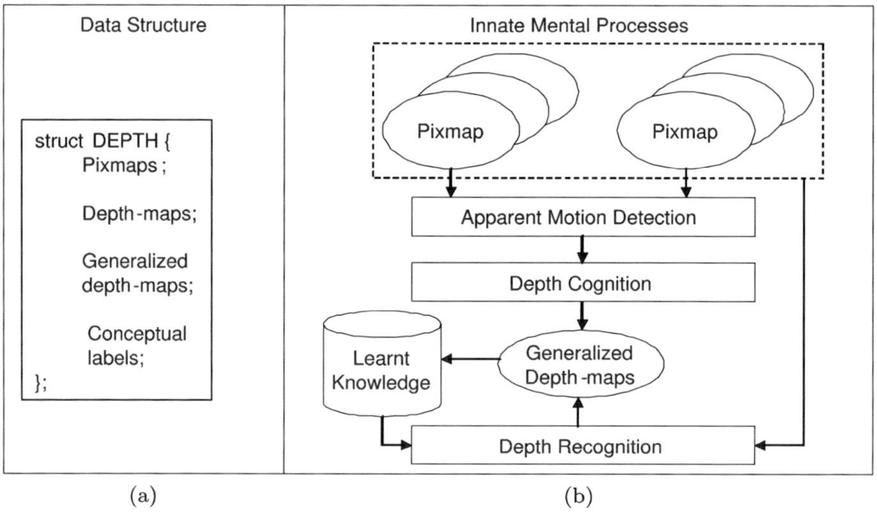

Fig. 9.24 Visual perception of depth must have an appropriately organized memory, which should properly store: pixmaps from the image planes of the cameras, depth-maps in the camera frames, generalized depth-maps in a viewer frame, and conceptual labels which describe the possible conceptual meanings of depth. With such an organized memory in place, relevant innate mental processes could act and interact in order to produce the outcome of generalized depth-maps.

Refer to Fig. 9.24. A machine must have a well-designed memory, which should properly store data such as:

- Pixmaps which are images taken by the two cameras in a machine's vision system.
- Depth-maps which are the outcomes of depths in the camera frames, respectively.
- Generalized depth-maps which are the outcomes of the generalized depths with respect to a viewer frame.
- Conceptual labels which describe the possible conceptual meanings of depth in a human language.

With such an organized memory in place, the next issue is to design the relevant mental processes, which could act and interact in order to achieve the final outcome of generalized depth-map with respect to a viewer frame. For instance, the apparent motions detected in the image plane of a camera are necessary for a machine to compute the depth, or generalized depth, of a scene. And, when a machine has gained enough visual experiences in perceiving depths, it is possible for a robot to recognize the generalized depth-map of a familiar scene from static images, or photos. Under the context of cognitive vision, this visual task is known as depth recognition.

In the next section, we will study the topic of depth cognition first. And, it is expected that the designed mental processes should be simple and generic enough for them to be implemented as innate algorithms, or solutions, inside a machine's mind.

9.5 Cognition of Depth

The visual task of depth cognition is to determine depths from apparent motions, which are detected from live images coming from scenes. As shown in Fig. 9.25, a machine with a well-designed body and head could achieve depth cognition in two complementary ways, namely: a) visual computing and b) visual learning. The former aims at directly determining depths from apparent motions. And, the latter is to indirectly determine depths through the autonomous interaction in a scene. The autonomous interaction in a scene, which may involve the use of a machine's body and hands, provides an advantageous way for a machine to autonomously validate the relationship between depths and apparent motions.

9.5.1 *Depth from Visual Computing*

Refer to Fig. 9.25. A machine with a well-designed head, which houses a binocular vision, should be able to determine the depth of a scene from apparent motions. In principle, the apparent motions could be detected from any pair of live images. And, a pair of live images coming from scsnes could be obtained in two different ways such as

- Two consecutive images are taken by a moving camera at two different poses in a scene.
- Two images are respectively taken at the same time by a pair of cameras, which are placed at two different poses in a scene.

Without loss of generality, we consider here the case, in which two live images coming from a scene are captured at two different poses, which are named as pose

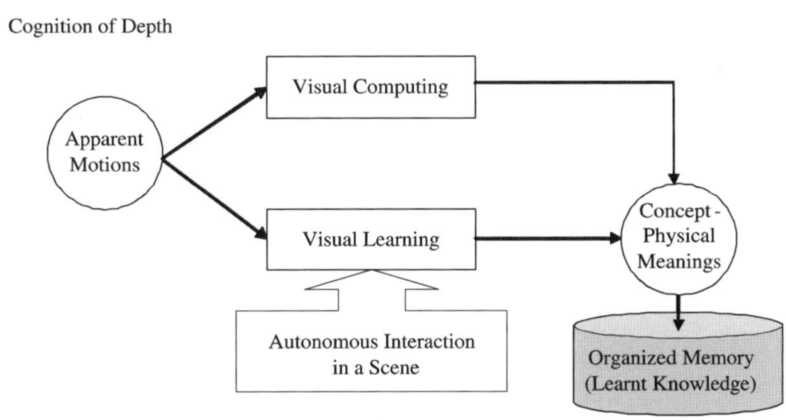

Fig. 9.25 Depth cognition consists of visual computing and visual learning. The former is to directly compute depths from apparent motions. The latter is to learn depths through the autonomous interaction in a scene, which may involve the use of machine's body and hands to measure depths.

1 and pose 2. In practice, the configuration of pose 1 and pose 2 may correspond to one of the following cases:

- Case 1:
  Pose 1 refers to the left-camera's pose at time instance t_1. And, pose 2 refers to the right-camera's pose at the same time instance (i.e. t_1).
- Case 2:
  Pose 1 refers to the left-camera's pose at time instance t_2. And, pose 2 refers to the right-camera's pose at the same time instance (i.e. t_2).
- Case 3:
  Pose 1 refers to the left-camera's pose at time instance t_1. And, pose 2 refers to the left-camera's pose at time instance t_2.
- Case 4:
  Pose 1 refers to the right-camera's pose at time instance t_1. And, pose 2 refers to the right-camera's pose at time instance t_2.

In the following sections, we present solutions which determine either the depth of a point in a scene with respect to a camera-frame,.or the generalized depth of a point in a scene with respect to a viewer frame.

9.5.1.1 *Depth Computing in Camera Frame*

We assume that two images are taken at two different poses as shown in Fig. 9.26. Also, we assume that lights emanated from position A in the scene are captured at pixel location B in pose 1 and pixel location C in pose 2, respectively. Then, the displacement from pixel location B to pixel location C in the image plane is called an apparent motion, which is assumed to be detected in advance.

Now, in the camera frame at pose 1, let (X, Y, Z) denote the coordinates of position A. Then, the application of Eq. 9.56 will yield

$$\begin{pmatrix} X \\ Y \\ Z \end{pmatrix} = \begin{pmatrix} \frac{x_1}{f_1} \\ \frac{y_1}{f_1} \\ 1 \end{pmatrix} Z \tag{9.69}$$

where f_1 is the focal length of the camera at pose 1, and (x_1, y_1) the image coordinates at pixel location B in image 1.

It is worth noting that image coordinates (x_1, y_1) can be easily computed if index coordinates (r_1, c_1) at the same pixel location are known (see Eqs. 9.1, 9.2, 9.3, and 9.4). However, if a camera's parameters are not obtained by design, then we have to use the calibration model of a camera. For instance, from Eq. 9.1, we

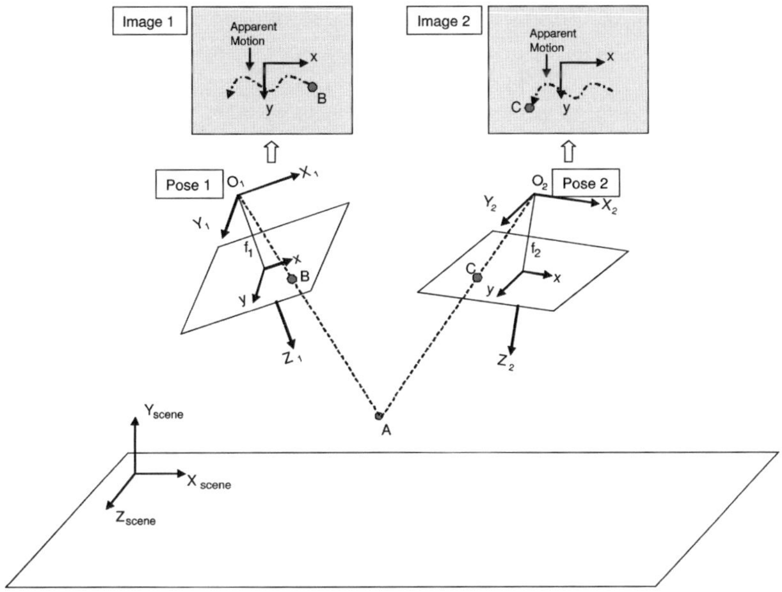

Fig. 9.26 Visual computing of depth with respect to a camera frame. Two images are taken at two different poses: pose 1 and pose 2. And, the apparent motion at pixel location B is given as an input.

can easily derive the following relationship with regard to the camera at pose 1:

$$\begin{cases} c_1 = c_{0,1} + \dfrac{x_1}{w} = c_{0,1} + \dfrac{f_1}{w} \bullet \dfrac{x_1}{f_1} \\ r_1 = r_{0,1} + \dfrac{y_1}{h} = r_{0,1} + \dfrac{f_1}{h} \bullet \dfrac{y_1}{f_1}. \end{cases} \tag{9.70}$$

If we define $f_{w,1} = f_1/w$ and $f_{h,1} = f_1/h$, Eq. 9.70 could be expressed as

$$\begin{cases} c_1 = c_{0,1} + f_{w,1} \bullet \dfrac{x_1}{f_1} \\ r_1 = r_{0,1} + f_{h,1} \bullet \dfrac{y_1}{f_1}. \end{cases} \tag{9.71}$$

By definition, parameters $(r_{0,1},\ c_{0,1},\ f_{h,1},\ f_{w,1})$ are called the *intrinsic parameters* of the camera at pose 1.

In general, parameters $(r_0,\ c_0,\ f_h,\ f_w)$ are called the intrinsic parameters of a camera (NOTE: $f_w = f/w$ and $f_h = f/h$).

Now, if we define $x_1^* = \frac{x_1}{f_1}$ and $y_1^* = \frac{y_1}{f_1}$, Eq. 9.69 will become

$$\begin{pmatrix} X \\ Y \\ Z \end{pmatrix} = \begin{pmatrix} x_1^* \\ y_1^* \\ 1 \end{pmatrix} Z. \tag{9.72}$$

And, coordinates $(x_1^*,\ y_1^*)$ could be computed from

$$\begin{cases} x_1^* = (c_1 - c_{0,1})/f_{w,1} \\ y_1^* = (r_1 - r_{0,1})/f_{h,1} \end{cases} \tag{9.73}$$

which can be easily derived from Eq. 9.71.

Similarly, let (X', Y', Z') denote the coordinates of position A with respect to the camera-frame at pose 2. By applying Eq. 9.56, we have

$$\begin{pmatrix} X' \\ Y' \\ Z' \end{pmatrix} = \begin{pmatrix} \frac{x_2}{f_2} \\ \frac{y_2}{f_2} \\ 1 \end{pmatrix} Z' \tag{9.74}$$

where f_2 is the focal length of the camera at pose 2, and (x_2, y_2) the image coordinates at pixel location C in image 2.

If the intrinsic parameters of the camera at pose 2 are $(r_{0,2},\ c_{0,2},\ f_{h,2},\ f_{w,2})$, Eq. 9.74 could be expressed as

$$\begin{pmatrix} X' \\ Y' \\ Z' \end{pmatrix} = \begin{pmatrix} x_2^* \\ y_2^* \\ 1 \end{pmatrix} Z' \tag{9.75}$$

with

$$\begin{cases} x_2^* = (c_2 - c_{0,2})/f_{w,2} \\ y_2^* = (r_2 - r_{0,2})/f_{h,2} \end{cases} \tag{9.76}$$

where (r_2, c_2) are the index coordinates at the same pixel location (x_2, y_2).

Now, let us examine how Eq. 9.75 is related to Eq. 9.72. In a machine's binocular vision, it is reasonable to assume that the relative pose between the left camera's coordinate system and the right camera's coordinates is known in advance. This can be the result of precise design and control of a binocular vision, or the result of an elaborated off-line calibration.

On the other hand, within a short time interval denoted by $[t_1, t_2]$, the relative pose between a camera's coordinate system at time instance t_1 and the same camera's coordinate system at time instance t_2 could also be assumed to be known in advance. This is because the well-designed body and head of a machine could undergo precise motions. Therefore, if we denote

$$T = \begin{pmatrix} t_x \\ t_y \\ t_z \end{pmatrix}$$

the position of the origin of the camera-frame at pose 2 with respect to the camera-frame at pose 1, and

$$R = \begin{pmatrix} r_{11} & r_{12} & r_{13} \\ r_{21} & r_{22} & r_{23} \\ r_{31} & r_{32} & r_{33} \end{pmatrix}$$

the orientation of the camera-frame at pose 2 with respect to the camera-frame at pose 1, then we have

$$\begin{pmatrix} X \\ Y \\ Z \end{pmatrix} = T + R \bullet \begin{pmatrix} X' \\ Y' \\ Z' \end{pmatrix}. \tag{9.77}$$

And, the application of Eq. 9.72 and Eq. 9.75 into Eq. 9.77 will yield

$$\begin{pmatrix} x_1^* \\ y_1^* \\ 1 \end{pmatrix} Z = T + R \bullet \begin{pmatrix} x_2^* \\ y_2^* \\ 1 \end{pmatrix} Z'. \tag{9.78}$$

Now, if we define a new vector, denoted by (r_x, r_y, r_z), as follows

$$\begin{pmatrix} r_x \\ r_y \\ r_z \end{pmatrix} = R \bullet \begin{pmatrix} x_2^* \\ y_2^* \\ 1 \end{pmatrix} \tag{9.79}$$

Eq. 9.78 will become

$$\begin{pmatrix} x_1^* \\ y_1^* \\ 1 \end{pmatrix} Z = \begin{pmatrix} t_x \\ t_y \\ t_z \end{pmatrix} + \begin{pmatrix} r_x \\ r_y \\ r_z \end{pmatrix} Z'. \tag{9.80}$$

In Eq. 9.80, the third row allows us to derive the following equality

$$Z' = \frac{1}{r_z}(Z - t_z). \tag{9.81}$$

And, the sum of the first and second rows in Eq. 9.80 will be

$$(x_1^* + y_1^*)Z = (t_x + t_y) + (r_x + r_y)Z'. \tag{9.82}$$

Finally, it is easy to verify that the substitution of Eq. 9.81 into Eq. 9.82 will result in the following solution for depth Z

$$Z = \frac{(t_x + t_y)r_z - (r_x + r_y)t_z}{(x_1^* + y_1^*)r_z - (r_x + r_y)}. \tag{9.83}$$

And, coordinates X and Y will be

$$\begin{cases} X = x_1^* \bullet Z \\ Y = y_1^* \bullet Z. \end{cases} \tag{9.84}$$

Example 9.7. There is a binocular vision system, in which the intrinsic parameters of both the left-camera and the right-camera are

$$(r_{0,l}, c_{0,l}, f_{h,l}, f_{w,l}) = (272.1411, 255.7699, 944.2594, 662.1295)$$

and

$$(r_{0,r}, c_{0,r}, f_{h,r}, f_{w,r}) = (273.1110, 244.7648, 940.5618, 659.1966)$$

respectively. And, the relative pose of the right-camera with respect to the left-camera is described by the rotation matrix

$$R = \begin{pmatrix} 0.9989 & 0.0068 & -0.0455 \\ -0.0066 & 1.0000 & 0.0016 \\ 0.0455 & -0.0012 & 0.9989 \end{pmatrix}$$

and the translation vector

$$T = (61.5911 \quad -0.2889 \ 2.9516)^t$$

in millimeters. Now, the binocular vision system is placed inside an office. And, a pair of stereo images has been taken. These two images are shown in Fig. 9.27a and Fig. 9.27b, respectively.

Then, we manually select a pixel in the left image, and let the program to automatically find the match along the epipolar line in the right image, as shown in Fig. 9.27b. In this case, the index coordinates of the selected pixel in the left image are

$$(c_1, r_1) = (326, 284).$$

And, the index coordinates of the matched pixel in the right image are

$$(c_2, r_2) = (334, 284).$$

Now, by applying Eq. 9.73 and Eq. 9.76, we will obtain

$$(x_1^*, y_1^*) = (0.1061, 0.0126)$$

and

$$(x_2^*, y_2^*) = (0.1354, 0.0116).$$

(a) Left Image and Selected Point

(b) Right Image and Matched Point

(c) Conglomerated Appearance (Left)

(d) Conglomerated Appearance (Right)

Fig. 9.27 Stereo images and correspondence from a scene of an office: (a) the left image and the selected pixel, (b) the right image and the matched pixel, (c) the conglomerated appearance centered at the selected pixel, and (d) the conglomerated appearance centered at the matched pixel. In this example, the conglomerated appearances are computed at scale 4 of color index. And, in (b), the line across the image is the epipolar line, which contains the potential matches of the selected pixel in (a). (Stereo images and calibration data reproduced by courtesy of INRIA).

Subsequently, the application of Eq. 9.79 will produce the following intermediate result

$$\begin{pmatrix} r_x \\ r_y \\ r_z \end{pmatrix} = \begin{pmatrix} 0.0898 \\ 0.0123 \\ 1.0051 \end{pmatrix}.$$

Finally, by applying Eq. 9.83, the computed depth will be

$$Z = 3571.9\ (mm).$$

And, from Eq. 9.84, coordinates (X, Y) will be

$$(X, Y) = (378.9, 44.9)\ (mm)$$

<><><><><>

9.5.1.2 *Generalized Depth in Viewer Frame*

As we have mentioned earlier, the three-dimensional representation of a scene perceived by a human being is referenced to a viewer's frame, instead of a particular eye's frame. Most importantly, it is advantageous to have such a viewer-centric representation, because the failure of one eye will not cause the whole human vision to fail altogether. This property of visual robustness to the partial failure of one eye should inspire us to adopt a similar viewer-centric representation for the geometry of a scene perceived by a machine.

However, in order to obtain such a viewer-centric representation of a scene's geometry, it is necessary to first be able to perceive generalized depths in a viewer frame.

Refer to Fig. 9.28. A viewer frame is placed in a scene, and this viewer frame should be attached to a machine's head. In the viewer frame, two poses are defined, which are named as pose 1 and pose 2. In the scene, we assume that lights emanate from position A, and that a pair of images are taken at pose 1 and pose 2, respectively. In image 1, lights from position A are focused on pixel location B. And, in image 2, lights from position A are focused on pixel location C. If the apparent motion at pixel location B is given as input, then the question will be: What should be the coordinates of position A with respect to the viewer frame?

Refer to Fig. 9.28 again. Let $({}^vX,\ {}^vY,\ {}^vZ,\ 1)$ denote the homogeneous coordinates of position A with respect to the viewer frame. If ${}^1M_v$ describes the relative pose of the viewer frame with respect to pose 1, the homogeneous coordinates of position A with respect to pose 1, denoted by $({}^1X,\ {}^1Y,\ {}^1Z,\ 1)$, will be

$$\begin{pmatrix} {}^1X \\ {}^1Y \\ {}^1Z \\ 1 \end{pmatrix} = {}^1M_v \bullet \begin{pmatrix} {}^vX \\ {}^vY \\ {}^vZ \\ 1 \end{pmatrix}. \tag{9.85}$$

In the image plane of the camera placed at pose 1, the image of position A will be focused on the pixel at location B. If (x_1, y_1) denotes the image coordinates at

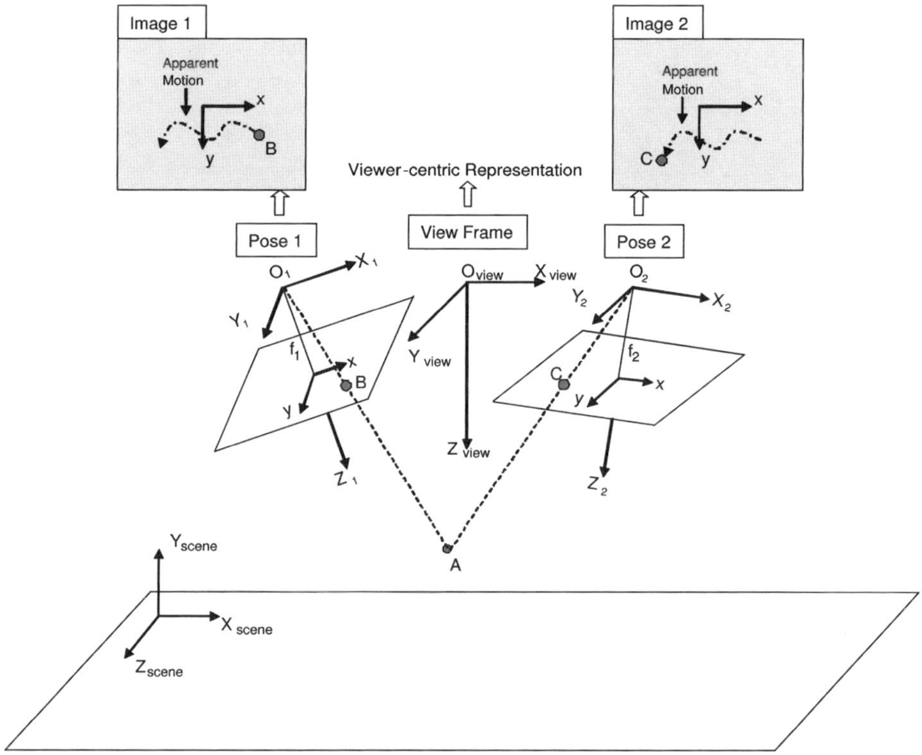

Fig. 9.28 Visual computing of generalized depth in a viewer frame. Poses 1 and 2 are defined with respect to a viewer frame. And, a pair of images are taken at pose 1 and pose 2, respectively. Within the pair of images, the apparent motion at pixel location B is given as an input.

pixel location B, we have

$$\begin{cases} \dfrac{x_1}{f_1} = \dfrac{{}^1X}{{}^1Z} \\ \dfrac{y_1}{f_1} = \dfrac{{}^1Y}{{}^1Z} \end{cases} \tag{9.86}$$

where f_1 is the focal length of the camera placed at pose 1. If we define

$$\begin{cases} x_1^* = \dfrac{x_1}{f_1} \\ y_1^* = \dfrac{y_1}{f_1} \end{cases} \tag{9.87}$$

Eq. 9.86 will become

$$\begin{cases} x_1^* = \dfrac{{}^1X}{{}^1Z} \\ y_1^* = \dfrac{{}^1Y}{{}^1Z} \end{cases} \tag{9.88}$$

Moreover, if we define $k_1 = {}^1Z$, then Eq. 9.88 could be expressed as

$$\begin{cases} k_1 \bullet x_1^* = {}^1X \\ k_1 \bullet y_1^* = {}^1Y \\ k_1 = {}^1Z. \end{cases} \tag{9.89}$$

or in a matrix form as follows:

$$k_1 \begin{pmatrix} x_1^* \\ y_1^* \\ 1 \end{pmatrix} = \begin{pmatrix} 1 & 0 & 0 & 0 \\ 0 & 1 & 0 & 0 \\ 0 & 0 & 1 & 0 \end{pmatrix} \begin{pmatrix} {}^1X \\ {}^1Y \\ {}^1Z \\ 1 \end{pmatrix}. \tag{9.90}$$

Now, the substitution of Eq. 9.85 into Eq. 9.90 will yield

$$k_1 \begin{pmatrix} x_1^* \\ y_1^* \\ 1 \end{pmatrix} = \begin{pmatrix} 1 & 0 & 0 & 0 \\ 0 & 1 & 0 & 0 \\ 0 & 0 & 1 & 0 \end{pmatrix} {}^1M_v \begin{pmatrix} {}^vX \\ {}^vY \\ {}^vZ \\ 1 \end{pmatrix}. \tag{9.91}$$

If we define

$${}^1H_v = \begin{pmatrix} 1 & 0 & 0 & 0 \\ 0 & 1 & 0 & 0 \\ 0 & 0 & 1 & 0 \end{pmatrix} {}^1M_v, \tag{9.92}$$

Eq. 9.91 will become

$$k_1 \begin{pmatrix} x_1^* \\ y_1^* \\ 1 \end{pmatrix} = {}^1H_v \begin{pmatrix} {}^vX \\ {}^vY \\ {}^vZ \\ 1 \end{pmatrix}. \tag{9.93}$$

In fact, Eq. 9.93 describes the mapping from the homogeneous coordinates in a viewer frame to the image coordinates in the image plane of a camera placed at pose 1. If we examine Eq. 9.92 and Eq. 9.93, it is easy to notice that the mapping from homogeneous coordinates to image coordinates depends not only on the focal length of a camera, but also on the relative pose between a camera's frame and the viewer frame. Most importantly, if a camera changes its focal length or its pose, such a mapping will be different. In other words, such a mapping is not unique, if the camera is allowed to dynamically change its parameters and pose. Therefore, when we design a robot's vision system under the context of dynamic vision, it is necessary to consider the requirement of automatic update of the mapping from homogeneous coordinates to image coordinates by the machine itself.

Similarly, if we denote f_2 the focal length of the camera placed at pose 2 and ${}^2M_v$ the relative pose of the viewer frame with respect to pose 2, then the mapping from the homogeneous coordinates in the viewer frame to the image coordinates in the image plane of a camera placed at pose 2 will be

$$k_2 \begin{pmatrix} x_2^* \\ y_2^* \\ 1 \end{pmatrix} = {}^2H_v \begin{pmatrix} {}^vX \\ {}^vY \\ {}^vZ \\ 1 \end{pmatrix} \tag{9.94}$$

with

$$\begin{cases} x_2^* = \dfrac{x_2}{f_2} \\ y_2^* = \dfrac{y_2}{f_2} \end{cases} \tag{9.95}$$

and

$${}^2H_v = \begin{pmatrix} 1 & 0 & 0 & 0 \\ 0 & 1 & 0 & 0 \\ 0 & 0 & 1 & 0 \end{pmatrix} {}^2M_v. \tag{9.96}$$

In Eq. 9.94, the mapping from homogeneous coordinates to image coordinates will not be unique, if the camera placed at pose 2 is allowed to change its parameters and pose.

Now, we are ready to derive the solution for generalized depth (${}^vX$, ${}^vY$, ${}^vZ$) at position A in the viewer frame. We start with Eq. 9.93, and express matrix ${}^1H_v$ as follows:

$${}^1H_v = \begin{pmatrix} a_{11} & a_{12} & a_{13} & a_{14} \\ a_{21} & a_{22} & a_{23} & a_{24} \\ a_{31} & a_{32} & a_{33} & a_{34} \end{pmatrix}. \tag{9.97}$$

In this way, the equality in the first row of Eq. 9.93 divided by the equality in the third row of Eq. 9.93 will yield the following equation:

$$x_1^* = \frac{a_{11}\ {}^vX + a_{12}\ {}^vY + a_{13}\ {}^vZ + a_{14}}{a_{31}\ {}^vX + a_{32}\ {}^vY + a_{33}\ {}^vZ + a_{34}}. \tag{9.98}$$

Eq. 9.98 could also be written as

$$(a_{11} - a_{31}x_1^*)\ {}^vX + (a_{12} - a_{32}x_1^*)\ {}^vY + (a_{13} - a_{33}x_1^*)\ {}^vZ = (a_{34}x_1^* - a_{14}). \tag{9.99}$$

Subsequently, we divide the equality in the second row of Eq. 9.93 by the equality in the third row of Eq. 9.93. The result will be

$$y_1^* = \frac{a_{21}\ {}^vX + a_{22}\ {}^vY + a_{23}\ {}^vZ + a_{24}}{a_{31}\ {}^vX + a_{32}\ {}^vY + a_{33}\ {}^vZ + a_{34}}. \tag{9.100}$$

or

$$(a_{21} - a_{31}y_1^*)\ {}^vX + (a_{22} - a_{32}y_1^*)\ {}^vY + (a_{23} - a_{33}y_1^*)\ {}^vZ = (a_{34}y_1^* - a_{24}). \tag{9.101}$$

Similarly, if we express matrix ${}^2H_v$ as follows

$$^2H_v = \begin{pmatrix} b_{11} & b_{12} & b_{13} & b_{14} \\ b_{21} & b_{22} & b_{23} & b_{24} \\ b_{31} & b_{32} & b_{33} & b_{34} \end{pmatrix} \tag{9.102}$$

Eq. 9.94 will allow us to obtain

$$(b_{11} - b_{31}x_2^*)\ {}^vX + (b_{12} - b_{32}x_2^*)\ {}^vY + (b_{13} - b_{33}x_2^*)\ {}^vZ = (b_{34}x_2^* - b_{14}) \tag{9.103}$$

and

$$(b_{21} - b_{31}y_2^*)\ {}^vX + (b_{22} - b_{32}y_2^*)\ {}^vY + (b_{23} - b_{33}y_2^*)\ {}^vZ = (b_{34}y_2^* - b_{24}). \tag{9.104}$$

Finally, if we define

$$L = \begin{pmatrix} a_{11} - a_{31}x_1^* & a_{12} - a_{32}x_1^* & a_{13} - a_{33}x_1^* \\ a_{21} - a_{31}y_1^* & a_{22} - a_{32}y_1^* & a_{23} - a_{33}y_1^* \\ b_{11} - b_{31}x_2^* & b_{12} - b_{32}x_2^* & b_{13} - b_{33}x_2^* \\ b_{21} - b_{31}y_2^* & b_{22} - b_{32}y_2^* & b_{23} - b_{33}y_2^* \end{pmatrix} \tag{9.105}$$

and

$$N = \begin{pmatrix} a_{34}x_1^* - a_{14} \\ a_{34}y_1^* - a_{24} \\ b_{34}x_2^* - b_{14} \\ b_{34}y_2^* - b_{24} \end{pmatrix} \tag{9.106}$$

Eqs. 9.99, 9.101, 9.103 and 9.104 could be combined together into the following matrix form:

$$L \begin{pmatrix} {}^vX \\ {}^vY \\ {}^vZ \end{pmatrix} = N. \tag{9.107}$$

By a least-square estimation, the solution for generalized depth $({}^vX,\ {}^vY,\ {}^vZ)$ at position A in the view frame will be

$$\begin{pmatrix} {}^vX \\ {}^vY \\ {}^vZ \end{pmatrix} = (L^tL)^{-1}(L^tN). \tag{9.108}$$

Example 9.8. If a camera is not a product of precise design and assembly, a camera's intrinsic parameters can only be determined by a process called *camera calibration*. Similarly, if a binocular vision system is not a product of precise design and assembly, the extrinsic parameters (i.e. the position and orientation) of each of the two cameras also have to be determined by a process of camera calibration.

In principle, the input to camera calibration consists of a set of known coordinates in a viewer frame, or a scene's frame. And, the output will be a camera's intrinsic parameters as well as its extrinsic parameters. In practice, a set of known coordinates could be precisely designed on a grid, which is placed onto two orthogonal planes as shown in Fig. 9.29a, or Fig. 9.29b. In this example, a binocular vision system is positioned in front of the calibration grid. After the calibration, the intrinsic parameters of both the left-camera and the right-camera are

$$(r_{0,l},\ c_{0,l},\ f_{h,l},\ f_{w,l}) = (284.8092,\ 375.3034,\ 1142.8196,\ 1178.9570)$$

and

$$(r_{0,r},\ c_{0,r},\ f_{h,r},\ f_{w,r}) = (284.6906,\ 376.5882,\ 1146.5695,\ 1182.1732),$$

respectively. And, the relative poses of the viewer frame (or a scene's frame) with respect to both the left-camera and the right-camera are

$$^{1}M_v = \begin{pmatrix} 0.634840 & -0.007955 & -0.772603 & -307.253874 \\ -0.229037 & 0.953066 & -0.198011 & -218.242988 \\ 0.737916 & 0.302660 & 0.603222 & 1289.794819 \\ 0.0 & 0.0 & 0.0 & 1.0 \end{pmatrix}$$

and

$$^{2}M_v = \begin{pmatrix} 0.835805 & -0.001960 & -0.549023 & -238.622886 \\ -0.165249 & 0.952725 & -0.254969 & -249.173349 \\ 0.523567 & 0.303830 & 0.795968 & 1396.878782 \\ 0.0 & 0.0 & 0.0 & 1.0 \end{pmatrix}$$

respectively.

Now, a football and a box are placed in between the binocular vision and the calibration grid. And, a pair of images is captured, which is shown in Fig. 9.29a and Fig. 9.29b.

Subsequently, a pixel is selected in the left-image. And, the match in the right-image is automatically identified with the help of conglomerated appearances. In this case, the location of the selected pixel in the left-image is at

$$(c_1,\ r_1) = (302,\ 384).$$

And, the location of the match in the right-image is at

$$(c_2,\ r_2) = (308,\ 368).$$

Then, from Eq. 9.73 and Eq. 9.76, we will obtain

$$(x_1^*,\ y_1^*) = (-0.0622,\ 0.0868)$$

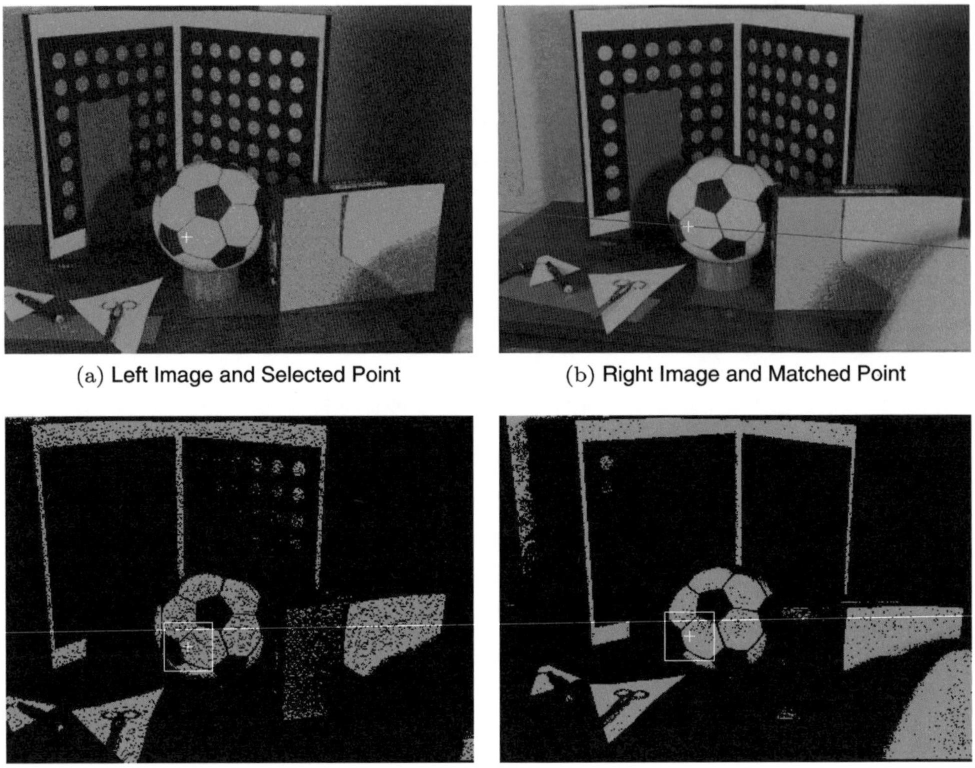

(a) Left Image and Selected Point (b) Right Image and Matched Point

(c) Conglomerated Appearance (Left) (d) Conglomerated Appearance (Right)

Fig. 9.29 Stereo images and correspondence from a scene of a calibration grid with a football in front: (a) the left image and the selected pixel, (b) the right image and the matched pixel, (c) the conglomerated appearance centered at the selected pixel, and (d) the conglomerated appearance centered at the matched pixel. In this example, the conglomerated appearances are computed at scale 4 of color index. And, in (b), the line across the image is the epipolar line, which contains the potential matches of the selected pixel in (a). (Stereo images and calibration data reproduced by courtesy of INRIA).

and

$$(x_2^*,\ y_2^*) = (-0.0580,\ 0.0727)$$

Now, from Eq. 9.92, matrix ${}^1H_v$ will be

$${}^1H_v = \begin{pmatrix} 0.634840 & -0.007955 & -0.772603 & -307.253874 \\ -0.229037 & 0.953066 & -0.198011 & -218.242988 \\ 0.737916 & 0.302660 & 0.603222 & 1289.794819 \end{pmatrix}.$$

And, from Eq. 9.96, matrix ${}^2H_v$ will be

$$
{}^2H_v = \begin{pmatrix} 0.835805 & -0.001960 & -0.549023 & -238.622886 \\ -0.165249 & 0.952725 & -0.254969 & -249.173349 \\ 0.523567 & 0.303830 & 0.795968 & 1396.878782 \end{pmatrix}
$$

With the above available data, matrix L could be computed from Eq. 9.105. And, the result will be

$$
L = \begin{pmatrix} 0.6807 & 0.0109 & -0.7351 \\ -0.2931 & 0.9268 & -0.2504 \\ 0.8662 & 0.0157 & -0.5028 \\ -0.2033 & 0.9306 & -0.3128 \end{pmatrix}.
$$

Also, from Eq. 9.106, vector N will be

$$
N = \begin{pmatrix} 227.0589 \\ 330.1904 \\ 157.5777 \\ 350.6701 \end{pmatrix}.
$$

Finally, the application of Eq. 9.108 will yield the following solution to $({}^vX,\ {}^vY,\ {}^vZ)$:

$$
({}^vX,\ {}^vY,\ {}^vZ) = (0.0323,\ 274.1493,\ -304.8090)\ (mm).
$$

<><><><><>

9.5.2 *Visual Learning of Depth*

We know that human vision is embedded onto human head which, in turn, is mounted on top of human body through human neck. Such a configuration not only enables human vision to have agile motion, but also makes human vision apt to visual learning.

Similarly, when a machine has a well-designed body and neck mechanism for the purpose of undertaking intended interactions in a scene, visual learning by a machine will become an important part of visual perception.

Generally speaking, visual learning is an effective way for a machine not only to experience the physical properties of the objects in a scene, but also to learn their conceptual meanings through the conversational dialogue with human masters (or other expert machines).

9.5.2.1 *Experiencing Depth by Hand*

Refer to Fig. 9.30. When a machine such as a humanoid robot is equipped with a binocular vision, it will have the advantage of readily coordinating the behavior of its hands and the task of visual perception.

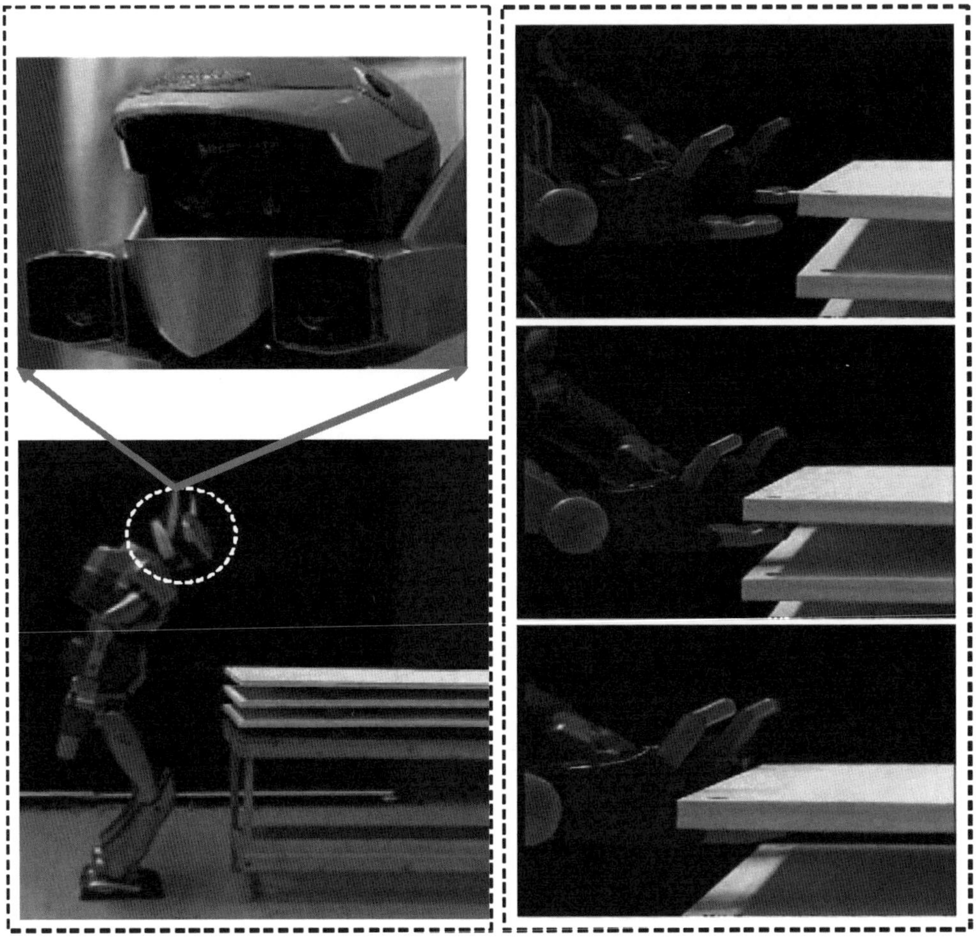

(a) A robot with binocular vision (b) Using hands to experience depths

Fig. 9.30 A machine such as a humanoid robot with well-designed body will make visual learning by hand an important skill, which is helpful not only for having the sense of depths, but also validating the relationship between depths and image coordinates. (photo courtesy of AIST).

For instance, under the context of visual learning of depth, two typical coordinations between hands and eyes (i.e. cameras) will be

- Image-guided Depth Measurement:
  In robotics, it is well-known that the forward kinematics of a robot's hand/arm mechanism permits the precise computation of the pose of a robot's fingertips. Therefore, given a pixel location in the image plane of a camera (e.g. pixel location B in Fig. 9.28), a robot could guide one of its fingertip to locate the generalized depth (e.g. position A in Fig. 9.28) in a scene, from which lights emanate toward the camera.

- Validation of Depth-to-Image Relationship:
  We know that Eq. 9.93 (or Eq. 9.94) describes the mapping from generalized depth in a viewer frame to image coordinates in the image plane of a camera. And, this mapping depends on the internal parameters of the camera. However, when a robot is able to precisely position its fingertips in the field of view of its vision, the image coordinates and generalized depths of the fingertips will serve as useful inputs to online validate the depth-to-image relationship described by Eq. 9.93 (or Eq. 9.94).

9.5.2.2 *Experiencing Depth by Body*

As we have discussed earlier, human vision do give us the impression that we are inside a scene, despite the fact that all the visual processes actually occur inside our brain. As one of the important purposes of visual perception is to enable autonomous behaviors for us to act or interact in a scene, it is advantageous to transform the three-dimensional representation of a scene perceived by human vision from a viewer frame into a body's frame. This is because a body's frame is placed at a contact point between a human being's body and a scene. In this way, a body's frame actually serves as a reference frame in a scene. As a result, the planning and control of behaviors to be executed by human body could be easily done in a body's frame. Moreover, when a person performs motions in a scene, his (or her) body actually helps him (or her) gain visual experience of depth.

In robotics, it is a well-known fact that the kinematics governing a robot's body is sufficient enough to fully determine the pose of any position on the body, with respect to a body's frame. This property shows that it is possible to involve a robot's body to undertake visual learning of depth, as long as a robot's body is well-designed in advance.

Refer to Fig. 9.31. When the kinematics of a robot's body is known precisely, it is possible to plan a trajectory or path, which will lead a position on the robot's body (e.g. a toe of one foot, or a fingertip of one hand) to reach an intended position in a scene (e.g. a particular point on a football on the ground). Therefore, under the context of visual learning of depth, a robot's body is particularly helpful for achieving image-guided depth measurement in a large space.

9.6 Recognition of Depth

There is no doubt that human vision is very skillful in deriving depth from apparent motion, in which relative motions between views are known. And, the visual perception of depth from apparent motion accomplishes the visual task of *cognition of depth.*

On the other hand, it is worth noting that visual perception of depth from apparent motion will accumulate visual experiences or outcomes, which will enable

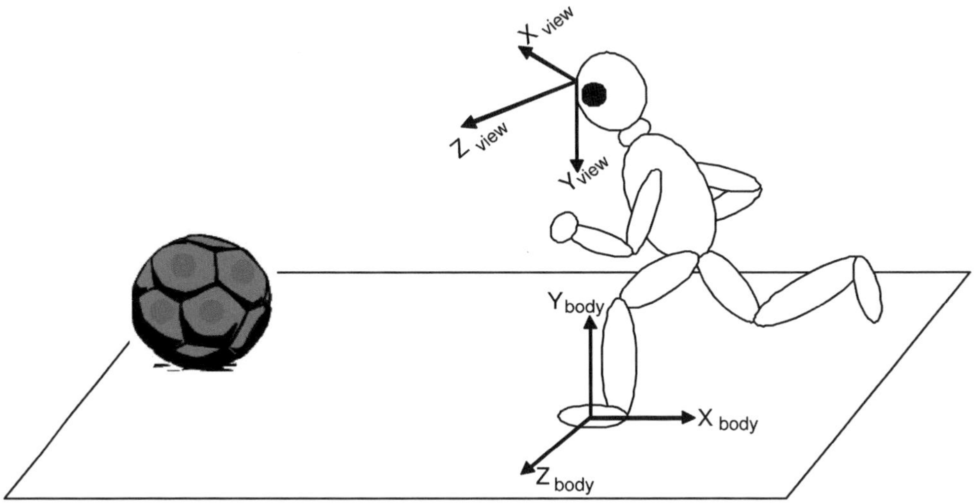

Fig. 9.31 A machine such as a humanoid robot with well-designed body could facilitate the visual experience of depth. In particular, a robot's body could help achieve image-guided depth measurement in a large space.

human vision to achieve another astonishing ability, which is the visual perception of depth from images or photos of monocular views of unknown poses. In other words, human vision is capable of doing photo interpretation or image understanding for the purpose of acquiring three-dimensional impressions from two-dimensional images (or photos), in which relative motions between views are unknown. For the sake of consistency, we name this mental ability of perceiving depth from images or photos of monocular views of unknown poses as the visual task of *recognition of depth*.

9.6.1 *Problem Statement*

In the process of depth cognition, two views with known motion are available. These two views may correspond to the left-camera view and the right-camera view in a binocular vision. Also, these two views may come from a monocular vision which moves from one pose to another pose. However, in the process of depth recognition, there are two scenarios. One scenario corresponds to the case of having individual photos as input. The second scenario corresponds to the case of having a single sequence of images (i.e. a video) as input. In both scenario, only single views of unknown poses are available. Here, we call such single views as *mono views*. Therefore, we face two challenges, which are:

- Depth from Photo: In this case, the input are photos taken from different monocular views, of which the poses are unknown. The question will be: What is the physical principle behind the interpretation of photos taken from different monocular views of unknown poses, which produces the three-dimensional impressions or depths in the mind of a machine?

- Depth from Video: In this case, the input are a sequence of images (i.e. video) taken from a same monocular view, of which the pose is unknown. Then, what is the physical principle behind the interpretation of the sequence of images taken from a same monocular view of unknown pose, which produces the three-dimensional impressions or depths in the mind of a machine?

9.6.2 *Use of Predication and Verification Loop*

To the best of our knowledge, photo interpretation or image understanding is still an open issue in computer vision as well as in artificial intelligence. Here, we propose a feasible solution which is outlined in Fig. 9.32.

The key modules inside the proposed solution include

- Image Composition: This module takes an image or photo as input. Here, the input image or photo has been taken at a view of unknown pose in the physical world which is outside the mind of a machine. The task of this module is to group input image's pixels into sub-images of shapes. The success of this module depends on a machine's mental capability of doing shape perception.
- Model Predication: For each shape detected from an input image or photo, this module predicts a suitable 3D model, which is chosen from a database of 3D models learnt by a machine. In practice, each 3D model may include a list of representative images synthesized from some key views. In this way, the predication of a 3D model could be guided by the sub-image of a given shape which is the input to this module.
- Synthesis of Model's Image: This module creates a mental world in which a global coordinate system is assigned to it. This global coordinate system is called *mental scene frame.* In mental scene frame, two mental views (i.e. virtual cameras) of known poses are looking at the 3D model predicated by the module of model prediction. These two mental views will synthesize two sub-image of shape as output.
- Match Verification: This module is to verify the truthfulness of model prediction. It undertakes the process of matching between the sub-image of shape extracted from input image (or photo) and the two sub-image of shape synthesized by the two mental views. If the feature-based matching (e.g. matching among line segments) among these three sub-images of shape is successful, the predicted model is confirmed to be true. Otherwise, it goes back to the module of model prediction which will predict a new 3D model.
- Estimation of Recognized View: This module is to estimate the pose inside the mental scene frame, in which the sub-image of shape (extracted from an input image) is supposed to be taken equivalently. Such pose is called as *recognized view.* And, each sub-image of shape extracted from an input image (or photo) will have its own recognized view which is associated to its recognized 3D model. The solution to this task will be described in the next section.

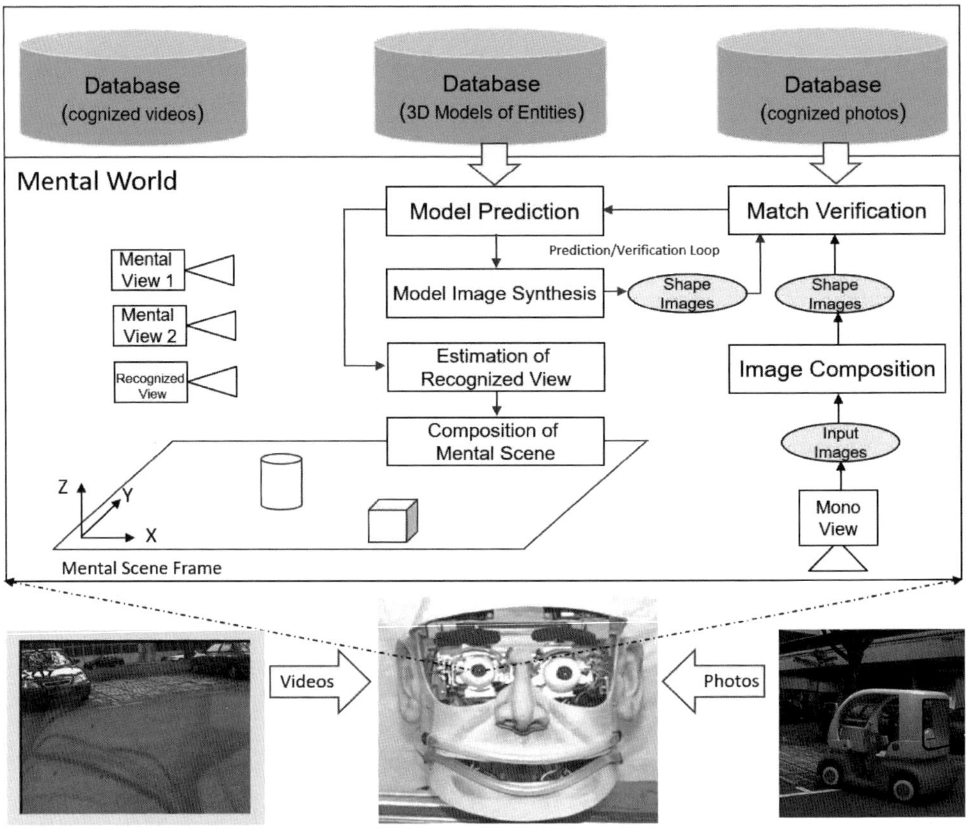

Fig. 9.32 Depth recognition with a closed-loop of predication and verification. The process starts with image composition, which groups input image's pixels into a set of shapes. For each shape, the process predicts its corresponding 3D model. Then, two mental images are synthesized at two mental views of known poses. If these two mental images match with the shape's sub-image extracted from the input image (or photo), the unknown pose of the input image's view could be determined. The result will be the pose of recognized view at which the input image (or photo) has been taken. After adding the 3D model of the shape into the recognized view, the process continues with the next shape detected from the input image (or photo) until all the shapes have been considered.

- Composition of Mental Scene: The closed loop of predication and verification allows the estimation of the recognized views for all the sub-images of shapes extracted from an input image or photo. And, each recognized view is associated with a recognized 3D model of a sub-image of shape. Hence, the task of this module is to transform all the recognized 3D models into a common coordinate system inside the mental world of a machine. In practice, this common coordinate system could be the mental scene frame or the coordinate system of the first recognized view. After such transformation, the output will be the recognized 3D models in a common coordinate system, which produce the three-dimensional impression or sensation of an image or photo inside the mind of a machine.

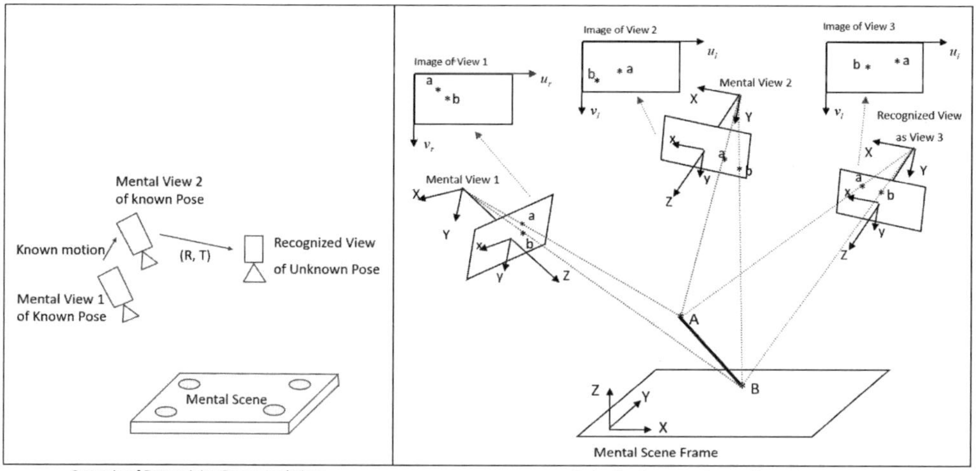

Scenario of Determining Recognized View Projective Geometry Among Three Views

(a) (b)

Fig. 9.33 Scenario of determining the pose of a third view from two views of known poses if three line segments are observed by these three views.

Clearly, the critical step toward the success of the above described methodology depends on the solution underlying the estimation of the unknown pose of a recognized view. In the next section, we present a deterministic solution, which allows the estimation of an unknown pose from two known poses.

9.6.3 *Estimation of Recognized View's Pose*

The problem here is to determine the unknown pose of a recognized view from the known poses of two mental views. This problem is illustrated in Fig. 9.33a, in which the poses of two mental views are known. Without loss of generality, we assume that the focal length of these three views are 1. For simplicity of notation, we rename mental view 1, mental view 2 and recognized view to be view 1, view 2 and view 3. In this way, let $({}^{v3}R_{v2}, {}^{v3}T_{v2})$ denote the motion transformation from view 2 to view 3. Then, the question will be: what are the equations which allow us to compute $({}^{v3}R_{v2}, {}^{v3}T_{v2})$?

Refer to Fig. 9.33b. Assume that the sub-image of a shape contains at least three line segments, one of which is line segment AB. Also, assume that end-points A and B can be seen by view 1, view 2 and view 3. In view 3's image plane, the image of line segment AB is a 2D line segment, which is described by

$$c_x \bullet x_{v3} + c_y \bullet y_{v3} + c_0 = 0 \tag{9.109}$$

where (c_x, c_y, c_0) are the coefficient of the 2D line segment, and (x_{v3}, y_{v3}) are the coordinates of an image point on the 2D line segment. According to perspective projection of each view (i.e. camera), we have $x_{v3} = \frac{X_{v3}}{Z_{v3}}$ and $y_{v3} = \frac{Y_{v3}}{Z_{v3}}$ where

(X_{v3}, Y_{v3}, Z_{v3}) are the coordinates of a point on line segment AB. Then, Eq. 9.109 could be rewritten as

$$c_x \bullet X_{v3} + c_y \bullet Y_{v3} + c_0 \bullet Z_{v3} = 0 \tag{9.110}$$

or in the following matrix form

$$C \bullet P^t_{v3} = 0 \tag{9.111}$$

where $C = (c_x, c_y, c_0)$ and $P_{v3} = (X_{v3}, Y_{v3}, Z_{v3})$.

Since P_{v3} is a point's position vector on line segment AB in view 3, the same point's position vector in view 2 will be P_{v2}. And, the relationship between P_{v2} and P_{v3} will be

$$P^t_{v3} =^{v3} R_{v2} \bullet P^t_{v2} +^{v3} T_{v2}. \tag{9.112}$$

Then, Eq. 9.111 becomes

$$C \bullet^{v3} R_{v2} \bullet P^t_{v2} + C \bullet^{v3} T_{v2} = 0 \tag{9.113}$$

Because the poses of view 1 and view 2 are known, the coordinates of points on line segment AB in view 2 could be calculated from Eq. 9.108. Given the position vectors of point A and point B in view 2, we have

$$C \bullet^{v3} R_{v2} \bullet A^t_{v2} + C \bullet^{v3} T_{v2} = 0 \tag{9.114}$$

$$C \bullet^{v3} R_{v2} \bullet B^t_{v2} + C \bullet^{v3} T_{v2} = 0 \tag{9.115}$$

Here, rotation matrix ${}^{v3}R_{v2}$ has three independent unknown and translation vector ${}^{v3}T_{v2}$ also has three unknown. Hence, we need to have six equations in order to determine these six unknown. Since one line segment in view 2 provides two independent equations. As a result, three line segments in view 2 will be sufficient enough to fully determine $({}^{v3}R_{v2}, {}^{v3}T_{v2})$ which is the motion transformation from view 2 to view 3. Since the pose of view 2 is known in advance, the pose of view 3 now becomes known. In practice, a triangle or rectangle on a shape contains at least three line segments.

9.7 Summary

Depth perception is an important step toward scene understanding (i.e. interpretation of images or photos of binocular views of known motion) as well as image understanding (i.e. interpretation of images or photos of monocular views of unknown motion) which are two important visual tasks underlying knowledge acquisition from visual signals.

In this chapter, we started with the study of the geometrical aspect of imaging, which governs the imaging of the geometry of a scene. For instance, we have learnt the important concepts such as: pixel, resolution, focal length, optical axis, image center, image coordinates, and imaging-related coordinate systems. Also, we have learnt the relationship between the coordinates in a scene and the coordinates in an image plane. Most importantly, this relationship reveals that the depth at a point in a scene could be computed from the so-called apparent motion, or disparity, if the relative geometry of a binocular vision is known from either the design or the off-line calibration.

Then, we continued with the study of apparent motion detection. In this section, we have learnt an important concept, which is called the *conglomerated appearance* at a pixel. After highlighting the drawbacks of template-based approach to apparent motion detection, we have learnt how to apply the concept of conglomerated appearance to detect apparent motion. In addition, we have learnt the epipolar constraint, which is useful to limit the search of match candidates to a line, instead of an area.

Subsequently, we moved on to the study of the concept-physical representation of the depth in a scene. In particular, we have learnt some interesting concepts such as depth-map, layered scene, discrete scene, and look-through scene.

In line with the principle of cognitive vision, we specifically consider that the depth of a scene is a kind of cognitive entity. Consequently, we have further studied the topics of depth cognition and depth recognition.

Under the topic of depth cognition, we have learnt that there are two ways of cognising depth. The first approach is visual computing, by which the depth at a point in a scene can be computed with respect to either the camera frame or the viewer frame. And, the second approach is visual learning, by which a machine could learn both the geometrical properties and the conceptual meanings of an object in a scene.

Finally, we have highlighted that the ability of cognising the depth of a scene will enable a machine to recognize the depth of a scene, from partial information such as images from multiple views of unknown poses, or videos from single views of unknown poses.

9.8 Exercises

(1) What is depth?
(2) What are the possible ways of determining depth?
(3) Explain the difference between the concept of shape and the concept of object.
(4) What are the important parameters inside a machine's vision system?
(5) What is apparent motion?
(6) What is disparity?
(7) Derive the geometrical relationship between the coordinates of a point in a scene and the coordinates of the point's image inside a camera.
(8) Derive the differential relationship between the coordinates of a point in a scene and coordinates the point's image inside a camera.
(9) What is a template?
(10) What are the drawbacks of employing a template-based approach to detecting apparent motion?

(11) What is the conglomerated appearance at a pixel?
(12) What are the advantages of applying the concept of conglomerated appearance to detect apparent motion?
(13) Explain various ways of representing a scene.
(14) Project 1: Write a program in MATLAB or C, which computes the conglomerated appearance at a manually-selected pixel.
(15) Project 2: Read relevant textbooks in order to find out the standard way of calibrating a camera. Implement a project, which aims at determining the parameters of a camera.
(16) Project 3: Write a program in MATLAB or C, which loads a pair of stereo images as input, and finds the match of a manually-selected pixel in any one of the two images.
(17) Project 4: Write a program in MATLAB or C, which computes the depth, or generalized depth, from a pair of matched pixels in a binocular vision (assume that the the parameters of the binocular vision are known).

Part 3
Knowledge Computation

The physical world consists of huge amount of entities, which are organized into a hierarchy of systems. These systems range from tiny-scale systems up to large-scale systems. Most importantly, the states of these systems are undergoing constant changes. Such changes result in the records of huge amount of data. Therefore, the knowledge in the physical world could be divided into two broad categories, which are (a) knowledge related to the statics of data and (b) knowledge related to the dynamics of systems. In science and engineering, all the disciplines are involved in the study of knowledge computation with data and from systems. It is not necessary for this book to repeat what have been achieved so far in the domain of knowledge computation. Instead, this part of the book puts the focus on two selected chapters. One addresses the methodologies underlying knowledge computation related to the statics of generic data, while the other addresses the principles behind knowledge computation related to the dynamics of mechanical systems.

Chapter 10

Computing Statics of Generic Data

10.1 Introduction

The physical world is full of systems, which constantly emit signals. All the signals coming from the physical could be captured by sensors, which could further convert signals into computer-readable versions of digital data. As a result, systems constantly produce data. For example, in a city or a country, the social systems produce data related to citizens, the ecosystems produce data related to environment, the transportation systems produce data related to networks of roads and vehicles, the school systems produce data related to pupils and students, the medical systems produce data related to patients and medical professionals, the financial systems produce data related to consumers and investors, the defense systems produce data related to weapons and armies, etc. Hence, big systems produce big data, which in return enable big intelligence at multiple scales. In other words, data is the source of knowledge. In this chapter, we are going to learn some basic methods underlying knowledge computation related to statics of generic data.

10.2 Computation of Data's Center

When we have a large amount of data coming from a similar category of sensors, there is a need to know the reference of these data. Such reference is called as a central value of data or a mean of data.

For example, if a city has a network of temperature sensors which are uniformly deployed, there is a need to know the city's temperature as a function of time in terms of minute or hour. In this case, a good indicator is the mean value of the temperatures measured by all the temperature sensors inside the network. These measurements could be taken at time intervals of minutes or hours.

Another example is the network of cameras which are deployed into a city's network of roads. Each camera could measure the traffic volume in terms of the number of detected vehicles per minute or hour. If we could compute the mean value of the traffic volumes inside a city's network of roads, we could have a good indicator about the traffic situation which is a function of time in terms of minute or hour.

Let $V = \{v_i,\ i = 1, 2, 3, \ldots, n\}$ denote a set of n digital values, which are the output from a set of similar sensors (e.g. temperature sensors). The mean value of such data set is as follows:

$$E(v) = \bar{v} = \frac{1}{n} \sum_{i=1}^{n} v_i \tag{10.1}$$

where $E(.)$ stands for expected true value.

10.3 Computation of Data's Spread

Due to the presence of noises and disturbances, sensory data are not crisp. For example, the repeated readings from a same sensor which sequentially measures a same physical quantity may not be exactly the same. Also, a set of sensors which measures a same physical quantity at the same time may not result in the same output. This type of phenomenon indicates that there must be a value which could tell us about the degree of crispness or the degree of spread of data which are obtained from repeated measurements by a single sensor or parallel measurements by multiple sensors.

Let $V = \{v_i,\ i = 1, 2, 3, \ldots, n\}$ denote a set of n digital values, which are the output from a set of similar sensors (e.g. temperature sensors). The degree of spread of such data set is described by a value called as *standard deviation* which is calculated as follows:

$$E(v - \bar{v}) = \sigma_v = \sqrt{\frac{1}{n-1} \sum_{i=1}^{n} (v_i - \bar{v})^2} \tag{10.2}$$

10.4 Computation of Data's Occurrence

A generic data may refer to many different properties in the physical world. For example, a data entry could be a weather condition (e.g. rainy, windy, sunny, cloudy), a working condition (e.g. normal, faulty, defective), a health condition (e.g. strong, weak, mild, sick), a color (e.g. red, green, blue, yellow), a face image (e.g. David, John, Smith, Sophie), a presence or absence of something or somebody, etc. For these types of data, knowledge about their occurrences or frequencies of occurrences are useful for decision-making or decision prediction.

Let e denote a data's entry. Its occurrence could be calculated as follows:

$$H(e) = H(e) + 1 \tag{10.3}$$

where $H()$ is a counter, which will be incremented by 1 when e occurs each time inside a data set.

If $E = \{e_i,\ i = 1, 2, 3, \ldots, n\}$ represents a whole population of a data set's entries, the frequency of occurrence of data entry e_i will be

$$P(e_i) = \frac{H(e_i)}{\sum_{i=1}^{n} H(e_i)} \tag{10.4}$$

where $P(e_i)$ denotes the frequency of occurrence or probability of data entry e_i.

10.5 Computation of Data's Similarity

There are many examples in which data's similarity plays an important role in knowledge-based decision-making. For example, the similarity of shapes within a sequence of images allow us to cognize or recognize objects. Also, the similarity of digital sound signals allow us to undertake speech cognition, speech recognition, and speaker verification, etc. In smart cities of future, the similarity of faces will allow us to track and monitor the daily activities of selected people.

In mathematics, the similarity among data entries under comparison refers to the similarity among the feature vectors of data entries under comparison. If (f_i, f_j) are the feature vectors of data entries i and j, a good indicator of similarity could be the value as follows:

$$S(f_i, f_j) = \frac{f_i}{\|f_i\|} \bullet \frac{f_j}{\|f_j\|} \tag{10.5}$$

where $\bullet$ denotes the dot-product of two vectors, and $S()$ varies between 0 and 1.

Example 10.1. A color image contains two pixels at locations A and B. The RGB values of color at pixel A are $(255, 20, 10)$, while the RGB values of color at pixel B are $(200, 10, 30)$. Let $A = (255, 20, 10)$ which represents the feature vector of pixel A, and $B = (200, 10, 30)$ which represents the feature vector of pixel B. Then, the similarity between feature vector A and feature vector B is calculated. The result is 0.9936, which means that the color at pixel A is very much similar to the color at pixel B.

<><><><><>

10.6 Computation of Data's Dissimilarity

Dissimilarity is an indicator which is a conjugate to similarity. For example, one may choose to measure the dissimilarity as follows:

$$D(f_i, f_j) = 1 - S(f_i, f_j) \tag{10.6}$$

where $D()$ denotes dissimilarity between two feature vectors.

Refer to Fig. 10.1. The calculation of Euclidean distance between two feature vectors also provides a good indicator about dissimilarity among data entries. In

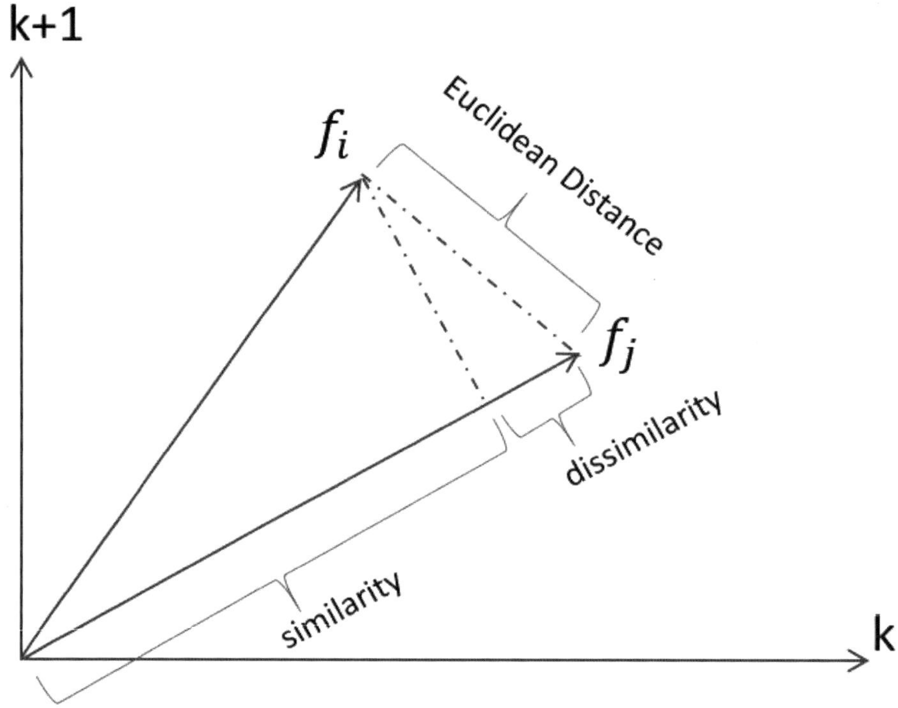

Fig. 10.1 Illustration of similarity, dissimilarity and Euclidean distance between two feature vectors f_i and f_j in the plane defined by their two adjacent elements k and $k+1$.

other words, the dissimilarity between two feature vectors could be calculated as follows:

$$D(f_i, f_j) = \|f_i - f_j\| = \sqrt{\sum_{k=1}^{m}(f_{i,k} - f_{j,k})^2} \tag{10.7}$$

where $f_i = \{f_{i,k},\ k = 1, 2, 3, \ldots, m\}$ and $f_j = \{f_{j,k},\ k = 1, 2, 3, \ldots, m\}$.

10.7 Computation of Data's Correlation

Data are the output of sensors, which measure the values of coefficients or parameters of systems in the physical world. Hence, it is interesting to know the level of dependence or independence among coefficients inside a system. Such level of dependence or independence among coefficients in a system is called as *correlation* in short. The knowledge about data's correlation are also very useful to knowledge-based decision-making.

For example, a vehicle's parameter of fuel consumption has a strong correlation with the vehicle's parameter of traveled distance. Also, a vehicle's parameter of air-drag coefficient has correlation with the vehicle's parameter of fuel consumption.

In addition, a vehicle's X coordinate has significant correlation with the vehicle's Y coordinate with respect to a global OXY coordinate system. This is because a vehicle is normally traveling on human-made roads which could be described by equations of curves or paths. Such equations impose a high level of dependence between a vehicle's X and Y coordinates.

Another important situation of computing data's correlation is to know the level of dependence or independence among the elements of feature vectors. In general, a good feature vector should consist of elements which are not strongly correlated. Feature vectors with weakly correlated elements are vital to the success of cognition and recognition in the process of knowledge acquisition.

Assume that we have a data set of n feature vectors as follows:

$$F = \{f_i,\ i = 1, 2, 3, \ldots, n\} \tag{10.8}$$

and each feature vector has m elements such as:

$$f_i = \{f_{i,k},\ k = 1, 2, 3, \ldots, m\} \tag{10.9}$$

Then, we could define vector of element x and vector of element y as follows:

$$c_x = \{f_{j,x},\ j = 1, 2, 3, \ldots, n\}, x \in [1, m] \tag{10.10}$$

$$c_y = \{f_{j,y},\ j = 1, 2, 3, \ldots, n\}, y \in [1, m] \tag{10.11}$$

Now, we could determine correlation ρ_{xy} between element x and element y as follows:

$$\rho_{xy} = \frac{\sigma_{xy}^2}{\sigma_x \sigma_y} \tag{10.12}$$

with

$$\sigma_x = \sqrt{\frac{1}{n-1}\sum_{j=1}^{n}(f_{j,x} - \bar{f}_x)^2} \tag{10.13}$$

$$\sigma_y = \sqrt{\frac{1}{n-1}\sum_{j=1}^{n}(f_{j,y} - \bar{f}_y)^2} \tag{10.14}$$

$$\sigma_{xy} = \sqrt{\frac{1}{n-1}\sum_{j=1}^{n}(f_{j,x} - \bar{f}_x)(f_{j,y} - \bar{f}_y)} \tag{10.15}$$

and

$$\bar{f}_x = \frac{1}{n}\sum_{j=1}^{n} f_{j,x} \tag{10.16}$$

$$\bar{f}_y = \frac{1}{n}\sum_{j=1}^{n} f_{j,y} \tag{10.17}$$

Example 10.2. Assume that a vehicle travels along a straight road. Its X and Y coordinates are recorded to be

$$(1.0\,km, 1.0\,km), (2.0\,km, 2.0\,km), (3.0\,km, 3.0\,km), (4.0\,km, 4.0\,km),$$
$$(5.0\,km, 5.0\,km), (6.0\,km, 6.0\,km), (7.0\,km, 7.0\,km).$$

We first calculate the mean values of X and Y coordinates. The results are $(4.0\,km, 4.0\,km)$. Then, we calculate the standard deviations (σ_x, σ_y) of X and Y coordinates. The results are $(2.1602\,km, 2.1602\,km)$. Also, we calculate the covariance σ_{xy}^2. The results is $4.6667\,km^2$. Finally, the value (i.e. coefficient) of correlation ρ_{xy} is 1.0, which means that the vehicle's X and Y coordinates are 100% dependent.

<><><><><>

10.8 Computation of Data's Principal Components

In the physical world, there are many simple data such as temperature, pressure, speed, force, etc. Simple data could be easily represented by values or low-dimensional vectors of values. However, the physical world contains also many complex entities which produce complex data. For example, a human being could produce complex data such as innate structure, innate mechanism, innate organs, appearance of face, appearance of silhouette, walking pattern, talking pattern, food preference, sport preference, entertainment preference, education, profession, etc. Also, a vehicle could produce complex data such as innate structure, innate mechanism, innate devices, innate sub-systems, appearance, dynamics, motion performance, etc.

In general, a complex data is represented by a feature vector of high dimension. Then, a question arises: is each element of a feature vector characteristic or sensitive enough? An element inside a feature vector will be characteristic or sensitive if the span or range of its variation is large. Hence, it is important to know how to measure the extent of variation of each element inside a feature vector. It is interesting to take note that some applications may require the elements of feature vector to be highly sensitive (i.e. large variation), while other applications may require the elements of feature vector to be highly stable (i.e. small variation).

The answer to the above question is to compute eigenvectors and eigenvalues from the data set of a feature vector. Here, the word "eigen" means characteristic. Most importantly, the eigenvectors indicate the directions in which a feature vector take most sensitive values. Such most sensitive values along eigenvectors are called *principal components* of feature vector. In addition, eigenvalues indicate the levels of variation of the principal components along their respective eigenvectors.

Assume that we have a data set of n feature vectors which are represented by $F = \{f_i,\ i = 1, 2, 3, \ldots, n\}$. And, each feature vector has m elements such as $f_i = \{f_{i,k},\ k = 1, 2, 3, \ldots, m\}$. Then, we could compute the covariance matrix which is described by

$$S = \{s_{xy}, x \in [1, m],\ y \in [1, m]\} \tag{10.18}$$

with

$$s_{xy} = \frac{1}{n-1} \sum_{i=1}^{n} (f_{i,x} - \bar{f}_x)(f_{i,y} - \bar{f}_y) \tag{10.19}$$

and

$$\bar{f}_x = \frac{1}{n} \sum_{i=1}^{n} f_{i,x} \tag{10.20}$$

$$\bar{f}_y = \frac{1}{n} \sum_{i=1}^{n} f_{i,y} \tag{10.21}$$

Then, the eigenvectors and eigenvalues of covariance matrix S could be calculated from the following equations:

$$S \bullet v = \lambda \bullet v \tag{10.22}$$

$$\|S - \lambda I\| = 0 \tag{10.23}$$

where I is a $n \times n$ identical matrix.

The above equations will produce n solutions to eigenvector v and n values to eigenvalue λ. We can order the list of eigenvalues from the largest one to the smallest one. Also, we can project the original set of feature vectors onto the eigenvectors so as to obtain a new set of feature vectors which take principal components as their elements.

Example 10.3. We take six samples of color from an image. The color vectors of these six samples are: (250, 50, 10), (245, 47, 12), (240, 44, 14), (235, 41, 16), (230, 38, 18), (225, 25, 20). We first calculate the covariance matrix of these six samples. The result is

$$\begin{matrix} 87.5000 & 77.5000 & -35.0000 \\ 77.5000 & 78.1667 & -31.0000 \\ -35.0000 & -31.0000 & 14.0000 \end{matrix}$$

Then, we calculate the eigenvectors and eigenvalues of the above covariance matrix. The results of eigenvectors are: (0.3714, −0.0000, 0.9285), (0.6094, −0.7545, −0.2438), (−0.7005, −0.6563, 0.2802). And, the results of eigenvalues are: 0.0000, 5.5519, 174.1148. This example shows that the principal components along eigenvector (−0.7005, −0.6563, 0.2802) is discriminative enough to represent this set of six samples.

<><><><><>

10.9 Summary

Data are source of knowledge. In this chapter, we have learnt some basic methods of computing knowledge from data. For simple data which are represented by values or low-dimensional vectors of values, we could be interested in computing knowledge about data's centers, data's spreads, data's occurrences, data's similarities, data's dissimilarities. For complex data which are represented by high-dimensional feature vectors, we could be interested in computing knowledge related to data's correlations and data's principal components.

10.10 Exercises

(1) What is the mean vector of the following data set: (250, 50, 10), (245, 47, 12), (240, 44, 14), (235, 41, 16), (230, 38, 18), (225, 25, 20)?
(2) What is the mean vector of the following data set: (10, 10, 250), (12, 13, 245), (14, 16, 240), (16, 19, 235), (18, 21, 230), (20, 24, 225)?
(3) What is the vector of standard deviations of the following data set: (250, 50, 10), (245, 47, 12), (240, 44, 14), (235, 41, 16), (230, 38, 18), (225, 25, 20)?
(4) What is the vector of standard deviations of the following data set: (10, 10, 250), (12, 13, 245), (14, 16, 240), (16, 19, 235), (18, 21, 230), (20, 24, 225)?
(5) What is the frequency of occurrence of value 20 in the following data set (2, 4, 6, 7, 8, 3, 15, 20, 50, 20, 4, 2, 7, 8, 20, 15, 4, 2, 20, 50, 2, 4, 8, 7, 3)?
(6) What is the similarity between the following two vectors: (230, 38, 18), (225, 25, 20)?
(7) What is the dissimilarity between the following two vectors: (245, 47, 12), (240, 44, 14)?
(8) What are the principal components of each feature vector from the following data set: (250, 50, 10), (245, 47, 12), (240, 44, 14), (235, 41, 16), (230, 38, 18), (225, 25, 20)?
(9) What are the principal components of each feature vector from the following data set: (10, 10, 250), (12, 13, 245), (14, 16, 240), (16, 19, 235), (18, 21, 230), (20, 24, 225)?

Chapter 11

Computing Dynamics of Mechanical Systems

11.1 Introduction

In general, data are passive, while systems are dynamic. This implies that knowledge computed from data basically represent static aspects of data, while knowledge computed from systems predict dynamic potentials of systems. The main reason behind the above statements is due to the fact that systems are able to respond to input and exhibit different patterns of output. In particular, the responses of systems are broadly categorized into steady-state responses and transient responses. Therefore, there are two general scenarios of computing knowledge about dynamics of systems. One scenario consists of computing the knowledge about systems' outputs if the knowledge about systems' inputs are given. The other scenario is to compute the knowledge about systems' inputs if the systems have to achieve given or expected outputs. Since there are many dynamic systems which are under active studies in different domains of science and engineering, this chapter simply focuses on computing knowledge about dynamics of mechanical systems (i.e. the bodies of machines), which are the major players of actions and interactions in the physical world.

11.2 Computation of Mechanism's Motion

All the physical entities have their existences in the physical world, which occupy spaces. The rate changes of their positions in the physical world result in what is called as motion. Hence, the study of motion concerns all the physical entities. Obviously, it is not possible or necessary for this book to cover the study of motion related to all the physical entities. Here, we focus on the computation of motion related to vehicles and robots which are the most popular categories of mechanical systems on earth.

11.2.1 *Equation of Path*

The first quantities, which describe motion, are positions. And, the positions of physical entities could be simply represented by coordinates in a coordinate system of spatial reference. Therefore, the first type of knowledge about motion is the spatial constraints among positions. For example, robots could undertake industrial tasks by repeatedly moving forward and backward through sets of waypoints. Also, vehicles could travel to some destinations by passing through sets of waypoints referenced by a coordinate system.

Without loss of generality, let $P = \{(x_i, y_i), i = 1, 2, 3, \ldots, n\}$ denote a set of n positions in a coordinate system OXY. Then, the question will be: how to determine the spatial constraints among a set of positions? The answer is to compute the equations of paths. By definition, a path is a spatial constraint imposed to a set of positions without consideration of time. For example, in Fig. 11.1, a network of roads could be represented by a set of waypoints, in which any sub-set of waypoints corresponds to a path.

In mathematics, the following equations provide good representations of paths:

- Equations of straight lines: If vehicles move along straight roads, the paths followed by vehicles could be described by

$$a_x x + b_y y + c_0 = 0$$

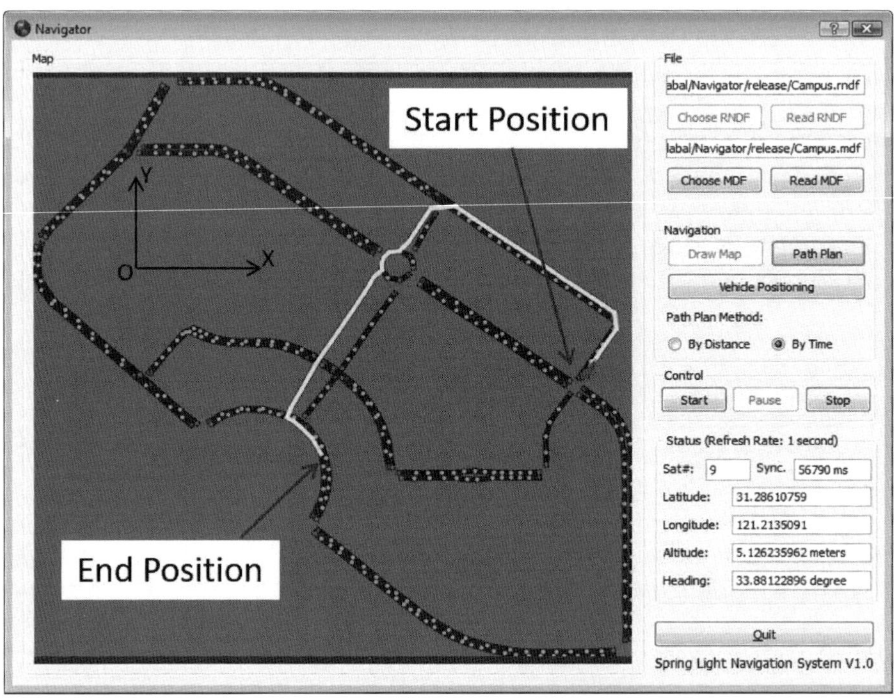

Fig. 11.1 Path refers to a spatial constraint imposed to a sub-set of waypoints which connect start-position to end-position.

where (a_x, b_y, c_0) are the coefficients of the equations of lines. If two waypoints are given, the coefficients (a_x, b_y, c_0) are fully determined. This is because only two of these three coefficients are independent.

- Equations of circular lines: If vehicles move along circular roads, the paths followed by vehicles could be described by

$$(x - x_0)^2 + (y - y_0)^2 = r^2$$

where (x_0, y_0) are the coordinates of a circle's center, while r is the radius of circle. If three waypoints are given, the parameters (x_0, y_0, r) are fully determined.

- Equations of curved lines: If vehicles move along curved roads, the paths followed by vehicles could be described by

$$y = a_n x^n + a_{n-1} x^{n-1} + \cdots + a_1 x + a_0$$

where n is an integer which indicates the order of the polynomial equation, while $(a_n, a_{n-1}, \ldots, a_1, a_0)$ are the coefficients of the polynomial equation. In practice, popular values for n are 2, 3, 4, and 5. If a set of $n+1$ waypoints are given, the coefficients $(a_n, a_{n-1}, \ldots, a_1, a_0)$ are fully determined.

11.2.2 *Equation of Trajectory*

It is interesting to take note that today's land vehicles could be equipped with the cruise control module, which regulates the speeds of vehicles at some values preset by vehicles' drivers. By introducing artificial self-intelligence into networks of roads and networks of vehicles, the cruise control module should be able to retrieve equations of trajectories from databases of smart roads. In addition, a vehicle could incorporate two motion control loops as shown in Fig. 11.2, which will enable it to follow a road with recommended time functions of driving speeds

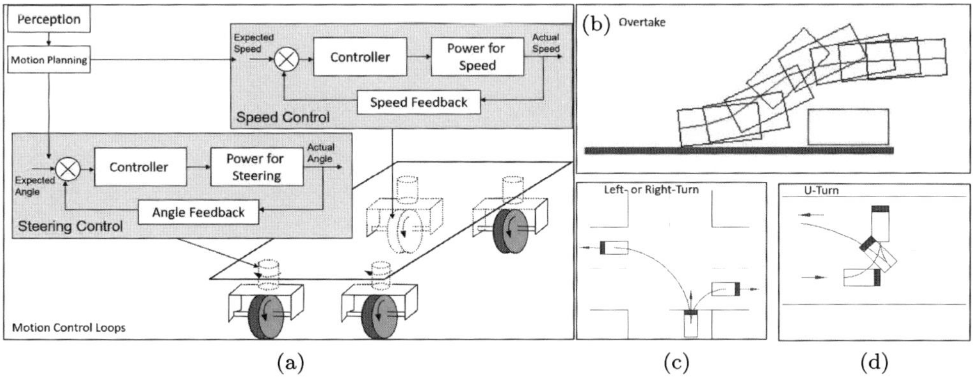

Fig. 11.2 Cruise control module's two loops: one for automatic control of driving speed and the other for automatic control of steering angle. The inputs (i.e. time functions of desired driving speeds and desired steering angles) to these two motion control loops may come from the databases of smart roads in which equations of trajectories (e.g. road following, vehicle overtaking, left-turn, right-turn, u-turn) are calculated and stored in advance.

and steering angles. In this way, the roles of future vehicles' drivers are to decide the switch between automatic mode (i.e. situations with low risk of collision) and manual mode (i.e. situations with high risk of collision).

By definition, a spatial constraint imposed to a set of positions with consideration of time is called as a trajectory. Since a network of roads (e.g. highways) has stable layouts or topology, the equations of path for each road could be calculated in advance and be stored into a database of smart road. Now, by adding time constraints, equations of path could be quickly converted into equations of corresponding trajectories.

In theory, the conversion from equations of paths into equations of trajectories is not unique. This is because there are many ways to satisfy a time constraint. For example, if a time constraint is a time duration for a vehicle to travel from position A to position B along a curved path, there are infinite number of solutions for us to derive equations of trajectories in this case. Therefore, this seams to be a problem of computing unique solutions to equations of trajectories in theory. However, it is actually an advantage from the viewpoint of designing equations of trajectories. This is because designers could enjoy the freedom of designing workable solutions to equations of trajectories.

In general, the popular design scheme is to divide a trajectory into three phases, which are

- Speed-up Phase: This phase starts from time instance t_0 and ends at time instance t_1.
- Cruising Phase: This phase starts from time instance t_1 and ends at time instance t_2.
- Slow-down Phase: This phase starts from time instance t_2 and ends at time instance t_3.

Now, the question will be: how to represent the time constraints imposed to trajectories? Here, we are going to present two scenarios of solutions. The first scenario corresponds to the cases of maintaining strict spatial constraints and time constraints, while the second scenario emphasizes on time constraints more than spatial constraints.

11.2.2.1 *Explicit Equation of Trajectory*

In robotics, the tool tip of a robotic arm has to have the capability of undergoing motions with strict spatial constraints and time constraints. In this case, we must have explicit equations to represent both spatial constraints and time constraints.

Refer to Fig. 11.3. Let $y = f(x)$ denote the equation of path, which represents the strict spatial constraints on (x, y) coordinates. Then, we can choose a time function of traveled distance to represent the time constraints imposed on (x, y) coordinates. Without loss of generality, let $q(t)$ denote a time function of traveled distance, which could be a traveled linear distance or a traveled angular distance. In this way, the solution to time function $q(t)$ within time interval $[t_i, t_j]$ ($i \in [0, 3]$,

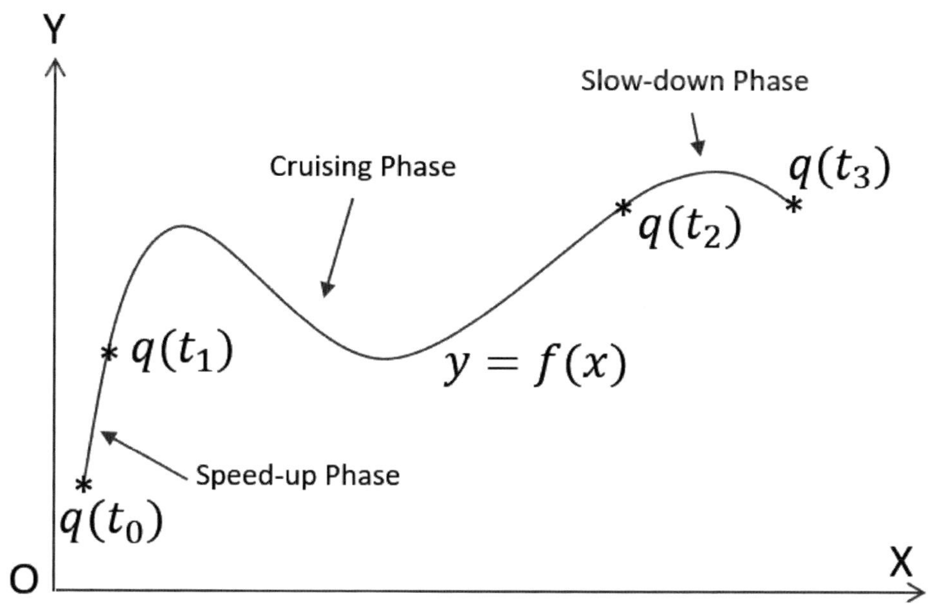

Fig. 11.3 Explicit equations for both spatial constraints and time constraints. Here, $y = f(x)$ represents the spatial constraints on (x, y) coordinates, while traveled distance $q(t)$ represents the time constraints imposed on (x, y) coordinates.

$j \in [0, 3]$) could be as follows

$$q(t) = k_5 t^5 + k_4 t^4 + k_3 t^3 + k_2 t^2 + k_1 t + k_0 \tag{11.1}$$

where $(k_5, k_4, k_3, k_2, k_1, k_0)$ are the coefficients which could be estimated from the knowledge about the following boundary conditions:

- Traveled distance at time instance t_i: $q(t_i)$
- Traveled distance at time instance t_j: $q(t_j)$
- Speed at time instance t_i: $\frac{dq(t_i)}{dt}$
- Speed at time instance t_j: $\frac{dq(t_j)}{dt}$
- Acceleration at time instance t_i: $\frac{d^2q(t_i)}{dt^2}$
- Acceleration at time instance t_j: $\frac{d^2q(t_j)}{dt^2}$

where

$$q(t_i) = \int_{x(t_0)}^{x(t_i)} \left[\sqrt{1 + (\frac{dy}{dx})^2} \right] dx$$

and

$$q(t_j) = \int_{x(t_0)}^{x(t_j)} \left[\sqrt{1 + (\frac{dy}{dx})^2} \right] dx$$

Knowing the time function of traveled distance, the velocity components of (x, y) coordinates could be calculated as follows:

$$\frac{dx}{dt} = \frac{1}{\sqrt{1 + (\frac{dy}{dx})^2}} \frac{dq(t)}{dt} \tag{11.2}$$

$$\frac{dy}{dt} = \frac{1}{\sqrt{1 + (\frac{dx}{dy})^2}} \frac{dq(t)}{dt} \tag{11.3}$$

And, the time functions of (x, y) coordinates will be

$$x(t) = \int_{t_0}^{t} \left(\frac{dx}{dt}\right) dt \tag{11.4}$$

$$y(t) = \int_{t_0}^{t} \left(\frac{dy}{dt}\right) dt \tag{11.5}$$

Example 11.1. A vehicle enters a circular path at point A and exits it from point B. The coordinates of point A are $(20.0\,m, 0.0\,m)$ and the coordinates of point B are $(0.0\,m, 0.0\,m)$. The circular velocities of the vehicle at points A and B are all zero. If the time constraint is for the vehicle to travel from A to B within 10 seconds, design equations of trajectory for accomplishing this task.

Solution: Let $t_0 = 0\,s$, $t_1 = 2\,s$, $t_2 = 8\,s$ and $t_3 = 10\,s$. We choose $\theta(t)$ to represent traveled angular distance by the vehicle. Hence, the time functions for the vehicle's (x, y) coordinates will be

$$x(t) = 10.0 + 10.0\cos(\theta(t))$$

and

$$y(t) = 0.0 + 10.0\sin(\theta(t))$$

Then, we design a trajectory which has three phases such as

- Speed-up phase between t_0 and t_1: In this phase, the vehicle moves with a constant tangential acceleration. The traveled angular distance will be:

$$\theta(t) = 0.0 + \frac{1}{2}\alpha(t - t_0)^2, t \in [t_0, t_1]$$

  where α is a constant tangential acceleration.
- Cruising phase between t_1 and t_2: In this phase, the vehicle moves with a constant circular velocity. The traveled angular distance will be

$$\theta(t) = \frac{1}{2}\alpha t_1^2 + \alpha t_1(t - t_1), t \in [t_1, t_2]$$

  where αt_1 is the constant circular velocity.

- Slow-down phase between t_2 and t_3: In this phase, the vehicle moves with a constant tangential deceleration. The traveled angular distance will be

$$\theta(t) = \theta(t_2) + \alpha t_1(t - t_2) - \frac{1}{2}\alpha(t - t_2)^2, t \in [t_1, t_2]$$

with

$$\theta(t_2) = \frac{1}{2}\alpha t_1^2 + \alpha t_1(t_2 - t_1)$$

Since the total traveled angular distance is π, we have $\theta(t_3) = \pi$. This condition allows us to determine the only unknown α. The result is $\alpha = \frac{\pi}{16}$.

<><><><><>

11.2.2.2 *Implicit Equation of Trajectory*

In some applications such as driving on a highway, the spatial constraints on coordinates are not so strict or strong. In other words, the spatial constraints on coordinates and the time constraints on velocities/accelerations could be mixed together. In this case, the solutions to equations of trajectories within time interval $[t_i, t_j]$ ($i \in [0, 3]$, $j \in [0, 3]$) could be as follows

$$x(t) = a_5 t^5 + a_4 t^4 + a_3 t^3 + a_2 t^2 + a_1 t + a_0 \tag{11.6}$$

$$y(t) = b_5 t^5 + b_4 t^4 + b_3 t^3 + b_2 t^2 + b_1 t + b_0 \tag{11.7}$$

where $(a_5, a_4, a_3, a_2, a_1, a_0)$ and $(b_5, b_4, b_3, b_2, b_1, b_0)$ are two sets of coefficients, which could be estimated from the knowledge about the following boundary conditions

- Coordinates at time instance t_i: $x(t_i)$ and $y(t_i)$
- Coordinates at time instance t_j: $x(t_j)$ and $y(t_j)$
- Velocity components at time instance t_i: $\frac{dx(t_i)}{dt}$ and $\frac{dy(t_i)}{dt}$
- Velocity components at time instance t_j: $\frac{dx(t_j)}{dt}$ and $\frac{dy(t_j)}{dt}$
- Acceleration components at time instance t_i: $\frac{d^2x(t_i)}{dt^2}$ and $\frac{d^2y(t_i)}{dt^2}$
- Acceleration components at time instance t_j: $\frac{d^2x(t_j)}{dt^2}$ and $\frac{d^2y(t_j)}{dt^2}$

By default, the above boundary conditions are specified by designers. These boundary conditions could also be determined from the specific requests of drivers.

Example 11.2. A vehicle approaches a traffic junction at a constant speed of 16.0 m/s. At a distance of 50.0 meters to the traffic junction, the traffic light changes to red color. Design equations of trajectory which make the vehicle to stop at the junction within 10.0 seconds.

Solution: When the traffic light changes to red color, we denote the time to be $t_0 = 0\,s$. At time t_0, we assign a coordinate system OY on the ground. Along OY axis, we design the following equations of trajectory:

$$\begin{cases} y(t) &= a_3t^3 + a_2t^2 + a_1t + a_0 \\ v(t) &= 3a_3t^2 + 2a_2t + a_1 \end{cases}$$

where

- $y(t)$ is the time function of vehicle's position,
- $v(t)$ is the time function of vehicle's speed, and
- (a_3, a_2, a_1, a_0) are the coefficients.

Then, two sets of boundary conditions could be written as follows: $[y(0) = 0.0\,m, v(0) = 16.0\,m/s]$ and $[y(10) = 50.0\,m, v(10) = 0.0\,m/s]$. Application of these boundary conditions yields the following equations

$$\begin{cases} y(0) &= 0 &= a_0 \\ y(10) &= 50 &= 10^3a_3 + 10^2a_2 + 10a_1 + a_0 \\ v(0) &= 16 &= a_1 \\ v(10) &= 0 &= 300a_3 + 20a_2 + a_1 \end{cases}$$

From the above equations, the values of coefficients (a_3, a_2, a_1, a_0) could be uniquely determined. And, the results are: $a_0 = 0$, $a_1 = 16$, $a_3 = -11/40$, and $a_3 = 3/50$.

<><><><><>

11.3 Computation of Mechanism's Kinematics

By definition, kinematics refers to the study of motion without consideration of force or torque. In general, the body of an entity, which could demonstrate motion, will include structure and mechanism. An entity's structure is a static part which provides space to house the entity's other components, while its mechanism is a dynamic part which has both input motion and output motion. If a mechanism's input motion is given, the study of its output motion is called as forward kinematics. Inversely, if a mechanism's output motion is given, the study of its input motion is called as inverse kinematics. Kinematics has been under intensive study in mechanical engineering. Here, we use simple examples to illustrate the concept of computing forward kinematics as well as inverse kinematics.

11.3.1 *Equation of Forward Kinematics*

The study of forward kinematics is to determine a mechanism's output motion from its input motion. For example, in Fig. 11.4, there is a planar robotic arm with two links, which are placed inside the XY plane of coordinate system $OXYZ$. The

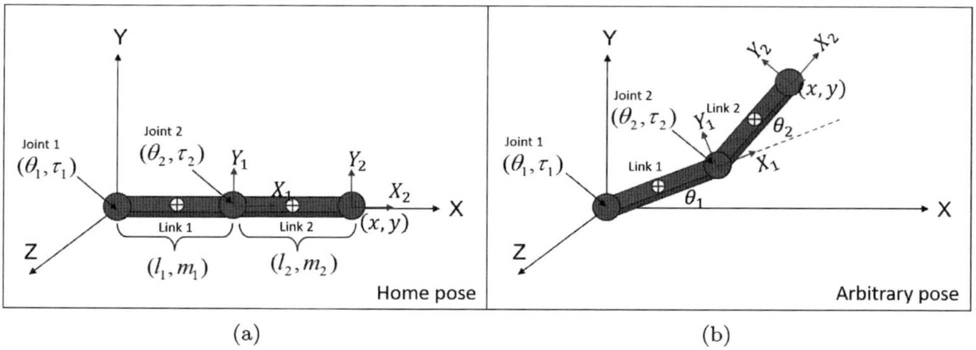

Fig. 11.4 A planar robotic arm with two links: (a) the links' poses when joint angles are zero, and (b) the links' poses when joint angles are not equal to zero.

robot's mechanism has two inputs of motion, which are the input of motion at joint 1 and the input of motion at joint 2.

Now, let's denote the parameters of a robotic arm as follows:

- Link length: $l_i, i \in [1, 2]$.
- Link mass: $m_i, i \in [1, 2]$.
- Joint Angle: $\theta_i, i \in [1, 2]$.
- Joint Torque: $\tau_i, i \in [1, 2]$.
- Link's coordinate system: $OX_iY_i, i \in [1, 2]$. In robotics, each link has two placements of joint. One is called the proximal placement of joint, while the other is called the distal placement of joint. A link's coordinate system is always attached to its distal placement of joint as shown in Fig. 11.4.

Then, the problem of forward kinematics in robotics is to determine the position and orientation of the last link's coordinate system with respect to the global coordinate system which is attached to the first link's proximal placement of joint. Without loss of generality, we could consider the rigid body, which supports link 1, to be link 0. In this way, the global coordinate system could be treated as the coordinate system attached to link 0.

In the example as shown in Fig. 11.4, the solutions to the computation of forward kinematics will be

- The position of the last link's coordinate system:

$$x(t) = l_1 \cos(\theta_1(t)) + l_2 \cos(\theta_1(t) + \theta_2(t)) \tag{11.8}$$

$$y(t) = l_1 \sin(\theta_1(t)) + l_2 \sin(\theta_1(t) + \theta_2(t)) \tag{11.9}$$

- The orientation of the last link's coordinate system:

$$\theta(t) = \theta_1(t) + \theta_2(t) \tag{11.10}$$

In general, if we have a robotic arm with n links, the solution to compute forward kinematics will be

$$^0M_n(\theta_1, \theta_2, \ldots, \theta_n) =^0 M_1(\theta_1) \bullet^1 M_2(\theta_2) \bullet \ldots \bullet^{n-1} M_n(\theta_n) \tag{11.11}$$

where ^iM_j is link j's matrix which represents the pose of link j's coordinate system with respect to link i's coordinate system. In particular, ^0M_n is the pose of the last link's coordinate system with respect to link 0's coordinate system.

11.3.2 *Equation of Inverse Kinematics*

In theory, a mechanism's output motion depends on its input motion. However, in practice, a mechanism's output motion is dictated by external tasks. For example, a task may require a tool to follow specific trajectories in workspace. In this case, the last link's motion of a robotic arm is known advance. As a result, an interesting question will be: how to determine a mechanism's input motion from its output motion? This problem is the focused study under the topic of inverse kinematics.

For example, in Fig. 11.4, if the robotic arm moves its tool tip to follow a specific trajectory, $(x(t), y(t))$ will be known in advance. In this case, the solutions to compute $(\theta_1(t), \theta_2(t))$ of the robot mechanism's input motion will be

$$\theta_1(t) = \arctan(\frac{y(t)}{x(t)}) - \alpha(t) \tag{11.12}$$

$$\theta_2(t) = \pi - \beta(t) \tag{11.13}$$

with the application of trigonometry's law of cosines such as

$$\alpha(t) = \arccos(\frac{l_2^2 - l_0^2 - l_1^2}{2l_0l_1}) \tag{11.14}$$

$$\beta(t) = \arccos(\frac{l_0^2 - l_1^2 - l_2^2}{2l_1l_2}) \tag{11.15}$$

and the virtual link length from the origin of OXY to position $(x(t), y(t))$ as follows

$$l_0 = \sqrt{x^2(t) + y^2(t)}$$

In general, if we have a robotic arm with n links, the solutions to compute a mechanism's input motion $(\theta_1, \theta_2, \ldots, \theta_n)$ could be derived from the following equation of constraints

$$^0M_n(\theta_1, \theta_2, \ldots, \theta_n) =^0 M_1(\theta_1) \bullet^1 M_2(\theta_2) \bullet \ldots \bullet^{n-1} M_n(\theta_n)$$

or its first-order derivatives. Here, we are not going to repeat the details which could be found in robotics textbooks.

11.4 Computation of Mechanism's Dynamics

When an entity demonstrates rate changes of positions in real-time, it actually manifests kinetic energy, which must come from somewhere. Hence, the final study of motion must consider the presence of energy in terms of force or torque. Not surprisingly, the study of motion in consideration of force or torque is under the topic of dynamics. In particular, if an entity has its own mechanism, the topic of dynamics refers to the study of the relationship between the motion at a mechanism's joints and the forces or torques applied to the mechanism's joints.

Refer to Fig. 11.4. Assume that the OXY plane is horizontal plane. In this case, the gravitational force will not contribute to the motion of the robotic arm. When the robotic arm's two joints output torques (τ_1, τ_2), the equations of dynamics will be

$$J_1 \frac{d^2\theta_1}{dt^2} = \tau_1 - \tau_2 \tag{11.16}$$

$$J_2 \frac{d^2\theta_2}{dt^2} = \tau_2 - \tau_1 \tag{11.17}$$

where (J_1, J_2) are the moments of inertia of link 1 and link 2, and the viscous frictions at the joints are ignored.

The above equations also indicate that we have two scenarios about the computation of dynamics. One scenario is to determine motion if force or torque is known in advance, while the second scenario is to determine force or torque if motion is known in advance. The former is the study of forward dynamics, and the later is the study of inverse dynamics. Similarly, dynamics has been under intensive study in mechanical engineering. Here, we use simple examples to illustrate the concept of computing forward dynamics as well as inverse dynamics.

11.4.1 *Equation of Forward Dynamics*

A land vehicle could be treated as a high-speed mobile robot. In mobile robotics, there are different types of wheeled platforms. In Fig. 11.5, we depict two types of wheeled platform. The first type is a single-wheeled platform in which the parameters are

- The wheel's mass: m.
- The wheel's radius: r.
- The wheel's angular velocity: ω.
- The wheel's linear position: (x, z).
- The wheel's linear velocity: v.
- The wheel's applied torque: τ.

If we do not consider the friction and air's drag force, we could establish the following equation:

$$J\frac{d\omega(t)}{dt} = \tau \tag{11.18}$$

where J is the moment of inertia of mass m when it rotates about the wheel's axis.

Since $v(t) = r\omega(t)$, we have

$$\frac{J}{r}\frac{dv(t)}{dt} = \tau \tag{11.19}$$

The application of Laplace transform will yield

$$V(s) = \frac{r}{Js}T(s) \tag{11.20}$$

where $(V(s), T(s))$ are the Laplace transforms of $(v(t), \tau(t))$. The above equation in frequency domain is the equation of forward dynamics, which determines velocity if torque is known in advance. Its Laplace inverse transform will yield the time domain function for the computation of velocity

$$v(t) = \mathcal{L}^{-1}\{V(s)\}$$

Refer to Fig. 11.5. The second type of wheeled platform is a car-like mobile robot in which the parameters are

- The mobile robot's mass: m.
- The mobile robot's center: (x, y) which is the center between the two rear wheels.
- The mobile robot's width: W which is the distance between the centers of the two rear wheels.
- The mobile robot's length: L which is the distance between the two front steering wheels and the two rear wheels.

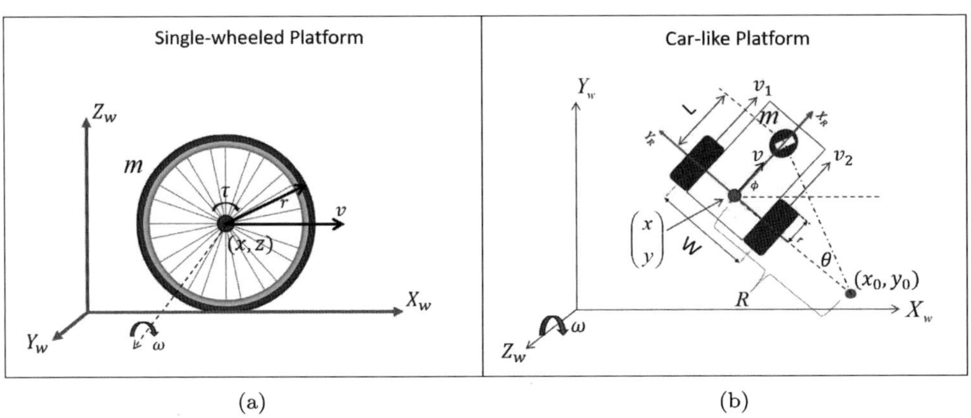

Fig. 11.5 Two wheeled platforms: (a) a single-wheeled platform, and (b) a car-like mobile platform.

- The mobile robot's linear velocity at the center: v.
- The mobile robot's linear velocity at the left wheel: v_1.
- The mobile robot's linear velocity at the right wheel: v_2.
- The mobile robot's angular velocity about OZ axis: ω.
- The mobile robot's steering angle: θ.
- The mobile robot's orientation: ϕ.
- The center of the mobile robot's circular path: (x_0, y_0).
- The radius of the mobile robot's circular path: R.
- The radius of the mobile robot's wheels: r.
- The applied torques at the mobile robot's two rear wheels: (τ_1, τ_2).

And, the knowledge about the mobile robot's dynamics could be calculated as follows:

- The equation of computing steering angle $\theta(t)$:

$$\theta(t) = \arctan(\frac{L}{R}) \tag{11.21}$$

with

$$R = \sqrt{(x(t) - x_0(t))^2 + (y(t) - y_0(t))^2}$$

- The equation of computing angular velocity $\omega(t)$:
Since the applied torques at the two rear wheels are (τ_1, τ_2), the equivalent forces acting on the two rear wheels will be $(\tau_1/r, \tau_2/r)$. These two forces will produce acting torque about the vertical axis which passes through (x_0, y_0). This acting torque will be

$$\tau = \frac{R + W/2}{r}\tau_1 + \frac{R - W/2}{r}\tau_2$$

which will produce the mobile robot's angular acceleration as follows

$$J\frac{d\omega(t)}{dt} = \tau$$

where J is the moment of inertia of the mobile robot's mass about the vertical axis passing through (x_0, y_0). Here, we ignore the friction and air's drag force. The Laplace transform of the above equation yields

$$\Omega(s) = \frac{1}{Js}T(s) = \frac{R + W/2}{rJs}T_1(s) + \frac{R - W/2}{rJs}T_2(s) \tag{11.22}$$

where $(\Omega(s), T_1(s), T_2(2))$ are the Laplace transforms of $(\omega(t), \tau_1(t), \tau_2(t))$. Its Laplace inverse transform will yield the time domain function for the computation of angular velocity

$$\omega(t) = \mathcal{L}^{-1}\{\Omega(s)\}$$

- The equation of computing linear velocity $v(t)$:
  Since $v(t) = R\omega(t)$, we have

$$V(s) = \frac{R^2 + RW/2}{rJs}T_1(s) + \frac{R^2 - RW/2}{rJs}T_2(s) \tag{11.23}$$

  where $V(s)$ is the Laplace transform of $v(t)$. Its Laplace inverse transform will yield the time domain function for the computation of linear velocity

$$v(t) = \mathcal{L}^{-1}\{V(s)\}$$

11.4.2 *Equation of Inverse Dynamics*

Inverse dynamics refers to the study of determining the applied torques to a robot's mechanism if the input motion imposed to the robot's mechanism are known in advance.

For example, in Fig. 11.5a, if the time function of the wheel's linear velocity $v(t)$ is known in advance, the wheel's angular velocity will be $v(t)/r$. Then, the time function of the applied torque $\tau(t)$ will be

$$\tau(t) = \frac{J}{r}\frac{dv(t)}{dt} \tag{11.24}$$

This is an equation of computing required torque if desired motion is known in advance.

Refer to Fig. 11.5b. If the time function of the mobile robot's angular velocity $\omega(t)$ is known in advance, the linear velocities of the two rear wheels will be $((R + W/2)\omega(t), (R - W/2)\omega(t))$. Then, the applied torques to the two rear wheels could be calculated as follows:

$$\tau_1(t) = \frac{J_1}{r}(R + W/2)\frac{d\omega(t)}{dt} \tag{11.25}$$

$$\tau_2(t) = \frac{J_2}{r}(R - W/2)\frac{d\omega(t)}{dt} \tag{11.26}$$

where (J_1, J_2) are the moments of inertia of the masses of the two rear wheels about the wheels' own axes of rotation. Here, we ignore the viscous frictions of the axes.

11.5 Summary

Knowledge could be computed from data. Also, knowledge could be computed from systems. In this chapter, we have studied typical scenarios of computing knowledge about the motion, kinematics and dynamics of mechanical systems. The principles studied in this chapter are equally applicable to other types of systems, which have been, or are still being, studied in different disciplines of science and engineering.

11.6 Exercises

(1) A road has two parallel lanes. The width of each lane is 4.0 m. Two vehicles are traveling along a same lane at a same speed of 15.0 m/s. The distance between these two vehicles is 30.0 m. At a time instance, the following vehicle decides to overtake the leading vehicle within 10.0 s. Design equations of trajectory which allow the following vehicle to achieve the task of overtaking the leading vehicle.

(2) In a coordinate system OXY, a vehicle enters a circular path at point A and exits it at point B. The coordinates of points A and B are $(20.0\,m, 0.0\,m)$ and $(30.0\,m, 10.0\,m)$. Assume that the radius of the circular path is 10.0m and the traveling time inside the circular is 10.0s. Also, assume that the circular speeds of the vehicle at points A and B are zero. Design equations of trajectory which enable the vehicle to accomplish the travel.

(3) A planar robotic arm has two links, which rotates in a horizontal plane. The lengths of link 1 and link 2 are $30.0\,cm$ and $20.0\,cm$, respectively. If the robotic arm's tool-tip has to reach the coordinates $(40.0\,cm, 30.0\,cm)$, what should be the values of the robot's two joint angles?

(4) A planar robotic arm has two links, which rotates in a horizontal plane. The lengths of link 1 and link 2 are $30.0\,cm$ and $20.0\,cm$, respectively. If the maximum errors of the robot's join 1 and joint 2 are 0.01^0 and 0.02^0, respectively, what will be the maximum errors of the coordinates at the robotic arm's tool-tip?

(5) A planar robotic arm has two links, which rotates in a horizontal plane. The lengths of link 1 and link 2 are $30.0\,cm$ and $20.0\,cm$, respectively. Assume that the uniformly distributed masses of link 1 and link 2 are $2.0\,kg$ and $1.0\,kg$, respectively. If the applied torques at the robotic arm's two joints are $0.5\,N.m$ and $0.3\,N.m$, respectively, what are the angular accelerations of the robotic arm's two links?

(6) A single-wheeled platform has a uniformly-distributed mass of 90.0kg (i.e. treat it as a disk). The radius of the wheel is 0.8 m. If a desired linear acceleration is $5.0\,m/s^2$, what will be the applied torque to the shaft of the wheel?

Part 4
Knowledge Application

There are many systems in the physical world. Most of them are nature-made creatures. However, the population of human-made systems is rapidly growing. It is interesting to take note that the primary purpose for human beings to invent tools and machines is to extend human beings' physical capabilities as well as mental capabilities. Therefore, the journey of inventing machines will inevitably consider the option of designing machines with self-intelligence, which could help human beings to undertake knowledge-based strategy-planning, knowledge-based decision-making, and knowledge-based action-taking. In this part of the book, we highlight some prominent applications in artificial intelligence.

Chapter 12

Knowledge-enabled Applications

12.1 Introduction

Human beings could demonstrate their intelligence in many different ways in terms of seeing, speaking, thinking, doing, analyzing, synthesizing, and designing, etc. Future machines with self-intelligence should possess human-like intelligence to a great extent. Although such goal is still a challenge for today's research and development in artificial intelligence, we believe that the new foundation presented in this book will help us to consolidate future efforts toward achieving significant breakthroughs in this domain. Therefore, we dedicate this chapter to highlight the promising directions in which artificial intelligence could make great progresses driven by impactful applications. In particular, this chapter will include insightful discussions about the following topics:

- Vision-centric Intelligence and Its Applications
- Speech-centric Intelligence and Its Applications
- Planning-centric Intelligence and Its Applications
- Control-centric Intelligence and Its Applications
- Data-centric Intelligence and Its Applications
- Design-centric Intelligence and Its Applications.

12.2 Vision-centric Intelligence and Its Applications

People say that seeing is believing. This implies that the cognitive state of knowing the meanings of visual signals depends very much on vision. People also say that an image is worth more than thousands of words. This witnesses the prowess of vision in attaining the awareness about a scene through images, or videos. There is no doubt that vision is important to many living-creatures because of their various needs such as acquiring knowledge, acquiring skills, avoiding dangers, and surviving in a changing environment. As for human-made creatures such as vehicles and robots, vision is also important because many tasks, which are to be undertaken by vehicles and robots, do require visual perception, visual feedback, or visual awareness.

In this section, we highlight the following typical scenarios of vision-centric intelligence and its applications:

- Vision-based Measurement
- Vision-based Guidance
- Visual Recognition
- Visual Understanding

12.2.1 *Vision-based Measurement*

Vision-based measurement is very important in industry and society. As shown in Fig. 12.1, the objective of visual measurement is to assess the conformity of observed characteristics with nominal characteristics of an object or a product.

In general, there are two ways to obtain the nominal characteristics of an object or a product. The first way is by design. For instance, the nominal characteristics can come from the designed features of products which are to be manufactured. During a process of manufacturing a products, it is often useful to use a vision system in order to monitor the variation of observed characteristics with respect to the nominal characteristics of a product to be manufactured. And, this type of vision applications is quite common for quality control in manufacturing industry.

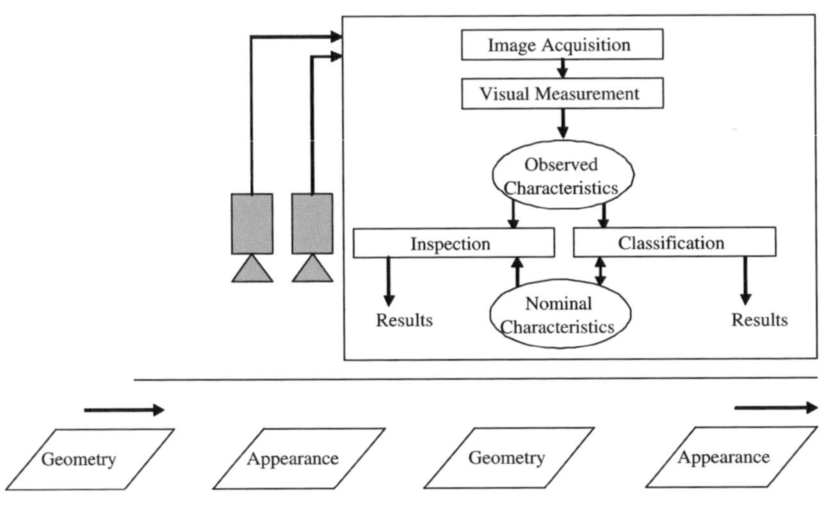

Fig. 12.1 Visual measurement, in which images of objects or products are taken as input in order to measure observed characteristics for the purpose of classification. Alternatively, the measured characteristics can be compared with the nominal characteristics in order to achieve the generic task of visual inspection.

The second way is by training or learning. For instance, the nominal characteristics of products can come from the visual measurement of samples selected from products which are purchased from suppliers. In this case, vision is useful for the generic task of classification.

12.2.1.1 *Visual Inspection*

In quality control, it may be necessary to make use of vision-based measurement in order to monitor and minimize the variation of observed characteristics of manufactured products, with respect to the designed characteristics.

We know that the input to a vision system are visual signals, which are the projection of light beams coming from a physical scene onto the image plane(s). Hence, visual signals contain not only information about the geometry of a product, but also its chromatic characteristics. As a result, vision is very useful for the measurement of geometry-related quantities such as

- Pose (i.e. position and orientation)
- Shape
- Dimension
- Distance
- Alignment
- Topology (i.e. spatial layout of several objects),

and appearance-related quantities such as

- Color
- Intensity
- Uniformity
- Texture

Example 12.1. Because of convenience, bottled (or canned) beverages are widely consumed in our modern society. In order to satisfy the demand by a very large population of consumers, beverages are filled into bottles by automated production lines. And, one important parameter for inspection is the quantity of beverage inside a filled bottle. In fact, the quantity of beverage inside a filled bottle could be visually verified by the liquid level. Therefore, a visual inspection station could be put in place for the purpose of indirectly inspecting the quantity of beverage inside every filled bottle in an automated production line.

As shown in Fig. 12.2a, a single camera could be placed beside a conveyor, which shifts forward the bottles which are filled with beverages. The camera is adjusted in such a way that the image plane of the camera is perpendicular to the surface of the conveyor. In addition, a sensor could be installed, which indicates to the camera when a filled bottle is within the field of view of the camera. During the process of inspection, the camera will capture an image when it receives a signal

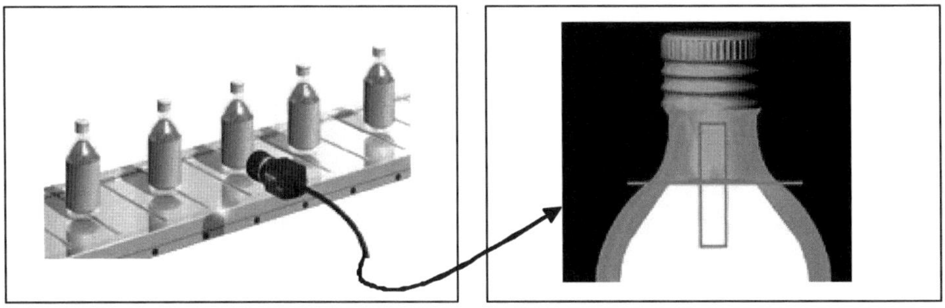

(a) System Setup (b) Sample Image

Fig. 12.2 Example of visual inspection of the liquid levels inside the bottles which are filled with beverages. A camera is placed at one side of a conveyor which carries forward filled bottles, and a sensor is installed, which detects the presence of a filled bottle within the field of the view of the camera. When the camera receives a signal from the sensor, it will capture an image. The processing unit of the visual inspection station will process the image in order to measure the liquid level of the bottle under inspection. (photo courtesy of KEYENCE Co.).

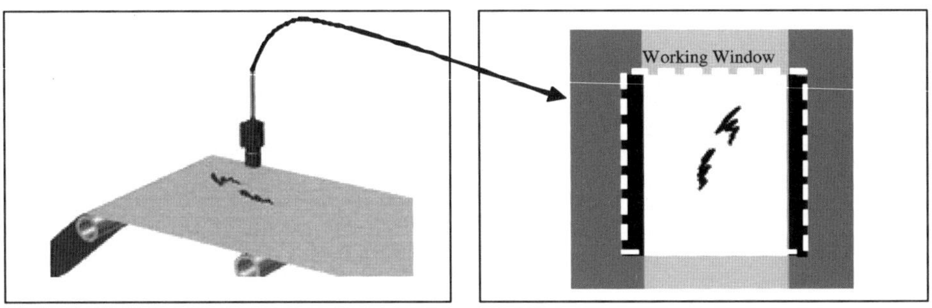

(a) System Setup (b) Sample Image

Fig. 12.3 Example of visual inspection of the stains in metal sheets (or fabric). The vision system consists of one camera, which outputs a series of images. For each image, a processing window is defined, in which the pixels are inspected in order to find those pixels which do not belong to the image of a metal sheet (or fabric). And, these pixels indicate the presence of stains on a metal sheet. (photo courtesy of KEYENCE Co.).

from the sensor. Then, the image will be processed in order to measure the level of the liquid as shown in Fig. 12.2b.

<><><><><>

Example 12.2. Metal sheets (or fabric) are common materials widely used in industry for the purpose of fabricating machines (or garment). Because of the demand in large volumes, metal sheets (or fabric) are normally produced in big quantities by automated production lines. In order to ensure the quality of metal sheets (or fabric), it is a common practice to use visual inspection station to detect stains on the surfaces of metal sheets (or fabric). As shown in Fig. 12.3a, a visual inspection station consists of one camera, which outputs a sequence of images. For

each image, the processing unit of the visual inspection station defines a working window, in which the pixels are inspected in order to verify whether they belong to a metal sheet or stains. When there are pixels which do not belong to a metal sheet, these pixels indicate the presence of stains on the metal sheet.

<><><><><>

12.2.1.2 *Visual Classification*

Visual classification is useful for security control, in which the nominal characteristics can come from entities such as fingers, faces, or irises. Besides security control, visual classification is also applicable to sort a set of entities (e.g. purchased products from a supplier) into different classes, each of which has its own definable characteristics in images. For instance, printed or hand-written characters, symbols, logos, or even specific types of defects are common candidates under consideration in visual classification.

Example 12.3. Connectors are common devices used inside electronic equipment, or machines, such as computers, hand-phones, video-players, and cars, etc. Because of the wide usage, connectors are manufactured by mass production. Therefore, it is necessary to automate the quality control process in order to maximize the throughput of a production line. And, a common solution to automate the quality control process is to set up a visual classification station. Alternatively, when a company has purchased a large quantity of connectors, it may be necessary to undertake the task of visual classification in order to reject those defective connectors, if any.

As shown in Fig. 12.4a, a visual classification station consists of two cameras, which work cooperatively for the purpose of detecting the warpage of pins

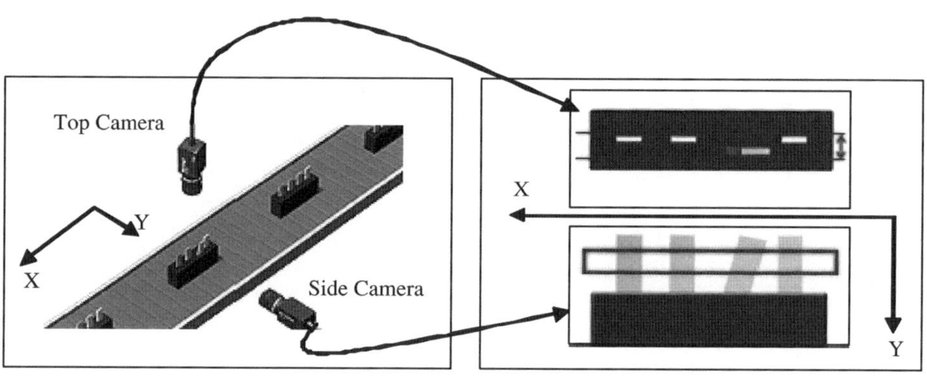

(a) System Setup (b) Sample Images

Fig. 12.4 Example of visual classification of the warpage of pins in a connector. The vision system consists of two cameras, which are named as the top-camera and the side-camera. In this example, the warpage of pin 3 is detected in both images from the top and side cameras. (photo courtesy of KEYENCE Co.).

in connectors being carried forward by a conveyor in a production line. In this example, the top camera is in charge of detecting the warpage in the direction parallel to axis OY, while the side camera helps detect the warpage in the direction parallel to axis OX. When a connector has a pin with warpage, it will then be classified as defective.

<><><><><>

12.2.2 *Vision-based Guidance*

One important purpose of a living-creature is to interact in a dynamically-changing environment for various needs including the self psychic- and physical-development. Moreover, any form of physical interaction will systematically result in what are called *behaviors.* In general, there are two types of behaviors, namely: a) instinctive behaviors, and b) learned or intended behaviors.

In control engineering, an instinctive behavior is the result of the dynamics of a system under control, while an intended behavior is produced by a feedback control system, in which the desired outcome could be determined by a motivation. In an organized, and disciplined, society and industry, what are important are learned, or intended, behaviors. Hence, when we use the term *behavior* in short, it normally refers to a learned, or intended, behavior. An instinctive behavior consists of an ordered sequence of actions as a result of interaction. However, a learned, or intended, behavior is influenced by motivation.

In general, most behaviors are requiring sophisticated sensory-feedbacks, as outlined in Fig. 12.5. Among various modalities of sensory feedback, vision is

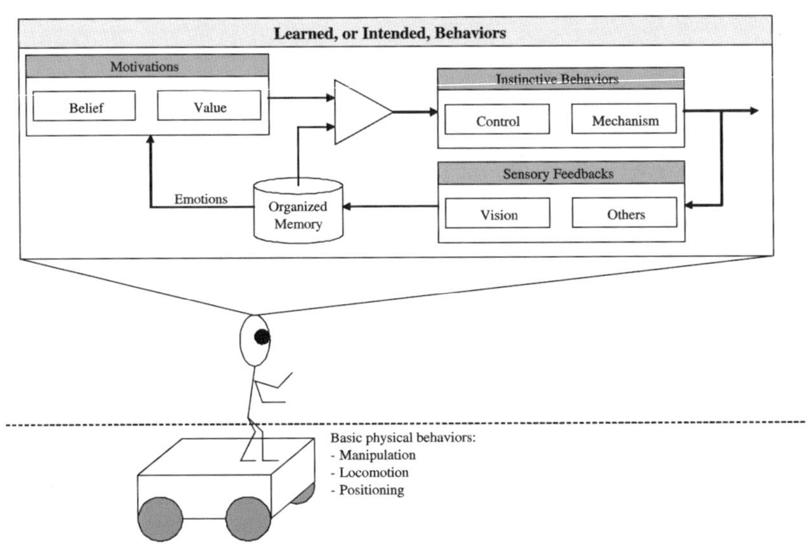

Fig. 12.5 Visual guidance of behaviors, where vision is indispensable for most of learned, or intended, behaviors such as manipulation, locomotion, or positioning.

undoubtedly the most important one. This is because vision enables the following three types of generic behaviors in the physical world:

- Vision-guided Manipulation
- Vision-guided Locomotion
- Vision-guided Positioning

12.2.2.1 *Vision-guided Manipulation*

Industrial robots are widely used for difficult, dirty and dangerous tasks such as welding, painting, material handling and assembly. By definition, an industrial robot is simply a re-programmable hand-arm system. With the visual feedback, a hand-arm system could achieve autonomous manipulation for generic tasks such as altering geometry (e.g. deformation), changing composition (e.g. insert, cut, pick, place), and imposing constrained (or unconstrained) motions (e.g. gesturing, writing).

Example 12.4. In automobile industry, manipulators are widely deployed to automate tasks such as assembly, painting, welding, and material handling, etc. By definition, a manipulator is a motion provider, which could carry various tools and execute re-programmable paths, or trajectories, in a workspace along a production line. By default, a manipulator repeatedly executes trajectories, which are designed by engineers. However, in order to increase the accuracy and flexibility in executing motions, it is desirable to equip a manipulator with a vision system.

As shown in Fig. 12.6, the output from a vision system could help plan new paths, or trajectories, for a manipulator to execute. Also, the output from a vision system could help do on-line alteration of a nominal path, or trajectory, so that a manipulator could adapt its actual motion to any unpredictable change in the workspace.

<><><><><>

12.2.2.2 *Vision-guided Locomotion*

When a physical system, such as a car, uses its own physical effort to displace its body along a surface such as a road, the event is called *locomotion*. The common means for a physical system to undertake locomotion are wheels, legs, or a combination of them. A typical example of vision-guided locomotion is the visual guidance of smart cars, in which vision plays the role of perceiving the road geometry and detecting road obstacles. As for legged locomotion such as the biped walking by a humanoid robot, vision is also indispensable if it wants to achieve autonomous walking in an unknown, or changing, environment.

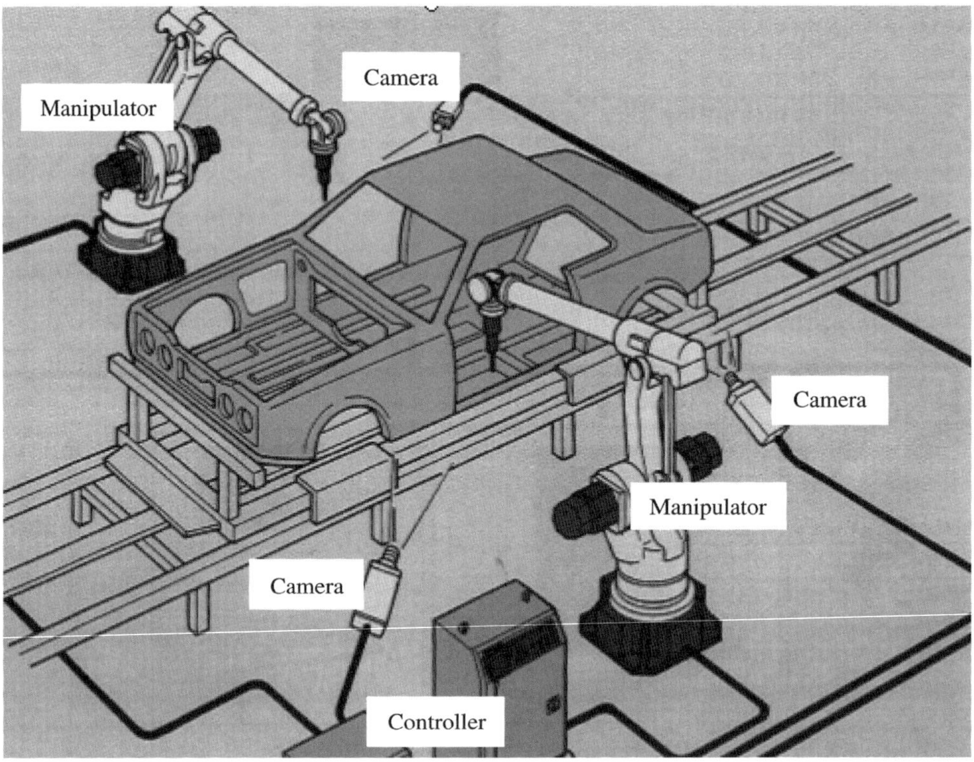

Fig. 12.6 Example of vision-guided manipulation, in which the output of vision helps plan new paths or trajectories for manipulators to execute.

Example 12.5. In our modern society, automobiles become an indispensable commodity. This is because they make us enjoy the freedom of going from places to places, and also boost our sense of social status. However, with the possession of automobiles by a large population of consumers, today's transportation faces three major problems, namely: a) congestion, b) pollution, and c) accident. In order to reduce the death rate due to car accidents, one solution is to make future automobiles sensible and intelligent.

For example, we can equip a car with a vision system together with a laser range finder, as shown in Fig. 12.7a. In this way, a car's vision system could help detect regions of interest in an image, which indicate the possible presence of obstacles in front of the car. Then, the laser range finder of the car will be activated in order to scan the areas of the scene, which may correspond to the detected regions of interest in an image. If a region of interest corresponds to an obstacle in a scene, the laser range finder will output a matrix of points, each of which carries an indication of depth between the laser range finder and the obstacle.

<><><><><>

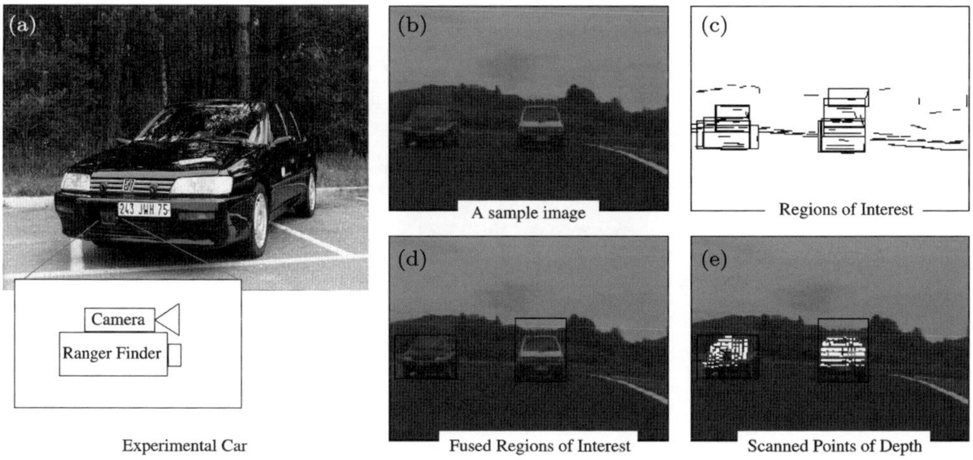

Fig. 12.7 Example of vision-guided locomotion, in which the vision system first detects the regions of interest from an image, and then uses them to guide the scanning done by the laser range finder. In this way, the response time of the laser range finder could be optimized.

12.2.2.3 *Vision-guided Positioning*

Positioning is an event which also involves the displacement of a physical system's body. However, in this case, it is not necessary to constrain the displacement to be along a specific surface. In general, positioning refers to the maneuver of a physical system's body so that it maintains a specific pose with respect to a fixed, or mobile, target. A typical scenario of positioning is the docking of a spacecraft, or a ship. Another example is when you approach a door in a room, you actually position your body with respect to the door.

In a road, you are also doing the maneuver of positioning when you drive your car behind a leading vehicle and maintain a fixed distance from the leading vehicle. Obviously, vision is indispensable for achieving autonomous positioning in an unknown, or changing, environment.

12.2.3 *Visual Recognition*

To a great extent, recognition means to match the input against something already known in advance in the mind of a natural, or artificial, creature. Hence, a prerequisite of recognition is that there must be something already known in advance. And, the question of how to cognise something, which is unknown, will fall into the topic of visual cognition, or understanding.

It is important to take note that in the mind of human beings, a social environment consists of two types of world, namely: a) the physical world where the nature shows its existence, and b) the conceptual worlds which are the representations of the physical world by human languages together with the technical language

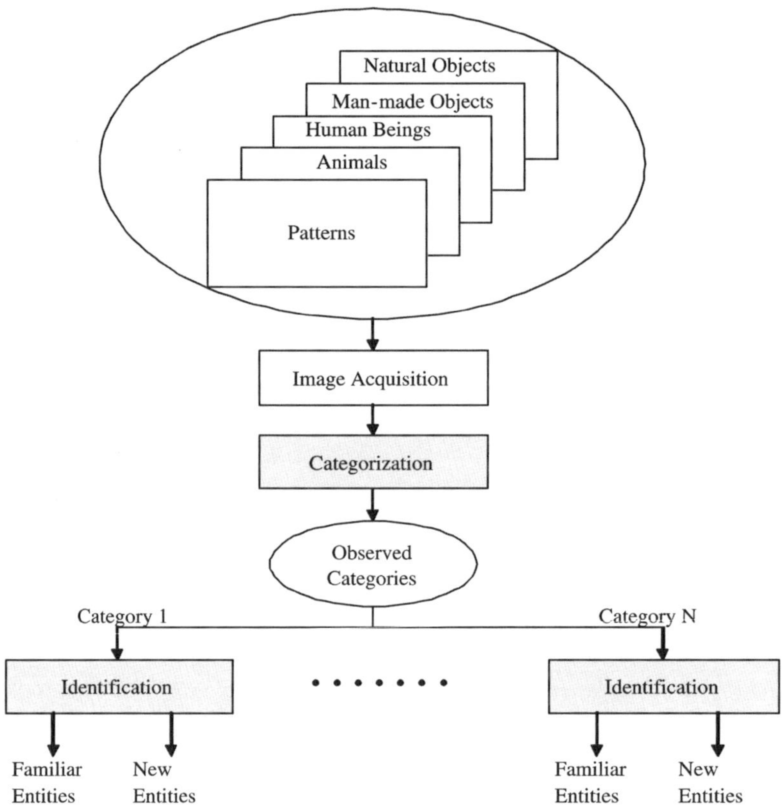

Fig. 12.8 Visual recognition of entities such as natural objects, human-made objects, human beings, animals, or patterns in general, in which vision plays two prioritized roles: a) categorization of observed entities, and b) identification of individual entities in each recognized category.

dedicated to sciences and engineering disciplines. Thus, it is useful to postulate the following definition:

In the physical, or a conceptual, world, an entity will have its own properties in order to manifest its identity. However, some entities may have a certain degree of similarity in their properties. This phenomenon gives rise to the notion of category. As a result, an entity will not only have its own identity, but also a hierarchy of categories.

Therefore, visual recognition literally means to recognize an entity so as to know its category and identify, as outlined in Fig. 12.8. In a more general term, visual recognition is a process, which categorizes and identifies entities in a scene, or image.

12.2.3.1 *Visual Categorization*

Categorization is a necessary step before identification. Without any prior knowledge, human vision is able to categorize observed entities such as cars, trees,

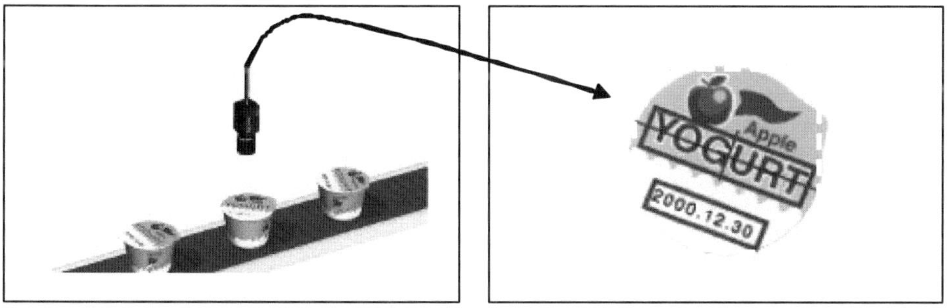

(a) System Setup (b) Sample Image

Fig. 12.9 Example of visual identification, in which the vision system is responsible for identifying the identity of the product and other information such as the expiry (or production) date of the product. (photo courtesy of KEYENCE Co.).

buildings, animals, etc. And, it is expected that machine vision could also achieve a similar level of success in future.

Interestingly enough, for many industrial applications, the category of entities in consideration may be known in advance. For instance, a production line normally outputs similar products of a same category. Therefore, visual recognition in industry concerns more about the problem of identification.

12.2.3.2 *Visual Identification*

With a prior knowledge, human vision is able to classify entities in a same category into two groups: a) the set of familiar entities, and b) the set of new, or unfamiliar, entities. For industrial applications, a typical example of visual identification is to convert a printed, or written, document into a computer-readable version (i.e. electronic format). This application is very much similar to the mental process of "reading texts" by human beings.

Also, it is possible to rely on visual recognition to identify specific patterns such as specific types of defects on manufactured products, or particular faces among a population.

Example 12.6. We know that every consumer-product in the market must have a label, which indicates the product's identity and other information such as the expiry (or production) date of the product, etc. Therefore, after the process of labeling products in a production line, it is necessary to verify that each product carries a correct label with the correct information.

As shown in Fig. 12.9a, a vision system consisting of a single camera could be deployed in a production line. Its role is to identify the printed text on a label. For instance, the labeled products are carried forward by a conveyor in a production line. A dedicated sensor will send a signal to the vision system in order to indicate that a labeled product is present within the camera's field of view. Then, the camera captures an image, which will subsequently be processed by the processing unit the vision system. Finally, the output from the vision system will help confirm the identity and expiry date of the product under identification. Once a product has passed the identification, it will be accepted and sent to a packaging station.

<><><><><>

12.2.4 *Visual Understanding*

Visual measurement, visual guidance and visual recognition cover a wide range of possible vision-enabled applications in industry and society. So far, these applications achieve limited successes because of the assumption of relatively simple input (i.e. images), or environment. If the context of an application is a complex scene, or a dynamically-changing environment, it is necessary to invoke visual understanding, in which visual cognition plays a crucial role.

As we have mentioned earlier, the social environment in the mind of human beings will consist of two types of world: a) the physical world, and b) the conceptual worlds. Therefore, the ultimate purpose of visual understanding is to cognise and recognize the meanings of a (real or virtual) scene in both the physical and conceptual worlds from images or videos.

As we have studied in previous chapters, knowledge or meanings refer to the properties and constraints of entities in both the physical world and the conceptual worlds. For example, a physical car has its meanings in the physical world, and the word *car* has its meanings in a conceptual world (e.g. a noun, an element in the category of land vehicle, etc).

Therefore, if a video, or image, is given as input, an important question will be: How to cognise and recognize the meanings recorded by the video, or image? And, this problem is known as *visual understanding.*

As shown in Fig. 12.10, the generic processes underlying visual understanding consist of: a) cognition, b) recognition, and c) synthesis. And, there is a fourth hidden process called *visual learning*, which is part of cognition. With these processes in place, a machine should be able to understand images and videos. Also, it should be able to read texts and visualize the physical meanings described by a text.

In practice, two typical applications of visual understanding are: a) surveillance (or reconnaissance) (i.e. to cognise and recognize the meanings of a scene from videos, or images which are captured at views of known poses), and b) interpretation (i.e. to cognise and recognize the meanings of a scene from videos, or images which are captured at views of unknown poses).

12.2.4.1 *Visual Surveillance*

Today, terrorism, violence and break-in are real threats in our modern society and industry. The most effective way to dissuade, prevent, and attenuate these threats

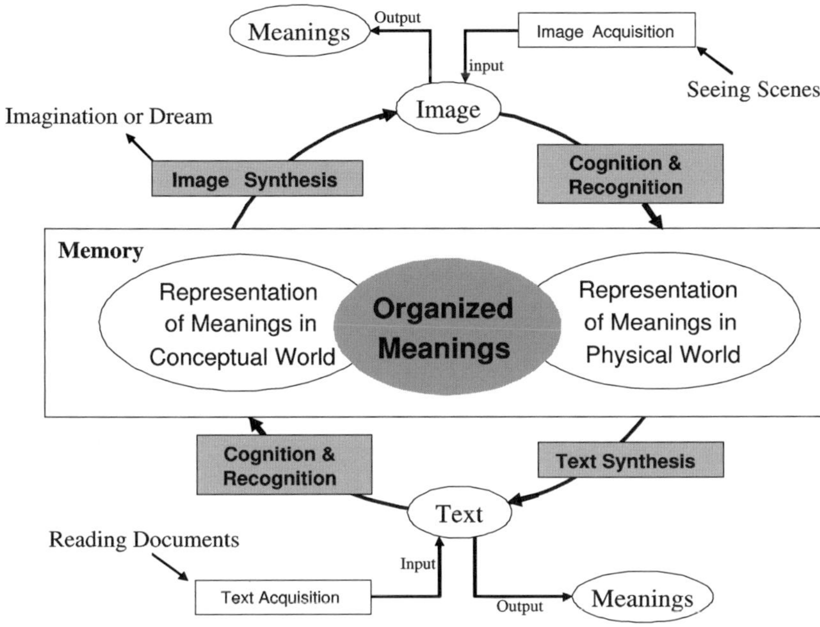

Fig. 12.10 Visual understanding of a scene from videos, or images, which are captured at views of known, or unknown, poses. The ultimate goal is to cognise and recognize the meanings in both the physical world and a conceptual worlds. In engineering terms, knowledge or meanings consist of properties and constraints of entities. Most importantly, knowledge or meanings are well organized, and are stored in the memory inside the brain of a natural, or artificial, creature. And, the mental processes, such as cognition, recognition, synthesis, and learning, concurrently make use of the organized meanings.

is to deploy a networked surveillance system, in which distributed "eyes" actively watch events and episodes occurring in some strategic lieu. With the ability of visual understanding, a surveillance system will be able to send advance warning or alarm, should there be any sign of anomaly detected.

Another example is military intervention, in which it is necessary to gather all kinds of useful information. One type of such information is the exact geography of an unauthorized region, or zone. And, the effective way to measure the geography of an unauthorized region is to fly an unmanned aerial vehicle over such a region and take video or photograph, which will then be analyzed in order to derive all the needed information. This application is usually called *reconnaissance*.

In a perfectly peaceful society where there is no any form of threat, surveillance is still useful. For instance, distributed "eyes" could be deployed in a shopping mall, which analyzes, in real-time, consumers' behaviors in order to measure interesting statistics such as product preference, age group, and gender, etc. These statistics may be useful for marketing purpose.

For other applications such as space exploration, it is desirable to send human-like robots to other planets for various tasks. And, human experts could remain in a

(a) Image at t1

(b) Image at t2

Fig. 12.11 Example of visual understanding, in which the vision system is expected to be able to understand the meanings of a scene, and also to describe the understood meanings with the help of a human language.

ground station on earth. In this scenario, it will be interesting if the robots could see and understand a visual scene, and communicate the description of the understood meanings of a scene, instead of images, back to the ground station. In this way, the delay in communication could be minimized and the ground station could rapidly reconstruct the scene from the received description. In fact, this application could also be extended to the simpler case of tele-operation on earth, if visual feedback is necessary.

Example 12.7. An image is worth more than thousands of words. Indeed, the most exciting application of vision waiting ahead is to understand the meanings of a scene, and also to describe the understood meanings of a scene with the help of a natural language. For instance, we may want to automate the surveillance of traffic condition in a city. In this case, the expected role of a vision system is to understand the traffic condition from a sequence of images taken by a camera fixed above a road. As shown in Fig. 12.11, an image of a road may contain a lot of information such as the number of vehicles, the types of vehicles, and the density of vehicles, etc. If a sequence of images from a same camera is processed, additional information about the dynamic behavior of each individual vehicle could also be cognised.

<><><><><>

12.2.4.2 *Visual Interpretation*

If the physical, or conceptual, meanings of an event are transcribed onto printed materials such as photos or documents, it is necessary to make use of vision to transform the visual signals coming from printed materials into descriptive knowledge.

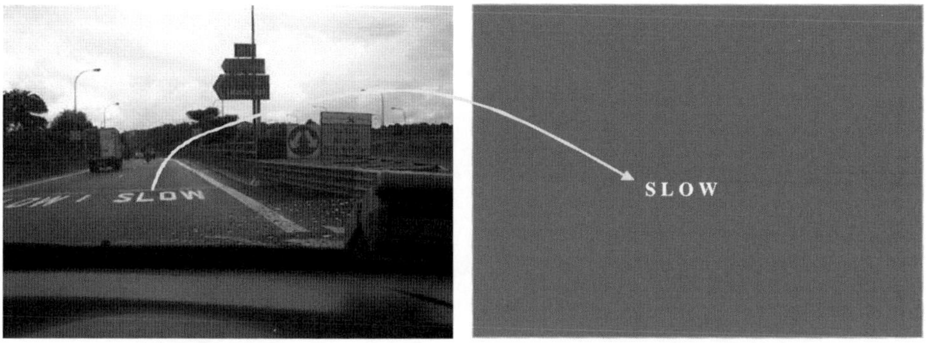

(a) A Road -image (b) Recognized Symbols

Fig. 12.12 Example of visual interpretation, in which the vision system is expected to be able to understand the meanings behind a text present in a road image.

For instance, reading a document is a typical example of visual interpretation. Another example is the visual appreciation of artworks such as paintings.

Example 12.8. In an intelligent transportation system, the ability to interpret human-readable road-signs is a must if a smart vehicle intends to autonomously navigate on roads, which are originally designed for human drivers.

As shown in Fig. 12.12, when a smart car travels on a human-made road, an image from its vision system will contain not only natural objects such as trees and hills, but also human-made objects such as the road, land-marks, and road-signs. If a smart car intends to achieve autonomous driving on a man-made road, it is necessary for it to gain the visual awareness of a traffic scene from an image, or video. In particular, a smart car's vision system must possess the ability to understand the meanings behind the texts present in a road-image. For instance, in this example, the in-vehicle vision system must not only recognize symbols such as "S", "L", "O", and "W", but also cognise the conceptual meanings behind the string of characters "SLOW".

<><><><><>

12.3 Speech-centric Intelligence and Its Applications

Knowledge are represented in the form of texts of human languages which are supplemented with extension to the technical language and programming languages. Texts could be printed on papers. Also, texts could be spoken out by speech sounds. In addition, speeches are the most convenient and popular way of communications among human beings. In this section, we highlight the following typical scenarios of speech-centric intelligence and its applications:

- Conversational Dialogue
- Voice-controlled Action

- Massive Sense-making
- Meaning-based Machine Translation
- Voice-based Inquiry

12.3.1 *Conversational Dialogue*

Human beings could easily talk to each other for various purposes which range from idea exchange, knowledge seeking, information communication, role playing, oral presentation, etc. In general, the process of conversational dialogue involves the following mental capabilities:

- Speech to text conversion: It converts signals of speech sounds into strings of symbolic characters in human languages.
- Text to meanings conversion: It recognizes the meanings behind strings of symbolic characters in human languages.
- Meanings to text synthesis: It uses texts to represent intended or understood meanings.
- Text to speech synthesis: It produces speech sounds from texts.
- Meanings to action synthesis: It plans actions according to intended or understood meanings.
- Meanings to scene & image synthesis: It interprets intended or understood meanings, and reconstructs a corresponding physical scene or image in a virtual space.
- Meanings cognition: It undergoes supervised learning or training in order to construct the inner database of organized meanings.

If future machines with self-intelligence could master the above-mentioned mental capabilities, it will be possible to achieve the scenario of conversational dialogue with the use of human languages between human beings and machines. Such scenario is illustrated in Fig. 12.13.

12.3.2 *Voice-controlled Action*

Today's human-machine interactions are mostly occurring at signal level. With the mental capability of understanding the meanings behind speech sound signals, human-machine interactions could occur at cognitive level, in which machines could respond to the meanings transmitted by speech sound signals.

For example, a vehicle could act according to the following sequence of voice commands:

- Continue to travel and turn into left road at the third junction.
- After that, continue to go straight and turn into right road at the fifth junction.
- Then, continue to travel about 300 meters and the destination will be reached.

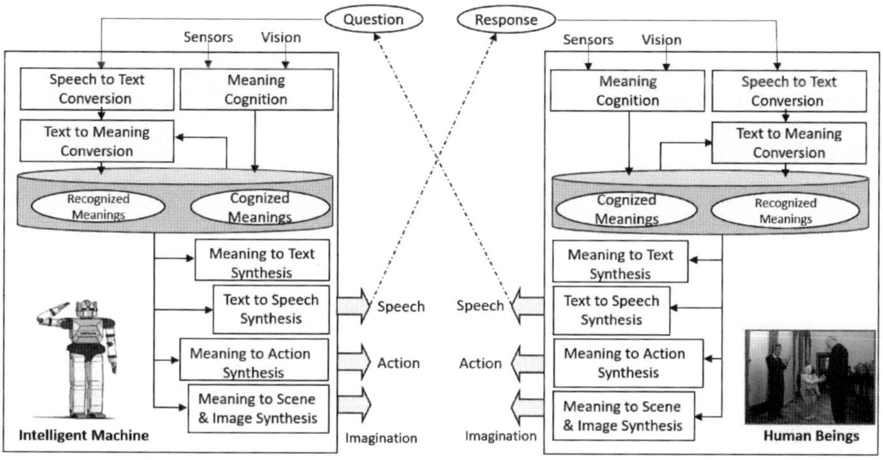

Fig. 12.13 Scenario of conversational dialogue between human beings and machines with self-intelligence.

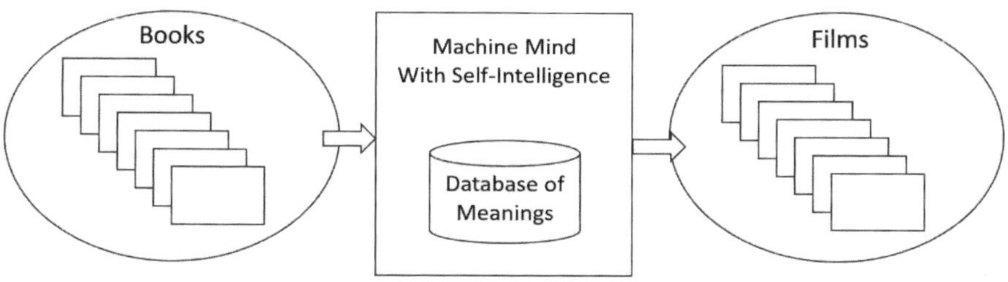

Fig. 12.14 Scenario of massive sense-making in which the input is a list of books and the output is a list of corresponding films.

12.3.3 *Massive Sense-making*

Today's cost of producing films or cinemas is very high. This is because the conversion of a novel or fiction into a film involving the complicated process of production. This could be an interesting domain of applications if the mind of a machine with self-intelligence could understand the meanings of a novel or fiction and convert the understood meanings into videos of films.

Fig.12.14 illustrates a scenario of massive sense-making, in which the input is a list of books and the output is a list of corresponding films.

12.3.4 *Meaning-based Machine Translation*

There are hundreds of human languages on earth. Hence, machines which could do language translation are important products in the market. So far, such products do not offer satisfactory performance. Hence, there is a huge room for improvement.

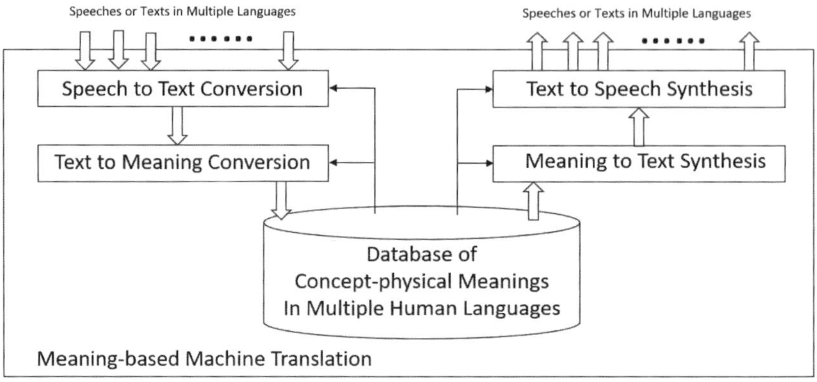

Fig. 12.15 Meaning-based Machine Translation.

One promising solution is to shift today's approach of look-up table based translation into future new approach of meaning-based translation.

Refer to Fig.12.15. The new approach to machine translation consists of the following steps:

- Speech to text conversion: The speech in one human language is recognized and converted into the text of the same language.
- Text to meaning conversion: The text in one human language is understood and converted into the meanings in the physical world, which are common to all human languages of all populations.
- Meaning to text synthesis: The understood meanings are described with the use of another human language.
- Text to speech synthesis: The text in another language is transformed into speech of that human language.

One advantage of the new approach is to achieve multiple outputs with a single input. In other words, the input of speeches or texts is in one human language, while the outputs could be speeches or texts in several other human languages.

12.3.5 *Voice-based Inquiry*

Today's society is largely driven by consumption. As a result, there are many needs for voice-based inquiry systems which could handle customers' demands in terms of inquiries about products, inquiries about services, inquiries about technical assistance, inquiries about bookings, and inquiries about deliveries, etc. Fig.12.16 shows an example of voice-based inquiry system, in which a human being makes an inquiry of "Is it lunch time now?", while a robot is making the effort of interpreting the inquiry and comparing the current time with a time interval for lunch in order to determine the answer of "yes" or "no".

Fig. 12.16 An example of voice-based inquiry.

In general, such voice-based inquiry systems could be implemented by a conversational dialogue module. Since a voice-based inquiry system is usually deployed to a specific commercial entity with limited sets of knowledge, the conversational dialogue could be led by a machine's mind. In this way, a machine's mind has the full control of conversations which will remain within the context of knowledge known to the machine in advance.

12.4 Planning-centric Intelligence and Its Applications

Planning is a mental activity which precedes control and management. Planning decides "what to do", while control and management take care of "how to do". For many intelligent activities such as transportation, manufacturing, shopping, surgical interventions, military interventions, sport competitions, travelings, just to name a few, planning always comes first. In general, planning involves the following mental procedures:

- Task planning: It decomposes a job or mission into a sequence of tasks.
- Action planning: It decomposes a task into a sequence of actions.
- Motion planning: It decomposes an action into a sequence of motions.

In this section, we highlight planning-centric intelligence and its applications under the context of intelligent transportation and smart manufacturing.

12.4.1 *Task Planning*

Refer to Fig. 12.17a. The mission of transportation is to move goods or people from one location to another location within a network of roads. Without loss of generality, a network of roads could be represented by a network of junctions. If a

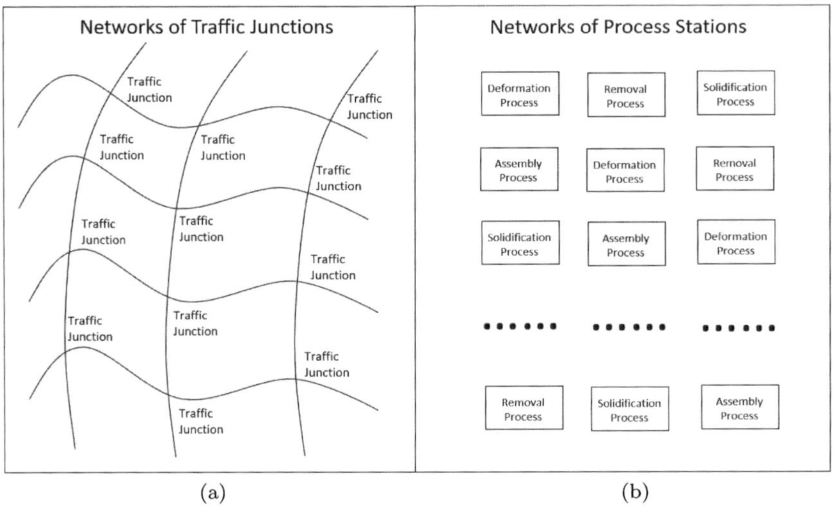

Fig. 12.17 Application contexts of intelligent transportation and smart manufacturing.

network of roads has n junctions, such network could be described by a matrix of connectivity as follows

$$C = \{J_{i,j}, i = 1, 2, \ldots, n;\ j = 1, 2, \ldots, n\} \tag{12.1}$$

with $J_{i,j} = 1$ if junction i is connected to junction j. Otherwise, $J_{i,j} = 0$. In this way, the objective of task planning is to determine subsets of junctions which allow vehicles to travel from one location to another location within a network of roads. Each subset of junctions actually tells a sequence of junctions for a vehicle to pass through. More specifically, the travel from the current location to the next junction is the next task of transportation.

Refer to Fig. 12.17b. A factory could deploy a network of process stations. These stations include the four typical manufacturing processes such as material removal process, deformation process, solidification process and assembly process. With such network of process stations in place, the task planning for achieving the job of manufacturing a specific product will be to determine possible sequences of process stations for materials and components to pass through in order to obtain the desired product as outcome. Within each sequence of process stations, a production task will consist of letting materials and components to pass through a corresponding manufacturing process.

12.4.2 *Action Planning*

Under the context of intelligent transportation, a vehicle's task consists of traveling from current location to the next junction indicated by the module of task planning. The execution of such task may involve a dynamic set of actions which could be road following, leading vehicle following, overtaking leading vehicle, stopping at the traffic light of the next junction, passing through the next junction, etc.

Under the context of smart manufacturing, a production task consists of letting materials and components to pass through the next process station indicated by the module of task planning. The execution of such task generally involves the actions of loading materials and components, processing materials and components, and unloading processed materials and components. In addition, the execution of a production task may invoke actions of process monitoring (e.g. to know whether the process is in control or out of control) and quality control (i.e. to know whether processed materials and components conform to desired standards or not).

12.4.3 *Motion Planning*

Under the context of intelligent transportation, a vehicle's action could be accomplished by executing a path and its corresponding trajectory. Therefore, the objective of motion planning is to determine the equations of path and the equations of trajectory for a vehicle to follow in order to accomplish the execution of a given action of transportation.

Under the context of smart manufacturing, an action at each process station may refer to loading materials and components, processing materials and components, or unloading processed materials and components. Clearly, the accomplishment of each action requires the execution of motion which may be simple or complex. As a result, the objective of motion planning is to determine the equations of path and the equations of trajectory, which will enable the execution of a production action.

12.5 Control-centric Intelligence and Its Applications

The physical world consists of entities which are organized into systems in one way or another. Systems with only steady-state responses are called as static systems, while systems with both steady-state and transient responses are called as dynamic systems. Most importantly, all intelligent systems must be controllable. In other words, all intelligent systems are closed-loop control systems. The beauty of closed-loop control scheme is to be able to make dynamic systems to behave like static systems as much as possible. Such objective is achievable with the following efforts

- Achieving stability: The purpose is to make all dynamic systems to be stable.
- Achieving smallest delay in time: The purpose is to all make dynamic systems to respond as fast as possible.
- Achieving smallest error in output: The purpose is to make all dynamic systems to produce zero errors as close as possible.

In this section, we highlight control-centric intelligence and its applications under the context of intelligent transportation.

12.5.1 *Achieving Stability*

A land vehicle has two independent control variables which are steering angle and driving speed. If these two variables are under the closed-loop controls, it is not a challenge to achieve the stability. Actually, the main cause behind a land vehicle's instability is the loss of friction between the vehicle and the ground. For example, if there is no such friction at all, the vehicle will totally loss its control. Therefore, knowledge about the friction between a land vehicle and the ground is an important parameter to monitor in real-time.

One solution to sense such knowledge is to determine the following difference

$$\triangle e_i = v_i - r_i \omega_i, \ i = 1, 2, 3, \ldots \tag{12.2}$$

where

- v_i is the linear or circular velocity of a land vehicle's wheel i.
- r_i is the radius of a land vehicle's wheel i.
- ω_i is the angular velocity of a land vehicle's wheel i.

By default, $\triangle e_i$ $(i = 1, 2, 3, \ldots)$ should remain to be zero. Any deviation from zero is an indication of danger or loss of stability.

12.5.2 *Achieving Smallest Delay in Time*

In general, a land vehicle has two closed-loop control systems. One is for the control of steering angle, while the other is for the control of driving speed. In normal situations, the outputs from these two closed-loop control systems vary smoothly. However, in situations of avoiding collisions or other dangers, a land vehicle must be able to execute the commands of sudden acceleration, sudden deceleration, and sudden change of lane, etc. Therefore, the response times of a land vehicle's two closed-loop control systems must be as short as possible.

For example, Fig. 12.18 shows a smart electric vehicle, in which the steering control is performed by the two front wheels, while the speed control is supported by the two rear wheels. In addition, each of the rear wheels has its own closed-loop control system. If the transfer function of a rear wheel's closed-loop control system is described by

$$G(s) = \frac{\omega_n^2}{s^2 + 2\zeta\omega_n s + \omega_n^2} \tag{12.3}$$

then the best value of ζ for achieving the smallest delay in time will be 0.7.

12.5.3 *Achieving Smallest Error in Output*

When a land vehicle travels on a road, the spatial clearances in front, rear and sides are usually quite large. Hence, it is not an issue for achieving smallest error in the

Fig. 12.18 Example of a smart electric vehicle in which the two front wheels support steering control, while the two rear wheels support speed control.

controlling a vehicle's pose. However, when a land vehicle undertakes the action of automatic parking into a constrained space of parking lot, the response accuracy of steering and driving is very important. In practice, PI or PID controllers should be the first choices for the design of a land vehicle's closed-loop control systems.

12.6 Data-centric Intelligence and Its Applications

The physical world consists of a huge amount of systems which compete against one another. For example, a nation usually has the following competing systems:

- Smart city systems: A nation may have many cities which provide space and convenience for people to live, to study, to work, to entertain, and to socialize, etc. Therefore, in generic terms, a city is a system which provides services supported by products to consumers who are the key players in a market.
- Smart building systems: A city may have many buildings which also provide space and convenience for people to live, to study, to work, to entertain, and to socialize, etc. Therefore, in generic terms, a building is a system which provides services supported by products to consumers who are the key players in a market.
- Smart hospital systems: A city may have several hospitals which provide space and medical services to patients who seek treatments and recoveries of their health conditions. Therefore, in generic terms, a hospital is a system which provides medical services supported by medical products to consumers (i.e. patients) who are the key players in a market.
- Smart school systems: A city may have many schools (including universities) which provide space and educational services to learners who seek knowledge and skills for betterment of lives and careers. Therefore, in generic terms, a school is a system which provides educational services supported by educational products (e.g. books, lecture videos) to consumers (i.e. learners) who are the key players in a market.
- Smart factory systems: A nation certainly has many companies who operate factories for the purpose of producing consumer products. Therefore, in generic terms, a factory is a system which makes use of equipment and machines to transform in-bound materials and components into outbound products for consumers (e.g. buyers) who are the key players in a market.

In addition, a nation will have many other big systems such as defense systems, homeland security systems, transportation systems, financial systems, etc. All these systems are the sources of production of big data. The wise use of big data will open the doors leading to the achievement of big intelligence which animates competitions at system levels.

Clearly, the presence of multiple systems with similar internal processes will automatically make them enter into competitions. Especially, in a free-market driven economy, the key forces, which propel the competitions among systems, are two-fold: one is to gain more market reward and the other is to gain more market acceptance. As shown in Fig. 12.19, a system must consume resources in order produce results. Most importantly, resources come from a market and results must be accepted by the market. Then, a question arises: how to win such competitions in a free-market driven economy?

Interestingly enough, all competing systems could exploit data in order to optimize internal processes and to enhance internal capabilities so as to be in better positions to gain market reward as well as market acceptance.

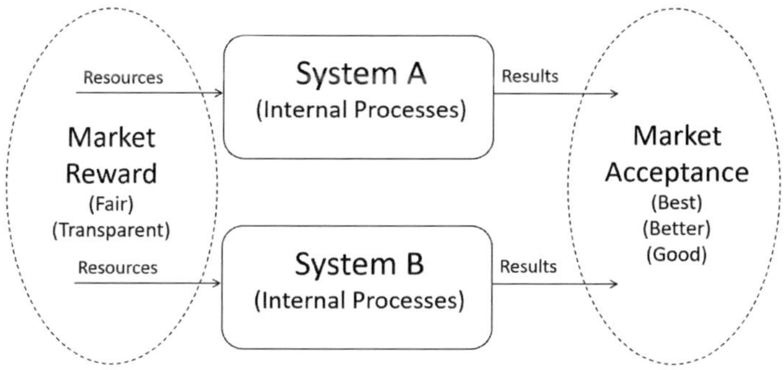

Fig. 12.19 Systems are competing for market reward and market acceptances in a free-market driven economy.

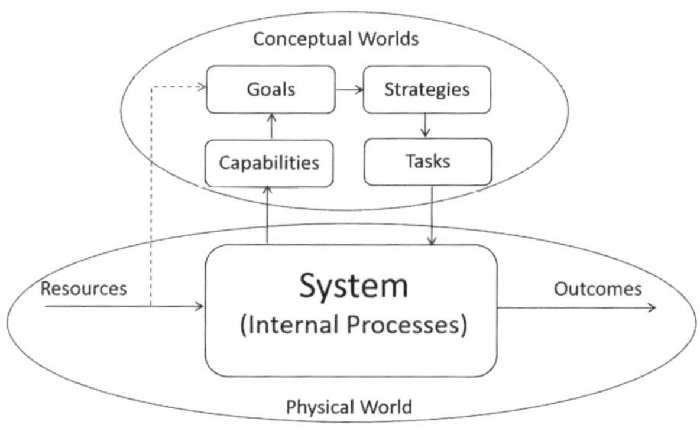

Fig. 12.20 Making systems to be competitive through strategic planning.

In this section, we highlight data-centric intelligence and its applications under the context of winning market competitions.

12.6.1 *Strategy for Market Reward and Market Acceptance*

The market has limited amount of resources in terms of capitals, manpower, and supplies. Hence, competitions for market reward are zero-sum games. Also, the market has a fixed size of population of consumers. Again, competitions for market acceptance are another zero-sum games. Therefore, the only way to win competitions is to have better internal capabilities and processes, and also to produce better external results and outcomes. These objectives could be achieved with the better use of data. More specifically, data could support strategic planning which will make a system competitive enough to gain market reward and market acceptance.

Refer to Fig. 12.20. The planning of goals for a system to achieve must consider the constraints of available resources and capabilities. Also, the achievements of planned goals by a system must lead to the enhancement of the system's internal capabilities and the increase of the system's external resources. This is the way for a system to remain competitive. The fulfillment of such scenario is possible if a proper strategic planning is put in place. Under the context of artificial self-intelligence, the purpose of strategic planning is for a system to autonomously make use of available data to design sequences of strategies at various levels.

For example, in a smart factory with self-intelligence and at the level of product design, strategies for design objectives could be:

- Strategy 1: To create new products for consumers to buy.
- Strategy 2: To create new products for consumers to adopt as their solutions within their products.
- Strategy 3: To create new products for consumers to depend on as parts of their internal capabilities.

At the level of production, strategies for manufacturing options could be:

- Strategy 1: To fabricate products of high quality.
- Strategy 2: To fabricate products of large quantity.
- Strategy 3: To fabricate products with least manpower.
- Strategy 4: To fabricate products with least wastes.
- Strategy 5: To fabricate products with least pollution.

At the level of product marketing, strategies for market penetrations could be:

- Strategy 1: To enter existing markets.
- Strategy 2: To grow existing markets.
- Strategy 3: To create new markets.

At the level of product sales, strategies for customer relations could be:

- Strategy 1: To have new buyers.
- Strategy 2: To keep existing buyers.
- Strategy 3: To grow existing buyers.

12.6.2 *Development of Internal Capabilities*

A system's internal capabilities depend on equipment and manpower. Strategies for developing better equipment could be:

- Strategy 1: To purchase better and new equipment.
- Strategy 2: To upgrade existing old equipment into better versions.
- Strategy 3: To design and develop alternative better replacements of existing equipments.

As for the development of a system's manpower, strategies could be:

- Strategy 1: To recruit new and competent manpower.
- Strategy 2: To train existing manpower so as to upgrade their knowledge and skill sets.

- Strategy 3: To reward existing manpower so as to unleash their creativity and job commitments.

12.6.3 *Optimization of Internal Processes*

Here, a system's internal processes refer to a system's internal operations. In general, there are two types of internal operations. One is product-oriented operations, while the other is service-oriented operations. For product-oriented operations, the generic steps include design (e.g. a new product), production, marketing, sales and customer feedback. For service-oriented operations, the generic steps include design (e.g. a new service), implementation, brain-storming, adoption, and customer feedback.

In general, the optimization of a system's internal processes involves the following two steps:

- Optimizing what is do: It is to define, and redefine, the strategic goals which are achievable on the basis of existing capabilities.
- Optimizing how to do: It is plan, and re-plan, the tasks which make the best use of existing capabilities in order to achieve thc goals.

12.7 Design-centric Intelligence and Its Applications

The mental capabilities at the highest level are design intelligence. It is a matter of fact that all human-made creatures are the outcomes of design works. Today, we have computer-aided design systems. Tomorrow, it will be certain that design by computer will become a reality. Hence, design-centric intelligence and its applications should be a big branch of research in artificial intelligence.

For example, in Fig. 12.21, we illustrate a general scenario of designing a smart product, in which a designer must be able to understand the design specifications coming from texts, images and scenes of gestures. Also, the designer must have the mental capabilities of applying learnt knowledge and skills to design a smart product's mechanical systems, control systems and software systems.

In this section, we highlight design-centric intelligence and its applications with a few examples which serve as possible directions for future research in artificial intelligence to shine.

12.7.1 *Designing Products*

Today's societies are organized around the chain of design, production, marketing, sales and consumption of products. Therefore, product design is an important pillar of our modern societies. In addition, product design is a highly intelligent activity which heavily depends on the use of knowledge and skills.

Refer to Fig. 12.22. In general, a complex product consists of a hierarchy of key systems (KS), in which each key system is another hierarchy of key modules (KM).

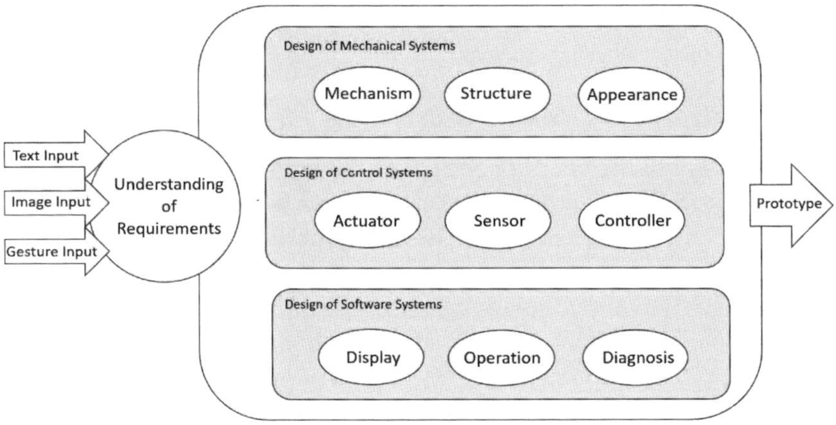

Fig. 12.21 Scenario of design by computer in which knowledge acquisition from texts, images and scenes of gestures plays a vital role.

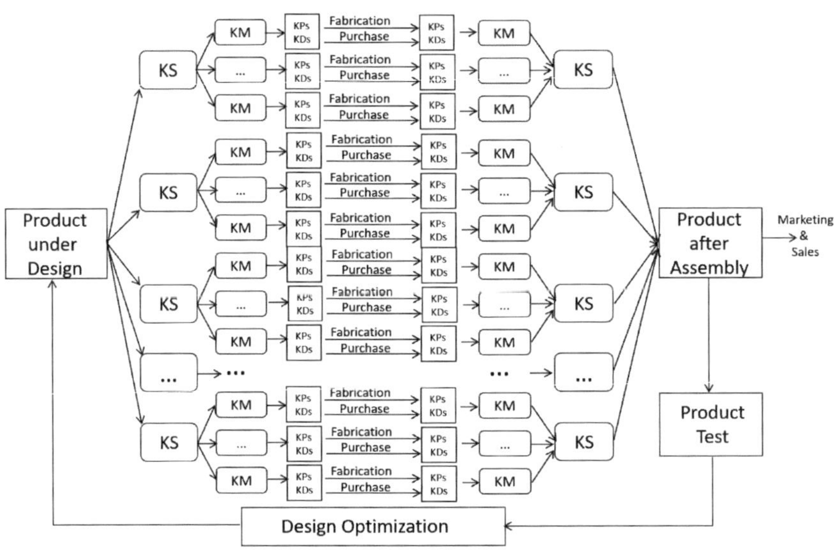

Fig. 12.22 Decomposition of product and cycle of design, prototyping, testing, and optimization.

In addition, a key module is composed of key parts (KP) and key devices (KD). Due to the complexity of a product, the procedure underlying product design may go through a few rounds of cycle, which consists of design, prototyping, testing, and optimization. The output from product design will be a set of part drawings and the procedure of doing assembly. Actually, the assembly procedure consists of putting key parts and key devices together to form key modules, which will be further integrated together to form key systems. The assembly of key systems will result in the final intended product.

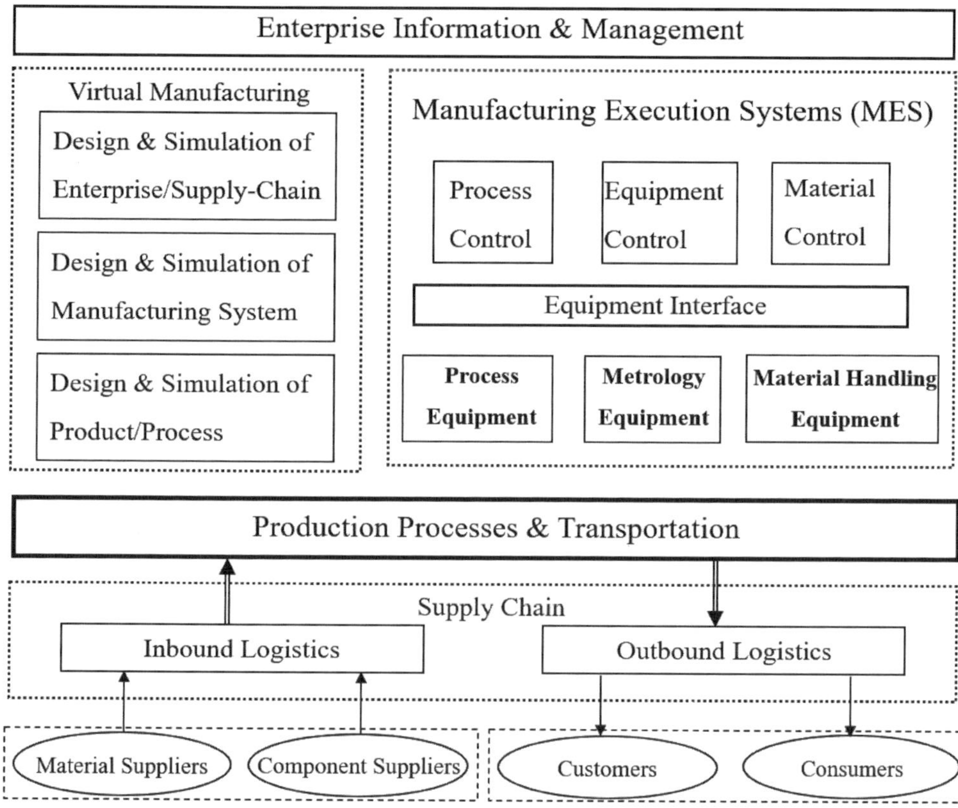

Fig. 12.23 Blueprint of intelligent manufacturing and production systems.

It is highly expected that machine's mind with self-intelligence could assist human beings in undertaking smart product design in future.

12.7.2 *Designing Factories*

Refer to Fig. 12.23. The input to a factory is a set of part drawings and the procedure of doing assembly. With such input, a factory will plan its activities which will include

- Design and Simulation of Production Processes: In manufacturing, the four typical manufacturing processes are material removal process, deformation process, solidification process and assembly process. In addition, there is a fifth process which is called as material handling or transportation.
- Design and Simulation of Manufacturing Systems: In manufacturing, production processes will be operated by machines. In addition, activities such as material handling and quality control will also be involved in production lines. When all

these machines work together, the scenario will be called as a manufacturing execution system, or MES in short.

- Design and Simulation of Enterprise and Supply-chain: A factory itself is a system which has input and produces output. Therefore, a factory's supply-chain will include both inbound logistics and outbound logistics.
- Actual Deployment of MES: Based on the results of design and simulation of manufacturing system, its actual implementation will take place. The result will be the actual deployment of MES in the form of production lines and production processes, which are interconnected and are networked with a factory's information and management systems.
- Actual Deployment of Supply-chain: Based on the results of design and simulation of supply-chain, the inbound logistics and outbound logistics will be implemented.

It is highly expected that machine's mind with self-intelligence could assist human beings in undertaking smart factory design in future.

12.7.3 *Designing Universities*

Universities are the major suppliers of human resources to the market. Although universities have housed most of the smartest people on earth, very few attention has been put on the issue of designing better universities.

Refer to Fig. 12.24. A university is actually founded on four key systems which are

- System for knowledge transfer: This is the educational system at the undergraduate level with the primary focus on knowledge transfer.
- System for knowledge application: This is the educational system which offers Master degrees and puts the primary focus on the applications of knowledge for the purpose of developing new technologies.

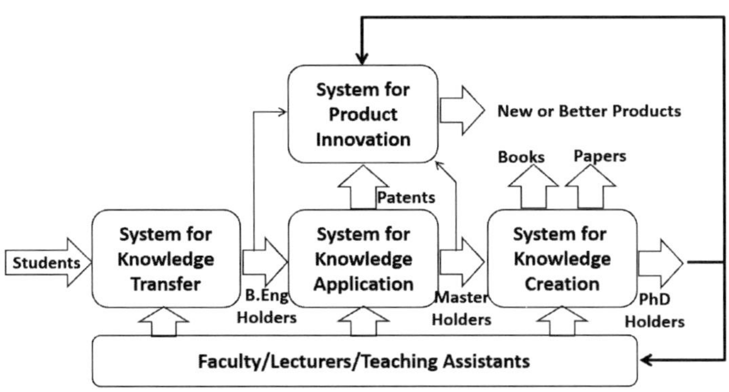

Fig. 12.24 Blueprint of modern university and its key systems.

- System for knowledge creation: This is the educational system which offers PhD degrees and puts the focus on the creation of new knowledge and new technologies.
- System for product innovation: Since universities are the primary locations of creating new technologies, universities will also be the primary places for students and faculties to venture into the creation of new products with new technologies inside.

It is highly expected that machine's mind with self-intelligence could assist human beings in undertaking smart university design in future.

12.7.4 *Designing Intelligent Transportation Systems*

The dynamics of a factory depends on transportation. Also, the dynamics of a society heavily depends on transportation. Therefore, the design and implementation of intelligent transportation system (ITS) are important activities which are relevant to artificial intelligence.

Refer to Fig. 12.25. An intelligent transportation system will consist of three parts which are

- Networked smart roads: Future smart roads will be equipped with massive networks of sensors which will produce huge amounts of data in order to support intelligent interactions between roads and vehicles.

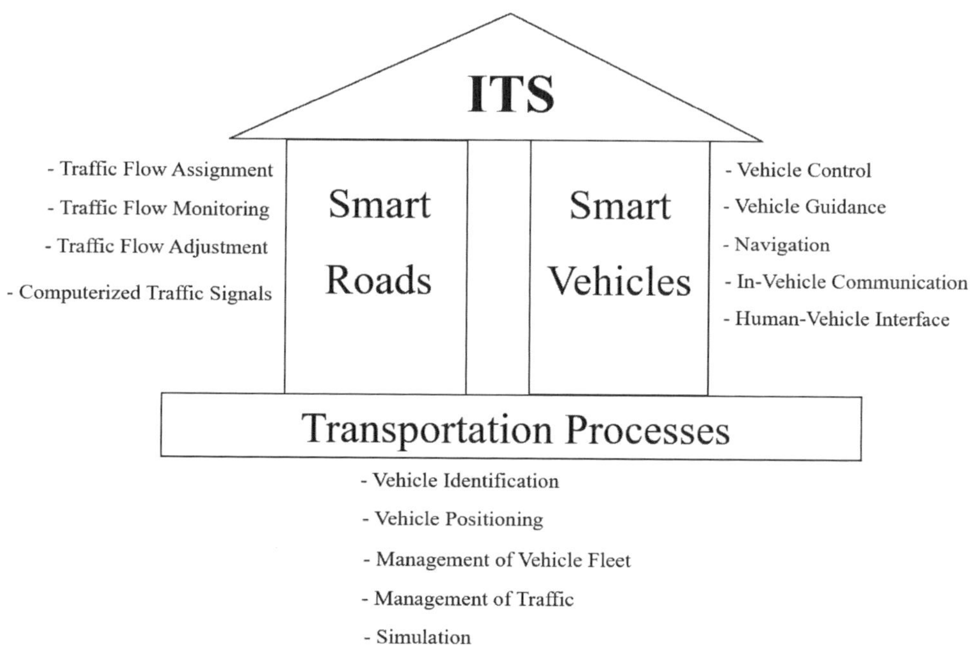

Fig. 12.25 Blueprint of intelligent transportation systems (ITS).

- Networked smart vehicles: Future smart vehicles on roads will have self-intelligence which could make good use of data in order to achieve future transportation with zero accident and zero traffic congestion.
- Knowledge-based transportation processes: These processes will collect data from networked roads and networked vehicles and make use of these data to support strategy-planning, decision-making and action-taking with the common goals of achieving safer, smarter and more efficient transportation in future.

It is highly expected that machine's mind with self-intelligence could assist human beings in designing intelligent transportation systems in future.

12.8 Summary

The physical world is full of entities, systems and signals. The properties and constraints make entities and systems to act and react. Most importantly, interactions among systems and entities could occur at signal level and also at cognitive level. However, it is the interactions at cognitive level which will make the physical world more pleasant for human beings to live in.

In this chapter, we have learnt that knowledge actually enable many systems to achieve interactions at cognitive level. For example, we have learnt examples such as vision-guided manipulation, vision-guided locomotion, speech-enabled dialogue, knowledge-based planning, and design intelligence. There are many other examples in which knowledge play the key and enabling roles. Hopefully, these promising knowledge-based applications will motivate people to continue the journey of making artificial self-intelligence to become a total success in a near future.

12.9 Exercises

(1) Write an essay about vision-centric intelligence and its possible applications.
(2) Write an essay about speech-centric intelligence and its possible applications.
(3) Write an essay about planning-centric intelligence and its possible applications.
(4) Write an essay about control-centric intelligence and its possible applications.
(5) Write an essay about data-centric intelligence and its possible applications.
(6) Write an essay about design centric intelligence and its possible applications.

Bibliography

Akerkar, Rajendra. (2014). *Introduction to Artificial Intelligence* (PHI Learning).

Bailey, James. (1996). *After Thought: The Computer Challenge to Human Intelligence* (Harper Collins Publishers).

Ballard, D. H. and Brown, C. (1982). *Computer Vision* (Prentice Hall).

Banks, S. (1990). *Signal Processing, Image Processing and Pattern Recognition* (Prentice Hall).

Barr, Avron. and Feigenbaum, Edward A. (1985). *The Handbook of Artificial Intelligence* (William Kaufmann Publishers Inc).

Berlin, B. and Kay, P. (1969). *Basic Color Terms: Their Universality and Evolution* (University of Calofornia Press).

Bond, Alan H. and Gasser, Les. (2014). *Readings in Distributed Artificial Intelligence* (Morgan Kaufmann Publishers Inc).

Bonabeau, Eric., Dorigo, Marco. and Théraulaz, Guy. (1999). *Swarm Intelligence: From Natural to Artificial Systems* (Oxford University Press).

Bratko, Ivan. (2001). *Prolog Programming for Artificial Intelligence* (Pearson Education).

Brooks, R. A. (1981). *Model-based Computer Vision.* (UMI Research Press).

A. de Callataÿ. (1992). *Natural and Artificial Intelligence* (Elsevier Science Publishers).

Castleman, K. R. (1996). *Digital Image Processing* (Prentice Hall).

Cawsey, Alison. (1997). *The Essence of Artificial Intelligence* (Prentice Hall).

Charniak, Eugene., Riesbeck, Christopher K., McDermott, Drew V. and Meehan, James R. (1987). *Artificial Intelligence Programming* (Lawrence Erlbaum Associates Publishers Inc).

Copeland, Jack. (1998). *Artificial Intelligence: A Philosophical Introduction* (Blackwell Publishers Inc).

Coppin, Ben. (2004). *Artificial Intelligence Illuminated* (Jones and Bartlett Publishers Inc).

Crick, F. (1994). *The Astonishing Hypothesis* (Simon & Schuster).

Dodigovic, Marina. (2005). *Artificial Intelligence in Second Language Learning: Raising Error Awareness* (Multilingual Matters Ltd).

Engelbrecht, Andries P. (2007). *Computational Intelligence: An Introduction* (John Wiley and Sons Ltd).

Faugeras, O. (1993). *Three-Dimensional Computer Vision* (The MIT Press).

Fetzer, J. H. (1988). *Aspects of Artificial Intelligence* (Kluwer Academic Publishers).

Flasiñski, Mariusz. (2016). *Introduction to Artificial Intelligence* (Springer).

Floreano, Dario. and Mattiussi, Claudio. (2008). *Bio-Inspired Artificial Intelligence: Theories, Methods, and Technologies* (The MIT Press).

Foley, J. D., Dam, A. V., Feiner, S. K., Hughes, J. F. and Phillips, R. L. (1994). *Introduction to Computer Graphics* (Addison-Wesley).

Gale, W. A. (1987). *Artificial Intelligence and Statistics* (United States: N. p).

Genesereth, Michael R. and Nilsson, Nils J. (2012). *Logical Foundations of Artificial Intelligence* (Morgan Kaufmann Publishers Inc).

Ginsberg, Matt. (2012). *Essentials of Artificial Intelligence* (Morgan Kaufmann Publishers Inc).

Goertzel, Ben. and Pennachin, Cassio. (2007). *Artificial General Intelligence* (Springer).

Goody, Esther N. (1995). *Social Intelligence and Interaction* (Cambridge University Press).

Haralick, R. M. and L. G. Shapiro (1993). *Computer and Robot Vision* (Addison-Wesley).

Hartley, R. I. and Zisseman, A. (1998). *Multiple View Geometry in Computer Vision* (Cambridge University Press).

Haugeland, John. (1989). *Artificial Intelligence: The Very Idea* (The MIT Press).

Haupt, Sue Ellen., Pasini, Antonello., and Marzban, Caren. (2009). *Artificial Intelligence Methods in the Environmental Sciences* (Springer).

Hunt, Earl. (2011). *Human Intelligence* (Cambridge University Press).

Hutter, Marcus. (2004). *Universal Artificial Intelligence: Sequential Decisions Based on Algorithmic Probability* (Springer).

Jackson, Philip C. (2019). *Introduction to Artificial Intelligence* (Dover Publications).

Jain, R., Kasturi, R. and Schunck, B. G. (1995). *Machine Vision* (McGRAW-Hill).

Johnson-Laird, P. N. (1983). *Mental Models* (Cambridge University Press).

Jayakumar, K. S. and Xie, M. (2010) *Natural Language Understanding by Robots* (LAP LAMBERT Academic Publishing Co.).

Kanal, L. N. and Lemmer, J. F. (1986). *Uncertainty in Artificial Intelligence* (Elsevier Science Publishers).

Kandel, E. R., Schwartz, J. H., and Jessel, T. M. (1995). *Essentials of Neural Science and Behavior* (McGRAW-Hill).

Korb, Kevin B. and Nicholson, Ann E. (2010). *Bayesian Artificial Intelligence* (CRC Press).

Marshall, A. D. and Martin, R. R. (1992). *Computer Vision, Models and Inspection* (World Scientific).

Marwala, Tshilidzi. and Hurwitz, Evan. (2017). *Artificial Intelligence and Economic Theory: Skynet in the Market* (Springer).

Minsky, Marvin. (1961). *Steps toward Artificial Intelligence*, Proceedings of the IEEE.

Moravec, Hans. (1988). *Mind Children: The Future of Robot and Human Intelligence* (Harvard University Press).

Munakata, Toshinori. (2008). *Fundamentals of the New Artificial Intelligence: Neural, Evolutionary, Fuzzy and More* (Springer).

Negnevitsky, Michael. (2005). *Artificial intelligence: a guide to intelligent systems* (Pearson Education).

Nilsson, Nils Johan. (1998). *Artificial Intelligence: A New Synthesis* (Morgan Kaufmann Publishers Inc).

Nilsson, Nils Johan. (2014). *Principles of Artificial Intelligence* (Morgan Kaufmann Publishers, Inc).

Nolfi, Stefano. and Floreano, Dario. (2001). *Evolutionary Robotics: The Biology, Intelligence, and Technology of Self Organizing Machines* (The MIT Press).

O'Hare, G. M. P. and Jennings, N. R. (1996). *Foundations of Distributed Artificial Intelligence* (John Wiley and Sons Publishers Inc).

Palmer, S. E. (1999). *Vision Science* (The MIT Press).

Paris, Cecile L., Swartout, William R. and Mann, William C. (1991). *Natural Language Generation in Artificial Intelligence and Computational linguistics* (Springer).

Rolf Pfeifer and Christian Scheier. (2001). *Understanding Intelligence* (The MIT Press).

Plataniotis, K. N. and Venetsanopoulos, A. N. (2000). *Color Image Processing and Applications* (Springer).

Poole, David L. and Mackworth, Alan K. (2010). *Artificial Intelligence: Foundations of Computational Agents* (Cambridge University Press).

Chrisley, Ronald. and Begeer, Sander. (2000). *Artificial Intelligence: Critical Concepts* (Routledge).

Russel, Stuart. and Norvig, Peter. (2013). *Artificial intelligence: a modern approach* (Pearson Education).

Sarfraz, M. (Editor). (2005). *Computer-Aided Intelligent Recognition Techniques and Applications (John Wiley & Sons)*.

Self, John. (1988). *Artificial Intelligence and Human Learning: Intelligent Computer-aided Instruction* (Chapman and Hall).

Shepard, Paul. (1978). *Thinking Animals: Animals and the Development of Human Intelligence* (The University of Georgia Press).

Shi, Zhongzhi. (2011). *Advanced Artificial Intelligence* (World Scientific).

Shoham, Yoav. (1994). *Artificial Intelligence Techniques in Prolog* (Morgan Kaufmann Publishers).

Simari, Guillermo. and Rahwan, Iyad. (2009). *Argumentation in Artificial Intelligence* (Springer).

Sternberg, Robert J. (1985). *Beyond IQ: A Triarchic Theory of Human Intelligence* (Cambridge University Press).

Sterne, Jim. (2017). *Artificial Intelligence for Marketing: Practical Applications* (Wiley and Sons Inc).

Szolovits, Peter. (1982). *Artificial Intelligence In Medicine* (Routledge Publishers).

Thompson, A. T. (2002). *Vision and Mind* (The MIT Press).

Ullman, S. (1996). *High-level Vision* (The MIT Press).

Weiss, Gerhard. (1999). *Multiagent Systems: A Modern Approach to Distributed Artificial Intelligence* (The MIT Press).

Weng, Juyang. (2012). *Natural and Artificial Intelligence: Introduction to Computational Brain-Mind* (BMI Press).

Wenger, Etienne. (1987). *Artificial Intelligence and Tutoring Systems* (Morgan Kaufmann Publishers Inc).

Xie, M. (2003). *Fundamentals of Robotics: Linking Perception to Action* (World Scientific).

Index

Acquisition, 75
Action, 364
Action-taking, 343
ADC, 91
Additive, 144
Apparent motion, 268
Appearance, 230
Appearance model, 210
Artificial intelligence, 9
Artificial vision, 124
Attentive, 119
Awareness, 118

Baseline, 261
Behavior network, 80
Binocular vision, 108
Block diagram, 52
Blueprint, 58, 90
Body frame, 111
Bottom-up optimization, 129
Boundary, 234
Brain, 116

Categorization, 354
Category, 166
Causality, 37
Center, 319
Circularity, 216
Class, 68
Classification, 238, 349
Closed-loop, 5, 240
Closed-loop control, 51
Cognition, 133, 163, 229, 293
Cognitive, 113
Cognitive vision, 184
Color, 97, 142
Color indices, 156
Color label, 161
Color space, 147
Competition, 41, 369
Composer, 118
Composite, 106
Composite meanings, 79
Composition, 95
Computation, 317
Computer-aided design, 371
Computing, 293
Concept network, 84
Conceptual world, 7, 10, 115
Confidence, 46
Consciousness, 118
Constraints, 6, 78, 82
Constructionism, 125
Control-centric, 365
Controller, 116
Conversational dialogue, 360
Correlation, 322
Cruise control, 329
Cruising, 330

Data, 319
Data structure, 59, 162
Data-centric, 367
Database, 89
Decision-making, 343
Depth, 247
Depth-map, 282
Derivative, 36
Design by computer, 371
Design intelligence, 371
Design-centric, 371
Developmental, 111
Diaphrame, 90
Digital computer, 61
Dissimilarity, 321
Dynamics, 317, 327, 337

Edge, 196, 197
Eigenvalue, 324
Eigenvector, 324

Enclosing rectangle, 214
Entity, 4
Episode network, 81
Euclidean, 322
Event network, 80
Experiencing, 307

Factory, 373
Feature model, 212, 233
Focal length, 252
Forward, 334, 337
Fourier transform, 48
Frame, 109
Frequency, 321
Function, 62
Fuzzy logic, 170

Gaussian function, 46
Generative, 227
Generator, 117
Gripper, 33
Grouping, 177
Guidance, 350

Histogram, 186
Human language, 1, 9, 15
Human vision, 105
Human-machine interface, 173

Identification, 355
Identity, 164
Image, 96
Image center, 251
Image coordinate system, 249
Image formation, 131, 145
Image plane, 248, 252
Image understanding, 132, 134
Imaging system, 120
Implicit knowledge, 29
Index coordinate system, 249
Ineraction, 57
Inquiry, 362
Inspection, 347
Inspector, 117
Integration, 35
Intelligence, 115, 123
Interaction network, 79
Internal capabilities, 370
Internal processes, 371
Interpretation, 358
Invention, 16
Inverse, 336, 340
ITS, 375

Kinematics, 334
Knowledge, 11

L*u*v, 148
Language model, 22, 85
Laplace transform, 50
Layered networks, 85
Layered scene, 286
Learning, 168, 307
Library, 67
Life, 124
Light, 140
Locomotion, 351
Logic, 37
Look-through scene, 289

Machine translation, 361
Machine vision, 105, 122
Machines, 343
Manipulation, 350
Market acceptance, 369
Market reward, 369
Matching, 275
Meanings, 78, 102, 123
Measurement, 346
Mechanism, 327
Membership, 37
Memory, 59
Microphone, 90
Mind system, 116
Mobile robot, 35
Model, 222
Monocular vision, 108
Motion, 365
Multi-scale, 157

Natural language, 1

Occurrence, 320
Ocean, 105
Open-loop, 237
Optical axis, 252
Organized meanings, 116, 228

Parallel, 106
Path, 328

Perspective projection, 121
Phase, 22
Phoneme, 89
Photo-detector, 143
Photo-receptor, 143
Phrase network, 82
Physical world, 3, 115
Pick and place, 33
Picture, 121
Pixel, 197, 248, 251
Pixmap, 146
Planning-centric, 363
Polynomials, 35
Pose, 208, 255
Positioning, 353
Possibility, 45
Prediction, 311
Principal axes, 216
Principal components, 324
Probability, 41, 321
Procedural knowledge, 16
Product, 371
Programming language, 57
Projection, 265
Properties, 5, 78, 82
Propositional knowledge, 15
Psycho-physical development, 123

Qualitative, 108
Quality control, 347
Quantitative, 107

Real world, 115
Recognition, 133, 173, 237, 309, 353
Rectangularity, 216
Redundancy, 289
Region, 189
Relationship, 37
Relative motion, 259
Representation, 77
Resolution, 249
Response accuracy, 366
Response time, 366
RGB, 147
Rigid body, 30
Road, 39, 329
Roadmap, 12
Robot, 33
Robustness, 126

S-plane, 52
Sampling, 91
Scale, 207
Scene, 121
Scene inference, 132
Script, 73
Search space, 278
Security control, 349
Segmentation, 98
Self-intelligence, 3, 8
Sense-making, 361
Sensort-motor mapping, 112
Sensory system, 29
Sentence, 22
Sentence network, 82
Shape, 183, 207
Signal, 75
Similarity, 177, 321
Sinusoids, 48
Slow-down, 330
Smart road, 375
Smart vehicle, 376
Sound signal, 91
Sound waves, 90
Speech, 75
Speech production, 87
Speech Signal, 105
Speech-centric, 359
Speed-up, 330
Spread, 320
Stability, 366
Standard deviation, 320
States, 317
Statics, 319
Stereopsis, 109
Strategy, 369
Strategy-planning, 343
Subdivision, 158
Subtractive, 145
Supervised learning, 75
Surveillance, 356
Symbol grounding, 94, 99
Symbol grounding network, 168
Symmetry, 220
Syntax, 62
Synthesis, 89, 144
Systems, 317, 327

Task, 363
Technical language, 29

Text, 86, 96
Texture, 200, 202
Thin-lens camera, 141
Tools, 343
Top-down design, 127
Topic network, 84
Trajectory, 329
Transformation, 75
Transformations, 12
Transportation, 375
Truthfulness, 38

Understanding, 356
Universe, 38
University, 374

Variable, 59
Vector, 31
Vehicle, 30, 329
Verification, 311
Viewer, 254
Viewer-centric, 110
Vision, 124
Vision-centric, 345
Visual attention, 120
Visual learning, 130, 235
Visual signal, 105
Vocabulary, 21
Voice coil, 91

Waypoints, 328

Zero-sum games, 369